T0335637

Biomarkers in Alzheimer's Disease

To
Blanchette Rockefeller
and
those who suffer from Alzheimer's disease

Biomarkers in Alzheimer's Disease

T.K. Khan, PhD
Blanchette Rockefeller Neurosciences
Institute, Morgantown, WV, United States

AMSTERDAM • BOSTON • HEIDELBERG • LONDON
NEW YORK • OXFORD • PARIS • SAN DIEGO
SAN FRANCISCO • SINGAPORE • SYDNEY • TOKYO
Academic Press is an imprint of Elsevier

Academic Press is an imprint of Elsevier
125 London Wall, London EC2Y 5AS, United Kingdom
525 B Street, Suite 1800, San Diego, CA 92101-4495, United States
50 Hampshire Street, 5th Floor, Cambridge, MA 02139, United States
The Boulevard, Langford Lane, Kidlington, Oxford OX5 1GB, United Kingdom

Copyright © 2016 Elsevier Inc. All rights reserved.

No part of this publication may be reproduced or transmitted in any form or by any means, electronic or mechanical, including photocopying, recording, or any information storage and retrieval system, without permission in writing from the publisher. Details on how to seek permission, further information about the Publisher's permissions policies and our arrangements with organizations such as the Copyright Clearance Center and the Copyright Licensing Agency, can be found at our website: www.elsevier.com/permissions.

This book and the individual contributions contained in it are protected under copyright by the Publisher (other than as may be noted herein).

Notices
Knowledge and best practice in this field are constantly changing. As new research and experience broaden our understanding, changes in research methods, professional practices, or medical treatment may become necessary.

Practitioners and researchers must always rely on their own experience and knowledge in evaluating and using any information, methods, compounds, or experiments described herein. In using such information or methods they should be mindful of their own safety and the safety of others, including parties for whom they have a professional responsibility.

To the fullest extent of the law, neither the Publisher nor the authors, contributors, or editors, assume any liability for any injury and/or damage to persons or property as a matter of products liability, negligence or otherwise, or from any use or operation of any methods, products, instructions, or ideas contained in the material herein.

Library of Congress Cataloging-in-Publication Data
A catalog record for this book is available from the Library of Congress

British Library Cataloguing-in-Publication Data
A catalogue record for this book is available from the British Library

ISBN: 978-0-12-804832-0

For information on all Academic Press publications
visit our website at https://www.elsevier.com/

  Working together
to grow libraries in
developing countries

www.elsevier.com • www.bookaid.org

Publisher: Mara Conner
Acquisition Editor: Melanie Tucker
Editorial Project Manager: Kathy Padilla
Production Project Manager: Lucía Pérez
Designer: Victoria Pearson

Typeset by Thomson Digital

Contents

Foreword

While monographs on Alzheimer's disease (AD) are the norm for family support books, it is a rare treasure to have a single voice, Tapan Khan, to skillfully lead us through the full spectrum of biomarkers. His authority puts imaging, genetics, cells, CSF, peripheral, and other markers into the immediate and long-term goals of supporting AD clinical therapeutics. If and when a disease-modifying drug is developed, it will be critical to intervene prior to major pathological changes. In the past decades, biomarkers have revealed diagnoses as specific as those confirmed by autopsy, by revealing that AD is a concurrent clinical and biological diagnosis. Biomarkers have revealed AD-like changes are seen in most middle-aged, apparently normal, individuals. Further study is required to determine if these AD-like changes are the beginning of AD or instead a novel feature of aging. Biomarkers are essential to sorting out these questions.

It is no surprise that biomarkers are one of the hottest areas of AD research and continue to be refined. This effort by Tapan Khan is a landmark in advancing the area by authoritatively reviewing, with a single objective voice, where we are and where we are going. The clarity of the book sets a path to single authored books as the new standard.

G. Perry
San Antonio, Texas

Preface

Research in the field of Alzheimer's disease (AD), and particularly in the area of AD biomarkers, continues its extraordinary growth with support from federal and nongovernmental funding sources, alongside stunning technological advances in imaging and diagnostics. Waves of neurobiological discovery in AD research have influenced new thinking on how to detect the disease earlier so it can be treated earlier to improve the lives of patients. In the last three decades, a rich array of new ideas, new discoveries, and new technologies have accelerated AD biomarker research. Even in the last few years, we have witnessed an explosive increase in the discovery, validation, and application of AD biomarkers for diagnosis, assessment of prognosis, longitudinal tracking, evaluation of therapeutic efficacy, and clinical trials. Despite this increase in our understanding about AD neurobiology and disease biomarkers, an integrated and cohesive summary of the field has been sorely lacking.

This book was written in an attempt to consolidate the history of AD biomarkers and new discoveries into a comprehensive collection, covering every testing modality and biomarker and their applications, and including a critical evaluation of each and the need for ongoing biomarker research. This book presents what we have learned about AD biomarkers from various disciplinary perspectives, including imaging, genetics, biochemistry, physiology, and neurology. In writing this book, I attempted to provide an unbiased and comprehensive discussion of all AD biomarkers, including neuroimaging biomarkers, cerebrospinal fluid (CSF) biomarkers (biochemical assays of $A\beta_{1-42}$, total tau [t-tau], and phosphorylated tau-181 [p-tau]), genetic biomarkers, and biomarkers in peripheral fluids and cells. The book contains 7 chapters, beginning with a brief introduction of AD in Chapter 1. Chapter 2 presents the different neuropsychological tests and their accuracies for clinical diagnosis of AD, the prodromal stages of Alzheimer's disease (preclinical, mild cognitive impairment, and dementia due to Alzheimer's disease), and the clinical overlap of AD with non-AD dementias [eg, vascular dementia (VaD), Lewy body dementia (LBD), frontotemporal dementia (FTD), Creutzfeldt–Jacob disease (CJD)]. The neuroimaging modalities and their use in the detection of specific AD imaging biomarkers are covered in Chapter 3. In Chapter 4, I discuss methodological approaches to identifying genetic AD biomarkers and provide a summary of AD genetic biomarker research. Chapter 5 includes a critical evaluation of the performance of AD biomarkers in the CSF in autopsy-validated cohorts, and discusses their crosscorrelation with other AD biomarkers and their clinical utility for early and preclinical diagnosis of AD, longitudinal assessment, differential diagnosis of AD and non-AD dementias, and assessment of drug efficacy in AD clinical trials. Because CSF can only be collected by lumbar puncture, the invasiveness of the procedure limits the clinical utility of the most widely researched CSF biomarkers in diagnostic screening of elderly patients. Detection of neuroimaging biomarkers is expensive and requires specialized training, which also limits their more widespread use in the clinical setting. For this reason, researchers have actively pursued less invasive, simple, and inexpensive biomarkers of AD, in particular, biomarkers in peripheral tissues. Chapter 6 summarizes recent findings of cross-sectional and longitudinal studies of AD markers in blood plasma, crosscorrelation of plasma AD biomarkers with other biomarkers, and the possible application of plasma AD biomarkers in clinical trials. Other peripheral sample sources of AD biomarkers, such as urine and saliva, have been explored and are also discussed in Chapter 6. Cell-based AD biomarkers are reviewed in the last chapter (Chapter 7).

The scope of this book is not limited to only a few well-studied AD biomarkers, but covers all modalities and biomarkers, from neuroimaging to cell-based bioassays. This book is intended to not only provide a comprehensive summary of the literature on all AD biomarkers, but also illustrate their current and potential applications and their utility in the detection of AD (preferably at the earliest, preclinical stages), prognosis, and guiding therapeutic intervention in the everyday care of patients. Each chapter presents the background and rationale for development of each type of AD biomarker, starting from a basic understanding of the biology to the advanced concepts applied in each field. I hope this book serves as a valuable resource for trainees, researchers, and clinicians in the fields of neurology and neurobiology, as well as investigators and supporters of clinical trials in AD, as they continue to push the field of AD research and medicine forward and make cutting-edge discoveries that will impact the lives of those affected by this devastating disease.

T.K. Khan, PhD
Associate Professor
Blanchette Rockefeller Neurosciences Institute,
Morgantown, WV, United States

Acknowledgments

I am grateful to all Alzheimer's disease researchers and the important work they have done and continue to do to understand and find treatments for this devastating disease. My sincere thanks go to those whose painstaking and brilliant work have contributed to the growth of research, discovery, and validation of potential Alzheimer's disease biomarkers. The main inspiration to write this book was to finally bring their unsung efforts into the light. I also wish to thank those researchers whose work has indirectly helped move the field forward but might not have been acknowledged in each chapter.

I would like to thank the manuscript proposal reviewers who provided thoughtful comments and suggestions on the organization of specific chapters. I have included many figures previously published in the literature and also want to express my gratitude to all the authors and journal staff members for their permission to use these images in the book. I am also incredibly grateful to my colleagues and collaborators for their contributions toward my own education and accomplishments in the field of Alzheimer's disease research.

I must also thank my mother and late father for their unconditional love and support throughout my life. Without moral support of my wife, Manabika K. Khan, I cannot imagine how this project would have been completed. I first started this work when our son, Soujatya K. Khan, came into our lives, and the manuscript grew along with him. My wife took care of raising our young son while I took care of the manuscript, and for this I am forever grateful to her. My elder brother and sister, and all my relatives (uncles, aunts, in-laws, cousins, nephews and nieces), also deserve my whole hearted thanks as well, supporting my efforts to their conclusion.

I am indebted to Dr. Stacey Tobin for assisting in the preparation of this manuscript. Ms. Kathy Padilla, the Editorial Project Manager and Ms. Melanie Tucker, the Senior Acquisitions Editor, Neuroscience of the Academic Press/Elsevier, have helped from the beginning. Last but not least, I would like to express my gratitude to the Academic press/Elsevier publishing group for their tireless support.

T.K. Khan, PhD
Associate Professor
Blanchette Rockefeller Neurosciences Institute,
Morgantown, WV, United States

CHAPTER 1

Introduction to Alzheimer's Disease Biomarkers

1.1 HISTORICAL BACKGROUND

Alzheimer disease (AD) was discovered in 1906 by Dr. Alois Alzheimer based on his observations and treatment of a 51-year-old patient, Miss Augusta "D" (Maurer et al., 1997; Strassnig and Ganguli, 2005), at the Frankfurt Psychiatric Hospital starting in Nov 1901. At the time, Dr. Alzheimer was a senior assistant at the institution. When Miss Augusta "D" died on Apr. 8, 1906, Dr. Alzheimer conducted the autopsy of her brain and described the morphological and histological changes in the brain that continue to stand as hallmarks of modern day AD neuropathology (Hippius and Neundörfer, 2003). Dr. Alzheimer presented his first report of Alzheimer's disease at a congress in Tübingen, Germany, in 1906. He described two neuropathological phenomena: amyloid plaques, which he referred to as "miliary bodies," and neurofibrillary tangles, which he described as "dense bundles of fibrils." In 1910, Dr. Emil Kraepelin (Dr. Alzheimer's supervisor) introduced the term "Alzheimer's disease" to recognize Dr. Alzheimer's contribution to

Biomarkers in Alzheimer's Disease. http://dx.doi.org/10.1016/B978-0-12-804832-0.00001-8
Copyright © 2016 Elsevier Inc. All rights reserved.

the discovery of this disorder. Dr. Alzheimer died in 1915 without knowing the outstanding contribution he made to the history of medical science.

1.2 IMPORTANT EVENTS IN ALZHEIMER'S DISEASE (AD) RESEARCH

Breakthroughs in AD research are presented chronologically in Table 1.1, along with historical perspectives. The first test used to quantify the cognitive and functional decline in individuals with dementia, including AD, was the Blessed Dementia Rating Scale (Blessed et al., 1968). In 1976, Dr. Robert Katzman was the first to identify AD as the most common cause of dementia among the elderly population, and predicted that AD would become one of the major public health problems in modern day society (Katzman, 1976). The purification, characterization, and amino acid sequence analysis of the amyloid protein seen in the post-mortem AD brain provided the first biochemical evidence of disease pathology and laid the foundation of the amyloid cascade hypothesis of AD (Glenner and Wong, 1984). The first clinical guidelines focused on the clinical diagnosis of AD were formulated in 1984 by the National Institute of Neurological and Communicative Disorders and Stroke and the Alzheimer's Disease and Related Disorders Association (NINCDS-ADRDA, McKhann et al., 1984). Based on available clinical diagnostic tools, the NINCSD-ADRDA guidelines categorized AD as either probable AD or possible AD and concluded that definitive diagnosis of AD was only possible through histological confirmation at autopsy. This NINCDS-ADRDA diagnostic guidelines successfully served AD clinicians and researchers worldwide for 27 long years until they were finally modified in 2011 (National Institute on Aging and Alzheimer's Association, 2011; McKhann et al., 2011), though there was an earlier attempt by the International Working Group 1 to revise them in 2007 (Dubois et al., 2007).

The "gold standard" of AD diagnosis is the presence of amyloid plaques and neurofibrillary tangles in the brain at autopsy. Amyloid plaques form as a result of the deposition of different truncated forms of amyloid beta peptides (Aβ) originating from the cleavage of amyloid precursor protein (APP). Grundke-Iqbal et al. discovered that hyper-phosphorylated tau protein is the main constituent of the neurofibrillary tangles in the AD brain in 1986 (Grundke-Iqbal et al., 1986). In the same year, a clinical trial of the cholinesterase inhibitor tetrahydroaminoacridine (tacrine [Cognex]) reported a temporary effect on improving clinical measures of cognitive function in AD patients (Summers et al., 1986). Tacrine was the first FDA-approved drug for the treatment of the memory impairment due to AD. In 1987, the first genetic evidence of an inherited form of familial AD (FAD) was reported, specifically mutations in the APP gene located on chromosome 21 (Robakis et al., 1987; Tanzi et al., 1987). This discovery further supported the amyloid cascade hypothesis of AD pathogenesis. Later in 1993, APOE4 was implicated as a genetic risk factor for late-onset AD (Saunders et al., 1993; Strittmatter et al., 1993). In 1995, a second AD gene, presenilin1 (PSEN1), was discovered (Sherrington et al., 1995). Most FAD patients carry a mutation in the PSEN1 gene (total of 185 mutations found

Table 1.1 Important Events and Their Impact in Alzheimer's Disease (AD)

Event	Impact to Alzheimer's Disease	References
The 37th Meeting of South-West German Psychiatrists (37 Versammlung Südwestdeutscher Irrenärzte) was held in Tübingen on Nov. 3, 1906.	Discovery of Alzheimer's disease: First presentation about Alzheimer's disease by Dr. Alzheimer.	Dr. Emil Kraepelin first names "Alzheimer's Disease" in the eighth edition of his book *Psychiatrie*, 1910.
Blessed Dementia Rating Scale	First cognitive test to quantify cognitive and functional decline in adults.	Blessed et al. (1968)
First evidence of public health problem	First to identify AD as the most common cause of dementia among elderly population and predict that it will be a major public health problem in the modern day.	Katzman (1976)
Isolation of amyloid protein; foundation of amyloid hypothesis	Amyloid protein was isolated from fibrils in cerebrovascular amyloidosis associated with AD.	Glenner and Wong (1984)
First guideline of clinical diagnosis of Alzheimer's disease by The National Institute of Neurological and Communicative Disorders and Stroke and the Alzheimer's Disease and Related Disorders Association (NINCDS-ADRDA, 1984).	Clinical diagnosis of AD can be categorized as probable AD or possible AD. Definitive diagnosis of AD is possible by histopathological confirmation at autopsy.	McKhann et al. (1984)
Discovery of phosphorylated tau related to AD	Tau is a major component of Alzheimer paired helical filaments, a hallmark of AD brain pathology.	Grundke-Iqbal et al. (1986)
First clinical trial for AD: Warner-Lambert Pharmaceutical Co. with the association of NIA and Alzheimer's Association.	Clinical trial of tacrine for targeting symptoms of AD. Tacrine (Cognex) was the first FDA-approved drug for the treatment of memory problems due to AD.	Summers et al. (1986)
Genetic mutation in APP (amyloid precursor protein) gene at chromosome 21q21.3	Foundation of amyloid β hypothesis.	Robakis et al. (1987), Tanzi et al. (1987)

(Continued)

Table 1.1   Important Events and Their Impact in Alzheimer's Disease (AD) (cont.)

Event	Impact to Alzheimer's Disease	References
Stages of neurodegeneration in AD	Neuropathological staging (I–IV) of AD-related changes in autopsy AD brains.	Braak and Braak (1991)
Discovery of genetic risk factor of LOAD	APOE (apolipoprotein E) at chromosome 19q13.2. The ε4 allele of APOE decreases Aβ clearance related to LOAD.	Strittmatter et al. (1993), Saunders et al. (1993)
Discovery of familial AD genes: Presinilin1 and Presinilin2	Familial AD gene related to production of Aβ.	Sherrington et al. (1995), Rogaev et al. (1995), Levy-Lahad et al. (1995)
National Alzheimer's Project Act into law by President Obama	This law was established to acknowledge the importance of personal, societal, and financial loss due to AD and creation of strategic plan to counter this escalating crisis.	US Public Law 111-375, 2011 http://aspe.hhs.gov/daltcp/napa/PLAW-111publ375.pdf
New guidelines for diagnosis of Alzheimer's disease	A working group of experts was established to revise and update the criteria and guidelines for diagnosis of AD by the Institute on Aging and Alzheimer's Association criteria.	Albert et al. (2011), Jack et al. (2011), McKhann et al. (2011), Sperling et al. (2011)
International working group 2 (IGW-2): New guidelines for diagnosis of Alzheimer's disease	IWG-2 criteria recognized the use of biomarkers as integral to the diagnosis of AD.	Dubois et al. (2014)

AD, Alzheimer's disease; EOAD, early-onset familial AD; FAD, familial AD; LOAD, late-onset (sporadic) AD.

to date). The third FAD gene (presenilin2 [PSEN2]) was discovered in the same year by Levy-Lahad et al. (1995), and confirmed by Rogaev et al. (1995).

Braak and Braak (1991) examined the amount of neurofibrillary tangles and neuropil threads in the brain at autopsy to describe disease severity in terms of stages I–VI. Stages I and II corresponded to very mild AD, stages II–IV corresponded to moderate AD, and stages V and VI corresponded to severe AD. However, the distribution of amyloid plaques was found to vary widely between individual patients and there was no correlation with clinical symptom severity.

As described previously, the 1984 NINCDS-ADRDA diagnostic guidelines were revised and updated in 2011 by a working group of experts convened by the National Institutes of Health (NIH) and the Alzheimer's Association (2011) (McKhann et al., 2011; Sperling et al., 2011; Jack Jr et al., 2011; Albert et al., 2011). Based on the significant discoveries made in the almost two decades since the original criteria were developed, the revised guidelines classified the clinical diagnosis of AD into three phases: a preclinical (asymptomatic) predementia AD phase, a predementia but symptomatic (mild decline in memory) phase; also referred to as mild cognitive impairment (MCI); and a symptomatic AD dementia phase. In the preclinical predementia phase, amyloid plaques may be developing but no neurodegeneration has occurred. The criteria used to identify preclinical AD are designated only for research purposes and are based on five AD biomarkers (brain atrophy measured by magnetic resonance imaging [MRI], high retention of amyloid tracers by positron emission tomography [PET], low fluorodeoxyglucose uptake by PET, low cerebrospinal fluid [CSF] $A\beta_{1-42}$ levels, and high CSF tau levels [total and phosphorylated]). In 2014, the International Working Group 2 recognized the use of AD biomarkers as integral part of the diagnosis of AD, signaling the first movement of AD biomarkers from the research setting to the clinical setting (Dubois et al., 2014).

1.3 SOME FACTS ABOUT ALZHEIMER'S DISEASE

According to the World Health Organization report on dementia (2012), there is a new case of dementia detected every 4 s somewhere in the world. Worldwide, a new diagnosis of AD occurs every 7 s (Cornutiu, 2015). AD was once considered to be a rare disease, but is now the most common dementia among the elderly population (Box 1.1). AD is a devastating disease: AD is the most common form of dementia in the elderly, representing 60–80% of all dementias in this population (Alzheimer's disease facts and figures 2012). AD affects approximately 3% of the total population aged 65–74 years, 10% aged 75–84 years, and 33% aged >85 years (Box 1.1). In the United States alone, more than 5.5 million people suffer from this irreversible neurodegenerative disorder. According to the World Alzheimer's report, approximately 40 million people worldwide are living with dementia, with an estimated cost of $604 billion in 2010. The mean life expectancy after a clinical diagnosis of AD is approximately 7 years, with only 3% of individuals living longer than 14 years after diagnosis (Alzheimer's Association fact sheet). A very recent estimate found that AD is the third leading cause

---

BOX 1.1 IMPACT OF ALZHEIMER'S DISEASE (AD) IN SOCIETY

1. AD accounts for >60% of all forms of dementia (source: 2014 Alzheimer's Disease Facts and figures).
2. The burden of dementia-related illness is growing. In 2012, the estimated number of people living with dementia was 35.6 million, and this number is expected to double by 2030. By 2050, it is estimated that 135 million people worldwide will have some form of dementia (source: Dementia: A Public Health Priority, The World Health Organization and Alzheimer's Disease International; 2012. This report was made to raise awareness of dementia and advocate for action on an international level.)
3. Globally, there were 35.6 million individuals living with AD in 2011. This figure is estimated to reach to 65.7 million by 2030 and 115.4 million by 2050. (http://www.alz.co.uk/research/files/WorldAlzheimerReportExecutiveSummary.pdf).
4. Between 2000 and 2010, there was a 68% increase in deaths due to AD in the United States (source: Alzheimer's Association 2014 and National Center for Health Statistics).
5. AD is the sixth leading cause of death in United States. By 2040, AD and other neurodegenerative diseases are expected to overtake cancer to become the second leading cause of death in developed countries, after cardiovascular disease (source: The World Health Organization).
6. The proportion of people over the age of 65 years will increase from 6.8% in 2000 to 16.2% by 2050 (source: The World Health Organization).
7. In the United States, members of the "baby boomer" generation are approaching the age of 65, and the number of people over 65 is expected to double by 2030.
8. In the United States, AD is the third leading cause of death, after heart disease and cancer, among people older than 65 years (James et al., 2014).
9. ~200,000 individuals with familial forms of AD in United States (Alzheimer's association 2011).
10. Each 7 s one AD case is being added in the world (Cornutiu, 2015). In United States, one new AD case is being added in every 68 s and this number will be 33 s in 2050 (Alzheimer's disease facts and figures 2013).

---

of death after heart disease and cancer among the elderly in the United States (James et al., 2014). The study was conducted using data from 2566 persons aged 65 years and older (mean 78.1 years) without dementia at baseline. The growing prevalence of AD presents a major challenge for the global health care system as it cares for an increasingly aged population (Dementia: A Public Health Priority, The World Health Organization and Alzheimer's Disease International; 2012). According to the National Center for Health Statistics, deaths due to AD have increased 68% between 2000 and 2010 (The National Vital Statistics Report. "Deaths: final data for 2010."National Center for Health Statistics. Available at: http://www.cdc.gov/nchs/data/dvs/deaths_2010_release.pdf. Accessed Jan. 30, 2013; National Center for Health Statistics. Deaths: final data for 2000. National Vital Statistics Reports. National Center for Health Statistics, Hyattsville, MD;

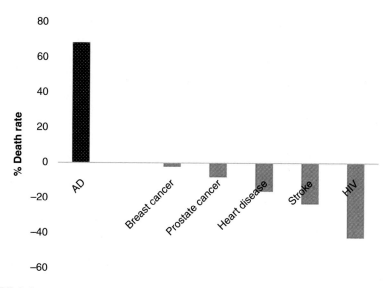

FIGURE 1.1

Percent change in death rate in a 10-year period (2000–10) for different diseases. *(This figure was created using data from the National Vital Statistics Report. "Deaths: final data for 2010." National Center for Health Statistics. Available at: http://www.cdc.gov/nchs/data/dvs/deaths_2010_release.pdf. Accessed Jan. 30, 2013; National Center for Health Statistics. Deaths: final data for 2000. National Vital Statistics Reports. National Center for Health Statistics, Hyattsville, MD; 2002; 2013 Alzheimer's disease facts and figures. Alzheimer's Association, 2013, Alzheimers & Dementia 9 (2), 208–245).*

2002; 2013 Alzheimer's disease facts and figures. Alzheimer's Association, 2013, Alzheimers & Dementia 9 (2), 208–245). During this same time period, the death rates due to human immunodeficiency virus (HIV) infection (−42%), stroke-related disease (−23%), heart disease (−16%), prostate cancer (−8%), and breast cancer (−2%) have decreased (Fig. 1.1). Taken together, these statistics reinforce the urgency for discovering biomarkers that can detect AD, preferably in its earliest stages, and the development of disease-modifying therapeutic strategies.

1.4 COMMON FEATURES OF ALZHEIMER'S DISEASE

As a neurodegenerative disease, the clinical symptom of AD is a loss of recent memory. The most common characteristic pathological features are synaptic loss, neuro-inflammation, neuronal cell death, cortical atrophy, and the presence of neurofibrillary tangles and senile plaques in the brain at autopsy. No standard method exists for diagnosing AD and no disease-modifying therapy has been developed, making this disease a major public health problem worldwide. Significant correlation between synaptic loss and cognitive decline suggests that synaptic dysfunction is the primary event in AD pathology (Terry et al., 1991; Masliah et al., 1991). Neuronal death and the formation of neurofibrillary tangles and senile plaques occur at later stages of the disease. AD is not caused by single factor, but instead develops over many years with multiple interacting causes. The most challenging aspect of AD diagnostic research is the pathological and clinical complexity and heterogeneity of the disease. AD often presents with other old-age neurodegenerative diseases such as vascular dementia (VaD), Lewy body disease (LBD), Parkinson's disease (PD), frontotemporal dementia (FTD), amyotrophic lateral sclerosis (ALS), and taupathy, making a clear diagnosis of AD difficult in many cases.

1.5   CLASSIFICATION OF ALZHEIMER'S DISEASE

The dominant hypothesis describing the pathogenesis of AD is the amyloid cascade hypothesis (Hardy and Selkoe, 2002). This theory is based on the observation that toxic oligomeric forms of Aβ are the main constituents of amyloid plaques seen in the AD brain at autopsy and patients with FAD produce excessive amounts of toxic Aβ peptides upon cleavage of APP. As described previously, two distinct kinds of AD dementia exist: (1) FAD or early-onset AD (EOAD), which has a well-defined genetic cause with autosomal dominant family history of AD, typically develops before the age 65, and is very rare (<1–5% of cases) and (2) late-onset AD (LOAD), which is sporadic and heterogeneous in nature, occurs later in life (onset age $\geq$65 years), and is the most common form (>95% of cases). EOAD is caused by mutation in familial AD genes such as APP, PSEN1, and PSEN2. Mutations in those genes cause abnormal toxic Aβ processing by β- and γ-secretase cleavage of APP. According to the amyloid cascade hypothesis, excess toxic Aβ causes synaptic dysfunction and neurodegeneration, and finally cognitive decline (Hardy and Selkoe, 2002). Aβ is the major protein component of the amyloid plaques in AD brains. The major Aβ peptide in AD brains is $A\beta_{1-40}$, followed by $A\beta_{1-42}$. $A\beta_{1-42}$ is hydrophobic, more toxic and fibrillogenic, and has a stronger mechanistic association with disease pathogenesis. Of note, the natural history of individuals with Down syndrome strongly supports the amyloid cascade hypothesis of AD pathogenesis. Down syndrome patients who have an extra copy of the APP gene also develop AD and produce excess Aβ.

The development of LOAD can also be described by the amyloid cascade hypothesis. LOAD is caused by a defect in the clearance of toxic Aβ from brain, again leading to toxic Aβ accumulation that causes synaptic loss and neurodegeneration. A growing body of literature suggests that soluble toxic Aβ oligomers impair long-term potentiation (Calabrese et al., 2007; Selkoe, 2008; Kittelberger et al., 2012), and induce tau hyperphosphorylation (Oddo et al., 2004; Jin et al., 2011). Therefore, deficiencies in the genes, genetic networks, and signaling pathways that regulate Aβ clearance provide a set of potential diagnostic markers and targets for therapeutic interventions for LOAD. Hypoxia is very common to aged brain and it is one of the main causes for sporadic AD. Hypoxia may also increase Aβ oligomers (Zetterberg et al., 2011). However, there are reports that at low physiological concentrations of Aβs are neuroprotective (Chan et al., 1999). Aβ also has antimicrobial properties against harmful organisms (Soscia et al., 2010).

1.6 RISK FACTORS FOR ALZHEIMER'S DISEASE

Age is the strongest risk factor for LOAD. Epidemiological studies have found that <1% of individuals aged 60–65 years have AD, and the prevalence increases exponentially to 24–33% of individuals aged 85 years (Alzheimer's Disease (AD)–World Health Organization). Another estimate found that among patients affected with AD, 4% are under age 65, 6% are 65–74, 44% are 75–84, and 46% are 85 or older (Alzheimer's disease facts and figures 2012). The incidence of AD increases exponentially with age, and females are disproportionally (2:1) more susceptible to AD compared to male.

Epidemiological studies have found that age, genetics, and environmental risk factors contribute to LOAD risk. Although researchers are still investigating the exact nature of several genes that have varying effects on conferring susceptibility to LOAD (ie, susceptibility genes), only apolipoprotein E-4 (APOE4) gene has been proven to be risk factor gene for LOAD. Identification of other genes will aid researchers and clinicians in identifying patients at risk of LOAD who would benefit from treatment to either prevent or slow the onset of the disease. Profiling genetic factors that affect the pathogenesis of LOAD may also identify a set of genetic diagnostic and prognostic biomarkers for LOAD. An intricate network of genes rather than specific individual genes might be responsible for the development of a complex and heterogeneous disease such as LOAD.

Genetics is the strongest risk factor for early-onset FAD. Three genes have been linked to early-onset FAD: APP on chromosome 21, PSEN1 on chromosome 14, and PSEN2 on chromosome 1. Individuals with mutations in these genes have higher levels of Aβ in the brain and peripheral tissues than those who do not carry the mutations. By contrast, sporadic AD (LOAD) is a multifactorial, heterogeneous neurodegenerative disorder resulting from the combined effects of genetics, age, and other nongenetic risk factors such as diet, life style, and incident of brain injury. The genetic component of LOAD is evident: the incidence of LOAD within a particular family with a history of AD is always higher than that within families with no history of the disease, with the risk of AD approximately twofold higher for first-degree relatives of family members with AD. Yet only a few low-penetrance genes have been linked to sporadic AD, including APOE4 (Green et al., 2009) and sortilin-related receptor (SORL1) (Rogaeva et al., 2007). More than 1000 articles have been published on candidate AD genetic susceptibility factors. Only the presence of the APOE4 allele has an established link to increased risk of LOAD; heterozygotes have an approximately 3-fold higher risk of sporadic AD and homozygotes have a 15-fold higher risk. However, the presence of the APOE4 allele alone is unable to predict AD, indicating that other factors are involved. One study found that the likelihood of developing AD is higher for monozygotic twins than for dizygotic twins if one twin already has AD (Gatz et al., 1997). Based on these observations, there is clearly a genetic component of LOAD. The genetics of AD and possible genetic biomarkers for AD are discussed in detail in Chapter 4.

Nongenetic risk factors for LOAD include brain injury, vascular disease, hypertension, high cholesterol, atherosclerosis, coronary heart disease, obesity, and diabetes, as well as an inactive lifestyle and poor diet. A growing body of literature and some clinical trials suggest that greater social connectivity/activity, cognitive stimulus activity, education, physical activity, and healthy diet can reduce the risk of AD. Traumatic brain injury (TBI) and posttraumatic stress disorder (PTSD) are two important risk factors for dementia, including AD, later in life. The military population has an increased risk of AD due to TBI and PTSD. A population-based prospective historical cohort study including World War II Navy and Marine veterans with head injury ($n = 548$) and without head injury ($n = 1228$) found that military veterans with moderate to severe TBI have two to four times higher risk of developing AD with advancing age (Plassman et al., 2000). Professional athletes with TBI also have a higher risk of developing AD, with one study reporting that PTSD increases the risk of dementia twofold (Yaffe et al., 2010). National Football League (USA) players have a threefold higher mortality rate from neurodegenerative diseases, including AD and ALS, than that of general population (Lehman et al., 2012).

1.7 DEFINITION OF A BIOMARKER

The NIH established the Biomarkers Consortium to accelerate discovery of new health-related technologies, medicines, and therapies for the early detection, prevention, diagnosis, prognosis, and treatment of diseases. A definition of a biomarker was first introduced by Hulka (1990) as "cellular, biochemical, or molecular alterations that are measurable in biological media such as human tissues, cells, or fluids." There are several ways to define biomarkers based on their applications, methods of development, and usefulness in disease diagnosis and prognosis. Updated definitions of a biomarker by different organizations and authors are presented in Table 1.2. NIH has suggested a definition for a biomarker as a characteristic that can be objectively measured and quantitatively evaluated as an indicator of a biologic process, pathogenic process, or pharmacologic response to a therapy (Biomarkers Definitions Working Group, 2001). According to the National Cancer Institute, a biomarker is defined as a quantitative measure of a signature biomolecule obtained from body fluids or tissues that shows abnormality in the disease condition and can be measured to examine treatment response (National Cancer Institute: Fact Sheet).

In general, a biomarker is a measurable parameter that reflects the physiological state and how it is affected by a disease condition or other pathological abnormality. It is a measure of the biochemical features of a disease that change with disease progression and with therapeutic intervention. Biomarkers can be various measurable chemical and biological entities, including proteins, lipids, RNA, and microRNA, as well as cellular characteristics, such as morphology or histology. The simplest definition of a biomarker is a characteristic (physical, chemical, biological, physicochemical, biochemical, or biophysical) of a disease state that can be measured and that changes in response to an intervention. An ideal biomarker reflects fundamental features of a specific disease, can differentiate

Table 1.2	Definition of Biomarker	
Organization/ authors	Definition	References
Hulka, 1990	Measure of cellular and biochemical alteration in human fluids, cells, or tissues.	Hulka, B.S., 1990. Overview of biological markers. In: Biological Markers in Epidemiology Hulka, B.S., Griffith, J.D., Wilcosky, T.C. (Eds.), Oxford University Press, New York, pp. 3–15.
National Institute of Health (2001)	Biomarker is an indicator of certain objective measure and evaluation of biological process, pathogenic process, or pharmacological evaluation of therapeutic efficacy.	Biomarkers Definitions Working Group, 2001. Clin. Pharmacol. Ther. 69, 89–95.
National Cancer Institute	Biomarker is quantitative measure of a signature biomolecule obtained from body fluids or tissues that shows abnormality in disease condition and a measure that can be used to examine the treatment response.	National Cancer Institute: Fact Sheet

that disease from other closely related diseases, and is measurable in early stages of the disease and throughout disease progression. The technological methods used to measure the biomarker should be highly reliable, easy to perform, and inexpensive, and sample sources should be easily collected (preferably from peripheral sources using noninvasive methods). Well-established biomarkers should be validated in trials that include a large number of patients, and should have high sensitivity, specificity, and diagnostic accuracy.

Clinicians, scientists, and health care professionals use biomarkers or biological markers as a measure of a person's present health condition or the response to intervention. The most clinically useful AD biomarkers should be able to diagnose disease at very early stages, clarify the risk of developing AD when combined with known risk factors, assess the efficacy of therapeutic agents, track disease progression through prodromal stages, distinguish AD from other dementias, and guide therapeutic decision making.

AD is an irreversible, progressive disease with long prodromal stages. The most recent classification system put forth by the National Institute on Aging and the Alzheimer's Association (NIA-AA) assigned working group (2011) describes three stages of AD: asymptomatic preclinical AD, dementia due to MCI, and symptomatic AD dementia (McKhann et al., 2011; Sperling et al., 2011; Jack Jr et al., 2011; Albert et al., 2011). The same working group also set up a revised set of diagnostic criteria that includes specific biomarkers, selected based on the current understanding of pathophysiological processes in AD.

There are three types of basic AD biomarkers according to Biomarker Definition Working Group. Type 0 are those biomarkers that measure the natural history of disease and correlate with progression of the disease over time. Neuroimaging biomarkers of AD such as atrophy of brain are Type 0 AD biomarkers. Type I are those biomarkers that can be used to evaluate the effect of therapeutic interventions. Cerebrospinal fluid biomarkers are considered Type 1 AD biomarkers. Type II biomarkers are surrogate biomarkers. There are no biomarkers for AD that are considered Type II.

1.7.1   A. Categorization of Biomarkers According to Their Role

Biomarkers can be classified by how they are used, for example, as a diagnostic biomarker, a prognostic biomarker, a pharmacological biomarker, and as a surrogate biomarker. In this way, the entire spectrum of valid biomarkers can be categorized into four groups: (1) detective biomarkers, (2) diagnostic biomarkers, (3) prognostic biomarkers, and (4) predictive biomarkers. Detective biomarkers detect certain biomolecule(s) specifically related to a particular disease, for example, identification of prostate specific antigen as a biomolecule that is associated with prostate cancer. Diagnostic biomarkers identify individuals who have a particular disease, for example, the presence of rheumatoid factors in the blood is a diagnostic marker for rheumatoid arthritis and can distinguish it from other types of arthritis. Prognostic biomarkers describe the likely progression of disease with or without therapeutic intervention. Predictive biomarkers focus on the risk factors for certain diseases and help identify those individuals who are in early stages of the disease and might benefit from therapeutic intervention. Bauer et al. (2006) also categorized biomarkers into several groups depending on their application, including disease burden biomarkers, diagnostic biomarkers, prognostic biomarkers, investigative biomarkers, and therapeutic efficacy biomarkers.

1.7.2   B. Definition of Common Statistical Parameters to Quantify the Performance of Biomarkers

The performance of any given biomarker is usually judged by parameters such as sensitivity, specificity, accuracy, positive predictive value, and negative predictive value [Box 1.2]. Sensitivity and specificity are used to determine the accuracy of the diagnostic biomarker against a gold standard of diagnosis; however, because a definitive diagnosis of AD occurs after development of dementia in late-stage disease and at autopsy with the presence of amyloid plaques and neurofibrillary tangles, these terms are not as useful as predictive values for clinicians who are trying to identify patients at risk of the disease or patients who may benefit from

---

BOX 1.2 STATISTICAL PARAMETERS TO QUANTIFY PERFORMANCE OF AD BIOMARKERS

Sensitivity = $a/(a + b)$ = probability that an abnormal biomarker test result will be seen in a patient with AD

Specificity = $d/(c + d)$ = probability that a normal biomarker test result will be seen in a patient without AD

Accuracy = $(a + d)/(a + b + c + d)$

Positive predictive value = $a/(a + c)$ = probability that a patient has AD if they have an abnormal biomarker test result

Negative predictive value = $d/(b + d)$ = probability that a patient does not have AD if they have a normal biomarker rest result

Where a = true positive, b = false negative, c = false positive, and d = true negative.

Positive likelihood ratio (LR +) = sensitivity/(1− specificity).

Negative likelihood ratio (LR−) = (1− sensitivity)/specificity.

Receiver operating characteristic (ROC) has been used to calculate the accuracy of a particular AD biomarker. When sensitivity vs (1 − specificity) is plotted as an ROC curve, the area under the curve is the accuracy of a specific diagnostic test.

Odds ratio (OR) = [sensitivity/(1− specificity)]/[(1−sensitivity)/specificity].

---

treatment. Furthermore, no single AD biomarker is useful as standalone test for diagnosis of the disease, and AD biomarkers can only be used as supportive measures in addition to clinical judgement. Therefore, positive and negative predictive values are more relevant to clinical decision making than sensitivity and specificity.

1.8 BIOMARKERS OF ALZHEIMER'S DISEASE

GE Healthcare sponsored a 10-country survey of 10,000 adults and found that three quarters of respondents would like to know whether they have a particular neurological disorder, even in the absence of a cure (S. Lawrence, Fierce Medical Device Aug. 19, 2014; http://www.fiercemedicaldevices.com). This survey also reported that 81% of the respondents would like to know whether their partner has a neurological disease. In addition, a majority of survey respondents indicated that diagnosis should be funded either by government or private health insurance companies. More than half of the people surveyed said that they would be willing to pay for diagnostic testing themselves, including in the most populous countries like China (83%) and India (71%). The size of the survey and results reinforce the urgent need for early diagnosis of AD. Currently available diagnostic tests have moved the field closer to early diagnosis of AD; however, a definitive diagnosis is made only with the development of clinical dementia later in the disease progression and the presence of amyloid plaques and neurofibrillary tangles at autopsy. At the present time, there are no biomarkers of early-stage AD that are ready for testing in clinical trials or application in primary care.

According to the 1998 Consensus Report of the Ronald and Nancy Reagan Research Institute of the Alzheimer's Association and the National Institute on Aging Working Group (1998) on "Molecular and Biochemical Markers of Alzheimer's Disease," the ideal biomarker for AD should have a sensitivity >80%, a specificity >80%, and ≥90% positive predictive value for detecting AD. Other criteria for the ideal AD biomarkers prescribed by this working group are: validation against "gold standard" AD pathology at autopsy, validation studies should be published in peer-reviewed journals and include specific non-demented controls (age-matched) and other non-AD dementia patients, and biomarkers should be validated against additional data. Furthermore, the ideal biomarker for AD should have a sensitivity >85% for detecting AD, and a specificity >75% for differentiating AD from non-AD dementias, according to the National Institute on Aging consensus criteria. At present, there are very few biomarkers that can diagnose AD with the specificity and sensitivity prescribed by the consensus criteria.

Because sporadic LOAD is often not diagnosed until later stages when cognitive deficits become clinically significant, in the past two decades, researchers have focused on the identification of biomarkers that can provide an earlier diagnosis of AD or assess the risk of developing AD. There are several biomarkers currently being investigated for the diagnosis of AD, including markers in the CSF, PET, and MRI neuroimaging markers, and markers detected in peripheral tissues such as blood and skin.

An ideal antemortem AD biomarker should meet the following criteria: (1) ability to detect fundamental features of AD neuropathology that can be validated at autopsy; (2) ability to differentiate AD from non-AD dementias; (3) ability to detect early stages of AD and differentiate between the stages of AD progression to guide therapy; (4) highly reliable, easy to perform, and inexpensive; and (5) use minimally invasive sample collection, such as from peripheral tissues, without requirement for lumbar puncture or other invasive sampling procedures.

Chapter 2 discusses the clinical diagnosis of AD using various neuropsychological tests, and discusses their accuracy in the clinical diagnosis of AD, detecting prodromal stages of AD (preclinical and MCI), and distinguishing between AD and non-AD dementias (VaD, LBD, FTD, CJD, etc). Since AD is a disorder of the central nervous system of the elderly population, noninvasive neuroimaging studies provide a powerful approach to AD diagnosis. The use of neuroimaging AD biomarkers is discussed in Chapter 3. Neuroimaging AD biomarkers include in vivo imaging of amyloid plaques, hyperphosphorylated tau, glucose hypometabolism, cerebral blood flow using modalities such as MRI and PET, and quantitative measurement of regional brain metabolites by magnetic resonance spectroscopy (MRS). Other neuroimaging modalities used to detect AD biomarkers in the brain imaging include single-photon emission computed tomography

(SPECT) and electroencephalogram (EEG). The most widely researched AD biomarkers in the CSF are levels of $A\beta_{1-42}$, total tau (t-tau), and phosphorylated tau-181 (p-tau). Chapter 4 describes genetic biomarkers of AD. Chapter 5 includes a critical evaluation of the performance of CSF AD biomarkers in autopsy-validated cohorts, and a discussion of their clinical utility for early and preclinical diagnosis of AD, longitudinal assessment, differential diagnosis of non-AD dementias, crosscorrelation with other AD biomarkers, and evaluation of drug efficacy clinical trials. CSF can only be collected by lumbar puncture, and given the invasiveness of the procedure, CSF biomarkers may not lend themselves to screening for AD or longitudinal assessment of AD in elderly patients. Neuroimaging biomarkers are expensive to perform and require specialized training and facilities. These limitations have led researchers to investigate peripheral AD biomarkers that require minimally invasive, simple, and inexpensive collection of blood, saliva, and urine samples. Chapter 6 summarizes recent findings of cross-sectional and longitudinal studies of blood plasma AD biomarkers, crosscorrelation with other AD biomarkers, and possible application of plasma AD biomarkers in clinical trials. Cell-based AD biomarkers are the focus of the final chapter (Chapter 7). This chapter also discusses the capacity for cell-based biomarkers to differentiate AD from non-AD dementias, and for early detection of the disease. Cell-based AD biomarker assays are also simple, inexpensive, noninvasive or minimally invasive in nature, and have potential to be useful for community-based screening and drug efficacy testing. More validation and large, multicenter cohort studies are needed before peripheral AD biomarkers are translated for clinical use.

1.9 FUTURE FOCUS: DETECTING PRECLINICAL STAGES OF ALZHEIMER'S DISEASE

There is a significant correlation between the decline in memory function and synaptic loss but no clear correlation between memory decline and the presence of amyloid plaques and neurofibrillary tangles at autopsy; this suggests that synaptic dysfunction is the primary indicator of AD pathology (Terry et al., 1991; Masliah et al., 1991). The newly revised criteria for AD emphasize that the AD pathophysiological process starts years or decades before clinical symptoms appear (Sperling et al., 2011; Alzheimer's Association and National Institute of Aging Working Group). Decades of fundamental research have generated substantial evidence that a silent pathophysiological process of AD starts before the clinical manifestations of memory and cognitive deficits appear. This idea is supported by evidence from studies of genetic risk factors of AD as well as by aging studies. A meaningful and actionable diagnosis of AD should be made during early stages of AD, when AD pathological changes start but clinical symptoms have not yet developed. Preclinical diagnosis of AD would allow therapeutic intervention to be initiated before neurons and connectomes are lost. The late-stage treatment timing, after the loss of synapses has already occurred, is one of the reasons why almost all therapeutic clinical trials in patients with severe AD failed. Early-stage biomarkers of AD will allow the identification of people still in the preclinical stage of AD and who will benefit from therapeutic intervention, prior to synaptic loss. Therapeutic intervention at later stages of AD (MCI or advanced AD) may be less effective, once synaptic loss has begun.

1.10   ECONOMIC IMPACT OF ALZHEIMER'S DISEASE BIOMARKERS

Economic impact of AD on society: A quantification of the actual cost and economic impact of AD in present day society is difficult. In Apr. 2012, the World Health Organization (WHO) identified dementia as a "public health priority." The medical cost of AD is expected to grow exponentially as the aging population increases and the disease progresses slowly in individuals. The cost of care for a patient with AD is directly related to the severity of the illness. The economic impact of AD is devastating to families, societies, and governments. The direct and indirect costs of AD have had an enormous impact on the US health care system, families, and state and federal budgets (Box 1.3). The direct costs of AD include medical treatment, hospital costs, and costs of caregivers. Indirect costs are loss of family income due to loss of working hours of the earning

---

BOX 1.3 ECONOMIC IMPACT OF ALZHEIMER'S DISEASE

1. The estimated cost of long-term care for AD and other related dementias was $214 billion in 2014, and is expected to reach $1.2 trillion in 2050 (source: Alzheimer's Association 2014; Alzheimer's disease facts and figures).
2. A clinical intervention that could delay AD by 1 year may reduce the number of AD cases by 9 million by 2050 (Brookmeyer et al., 2007). A therapy that delays AD progression by 5 years would save ~$3.97 trillion over 30 years (Alzheimer's Association Facts and Figures 2011). A treatment that delays AD onset by 5 years would save $935 billion over 10 years (in billions of 2015 dollars) (source: Alzheimer's Association 2015).
3. In the Asia-Pacific region, and estimated ~70 million people are expected to suffer from dementia by 2050, with an estimated cost of care of ~$185 billon by 2015 (source: Dementia in the Asia-Pacific Region by Alzheimer's Disease International Global Voice on Dementia 2014).
4. AD is the third most expensive disorder in the United States, after cancer and coronary heart disease (Meek et al., 1998).
5. Average per person Medicaid/Medicare costs for AD and other dementias are three times higher than those without dementia in the same age group (source: Testimony of Harry Johns, President and CEO of the Alzheimer's Association, 2014).
6. AD the leading cause of years lost due to disability in high-income countries and second highest cause of years lost due to disability worldwide (World Health Organization).
7. The Medicare cost of AD is projected to total $20 trillion (2010–50) if no effective therapeutic method is discovered (Alzheimer's Association Facts and Figures 2011).
8. The AD drug market was estimated to be worth ~$5.8 billion in 2011 and is expected to grow to $14.5 billion by 2020 (Datamonitor, Dec. 2011).

members of the family, increased health insurance premiums, the emotional cost, and the effect on professional careers of family members. The drug market in AD is estimated to be worth ~5.8 billion in 2011 and is expected to grow to 14.5 billion by 2020 (Datamonitor, Dec. 2011). Effective therapeutic intervention would reduce the financial burden of AD significantly. For example, one estimate showed that a clinical intervention that could delay AD by 1 year may reduce the number of AD cases by 9 million by 2050 (Brookmeyer et al., 2007). This epidemiological study employed a stochastic multistate model using data from the United Nations worldwide population forecasts and the risks of AD. A recent estimate by the Alzheimer's Association found that any treatment that delays AD onset by 5 years will save $935 billion over 10 years (in 2015 dollars) (source: Alzheimer's Association 2015).

The potential economic impact of Alzheimer's disease biomarkers: There are several prominent drivers of the steady increase in the global biomarker market, including AD biomarkers (Table 1.3). Biomarkers facilitate patient stratification

Table 1.3	Market Survey of Alzheimer's Disease Biomarkers Compared With All Other Biomarkers in an Expanding Global Market		
All other biomarkers in expanding global market			
Market survey	References	Current status (bn) (During the study)	Prediction (Year) (bn)
Piribo	http://www.piribo.com/publications/ biotechnology/biomarkers.htm	$5.6 (2007)	$12.8 (2012)
Visionain	http://www.visiongain.com	$14.18 (2010)	$19.86 (2015)
Aarkstore	http://www.aarkstore.com/ reports/Biomarkers-Advanced- Technologies-and-Global- Market-2009-2014--27515.html	$10.07 (2009)	$ 20.5 (2014)
Alzheimer's disease biomarker in expanding global market			
Study commissioned by Proteome Science	Alzheimer's biomarker market to reach $9bn in decade By Howard Lovy Jun. 8, 2011, FierceDiagnostics	—	$9.0 (2021)
Blood-based Alzheimer's disease biomarkers in United States	Market survey by Destum 2011	—	~275 million (2025)

bn, Billion.

and treatment selection, are useful in drug discovery, and are used by regulatory agencies to test the claims of pharmaceutical industries. AD biomarkers may have a huge impact on disease diagnosis, treatment decision making, and therapeutic efficacy testing. Table 1.3 shows the enormous future growth of the worldwide biomarker market. The worldwide in vitro diagnostic market accounts for only 1–2% of government healthcare expense; however, it influences 60–70% of healthcare decisions. An enormous amount of opportunity exists in the AD diagnostic market, despite the fact that there are few disease modifying treatment options. FierceDiagnostics forecasted in 2011 that the AD biomarker market is expected to reach $9 billion in the next decade (by 2021) (FierceDiagnostics: Howard Lovy Jun. 8, 2011). Blood-based AD biomarkers are most desirable for disease risk management, diagnosis, prognosis, and guiding therapeutic trials. In the United States, the estimated market for blood-based AD biomarkers is expected to reach ~$275 million by 2025 (Market survey by Destum 2011).

1.11 CONCLUSIONS

Future AD biomarker research and development will focus on (1) improving diagnostic specificity, sensitivity, and accuracy; (2) improving the ability to differentiate AD from non-AD dementias and MCI; (3) monitoring prodromal stages of AD; (4) developing tests to diagnose early-stage disease, (5) preclinical detection of AD, (6) developing surrogate markers, (7) use in AD risk management, and most importantly (8) their role in guiding therapeutic trials. AD is a multifactorial disease, therefore, combinations of AD biomarkers into a molecular signature or index may prove to be more accurate than any single biomarker. Less expensive and easily accessible blood-based AD biomarkers are urgently needed, particularly for early diagnosis. Loss of dendritic spines and synapses occurs before widespread neuronal loss and the clinical manifestation of AD. Thus, the ability to detect preclinical disease is the single most important goal in AD biomarker research, as it will allow for the treatment of AD before it progresses to widespread and irreversible neuronal loss.

Bibliography

Albert, M.S., DeKosky, S.T., Dickson, D., Dubois, B., Feldman, H.H., Fox, N.C., Gamst, A., Holtzman, D.M., Jagust, W.J., Petersen, R.C., Snyder, P.J., Carrillo, M.C., Thies, B., Phelps, C.H., 2011. The diagnosis of mild cognitive impairment due to Alzheimer's disease: recommendations from the National Institute on Aging-Alzheimer's Association workgroups on diagnostic guidelines for Alzheimer's disease. Alzheimers Dement. 7, 270–279.

Bauer, D.C., Hunter, D.J., Abramson, S.B., Attur, M., Corr, M., Felson, D., Heinegård, D., Jordan, J.M., Kepler, T.B., Lane, N.E., Saxne, T., Tyree, B., Kraus, V.B., Osteoarthritis Biomarkers Network, 2006. Classification of osteoarthritis biomarkers: a proposed approach. Osteoarthritis Cartilage 14, 723–727.

Biomarkers Definitions Working Group, 2001. Biomarkers and surrogate endpoints: preferred definitions and conceptual framework. Clin. Pharmacol. Ther. 69, 89–95.

Blessed, G., Tomlinson, B.E., Roth, M., 1968. The association between quantitative measures of dementia and of senile change in the cerebral grey matter of elderly subjects. Br. J. Psychiatry 114, 797–811.

Braak, H., Braak, E., 1991. Neuropathological staging of Alzheimer-related changes. Acta Neuropathol. 82, 239–259.

Brookmeyer, R., Johnson, E., Ziegler-Graham, K., Arrighi, H.M., 2007. Forecasting the global burden of Alzheimer's disease. Alzheimers Dement. 3, 186–191.

Calabrese, B., Shaked, G.M., Tabarean, I.V., Braga, J., Koo, E.H., Halpain, S., 2007. Rapid, concurrent alterations in pre- and postsynaptic structure induced by naturally-secreted amyloid-beta protein. Mol. Cell Neurosci. 35, 183–193.

Chan, C.-W., Dharmarajan, A., Atwood, C.S., Huang, X., Tanzi, R.E., Bush, A.I., Martins, R.N., 1999. Antiapoptotic action of Alzheimer Abeta. Alzheimers Rep. 2, 113–119.

Cornutiu, G., 2015. The epidemiological scale of Alzheimer's disease. J. Clin. Med. Res. 7, 657–666.

Dubois, B., Feldman, H.H., Jacova, C., Dekosky, S.T., Barberger-Gateau, P., Cummings, J., Delacourte, A., Galasko, D., Gauthier, S., Jicha, G., Meguro, K., O'brien, J., Pasquier, F., Robert, P., Rossor, M., Salloway, S., Stern, Y., Visser, P.J., Scheltens, P., 2007. Research criteria for the diagnosis of Alzheimer's disease: revising the NINCDS-ADRDA criteria. Lancet Neurol. 6, 734–746.

Dubois, B., Feldman, H.H., Jacova, C., Hampel, H., Molinuevo, J.L., Blennow, K., DeKosky, S.T., Gauthier, S., Selkoe, D., Bateman, R., Cappa, S., Crutch, S., Engelborghs, S., Frisoni, G.B., Fox, N.C., Galasko, D., Habert, M.O., Jicha, G.A., Nordberg, A., Pasquier, F., Rabinovici, G., Robert, P., Rowe, C., Salloway, S., Sarazin, M., Epelbaum, S., de Souza, L.C., Vellas, B., Visser, P.J., Schneider, L., Stern, Y., Scheltens, P., Cummings, J.L., 2014. Advancing research diagnostic criteria for Alzheimer's disease: the IWG-2 criteria. Lancet Neurol. 13, 614–629.

Gatz, M., Pedersen, N.L., Berg, S., Johansson, B., Johansson, K., Mortimer, J.A., Posner, S.F., Viitanen, M., Winblad, B., Ahlbom A, 1997. Heritability for Alzheimer's disease: the study of dementia in Swedish twins. J. Gerontol. A Biol. Sci. Med. Sci. 52, M117–M125.

Glenner, G.G., Wong, C.W., 1984. Alzheimer's disease: initial report of the purification and characterization of a novel cerebrovascular amyloid protein. Biochem. Biophys. Res. Commun. 120, 885–890.

Green, R.C., Roberts, J.S., Cupples, L.A., Relkin, N.R., Whitehouse, P.J., Brown, T., Eckert, S.L., Butson, M., Sadovnick, A.D., Quaid, K.A., Chen, C., Cook-Deegan, R., Farrer, L.A., REVEAL Study Group, 2009. Disclosure of APOE genotype for risk of Alzheimer's disease. N. Engl. J. Med. 361, 245–254.

Grundke-Iqbal, I., Iqbal, K., Quinlan, M., Tung, Y.C., Zaidi, M.S., Wisniewski, H.M., 1986. Microtubule-associated protein tau. A component of Alzheimer paired helical filaments. J. Biol. Chem. 261, 6084–6089.

Hardy, J., Selkoe, D.J., 2002. The amyloid hypothesis of Alzheimer's disease: progress and problems on the road to therapeutics. Science 297, 353–356.

Hippius, H., Neundörfer, G., 2003. The discovery of Alzheimer's disease. Dialogues Clin. Neurosci. 5, 101–108.

Hulka, B.S., 1990. Overview of biological markers. In: Hulka, B.S., Griffith, J.D., Wilcosky, T.C. (Eds.), Biological Markers in Epidemiology. Oxford University Press, New York, pp. 3–15.

Jack, Jr., C.R., Albert, M.S., Knopman, D.S., McKhann, G.M., Sperling, R.A., Carrillo, M.C., Thies, B., Phelps, C.H., 2011. Introduction to the recommendations from the National Institute on Aging-Alzheimer's Association workgroups on diagnostic guidelines for Alzheimer's disease. Alzheimers Dement. 7, 257–262.

James, B.D., Leurgans, S.E., Hebert, L.E., Scherr, P.A., Yaffe, K., Bennett, D.A., 2014. Contribution of Alzheimer disease to mortality in the United States. Neurology 82, 1045–1050.

Jin, M., Shepardson, N., Yang, T., Chen, G., Walsh, D., Selkoe, D.J., 2011. Soluble amyloid beta-protein dimers isolated from Alzheimer cortex directly induce Tau hyperphosphorylation and neuritic degeneration. Proc. Natl. Acad. Sci. USA 108, 5819–5824.

Katzman, R., 1976. The Prevalence and malignancy of Alzheimer disease: a major killer. Arch. Neurol. 33, 217–218.

Kittelberger, K.A., Piazza, F., Tesco, G., Reijmers, L.G., 2012. Natural amyloid-β oligomers acutely impair the formation of a contextual fear memory in mice. PLoS One 7, e29940.

Lehman, E.J., Hein, M.J., Baron, S.L., Gersic, C.M., 2012. Neurodegenerative causes of death among retired National Football League players. Neurology 79, 1970–1974.

Levy-Lahad, E., Wasco, W., Poorkaj, P., Romano, D.M., Oshima, J., Pettingell, W.H., Yu, C.E., Jondro, P.D., Schmidt, S.D., Wang, K., et al., 1995. Candidate gene for the chromosome 1 familial Alzheimer's disease locus. Science 269, 973–977.

Masliah, E., Hansen, L., Albright, T., Mallory, M., Terry, R.D., 1991. Immunoelectron microscopic study of synaptic pathology in Alzheimer's disease. Acta Neuropathol. 81, 428–433.

Maurer, K., Volk, S., Gerbaldo, H., 1997. Auguste D and Alzheimer's disease. Lancet 349, 1546–1549.

McKhann, G., Drachman, D., Folstein, M., Katzman, R., Price, D., Stadlan, E.M., 1984. Clinical diagnosis of Alzheimer's disease: report of the NINCDS-ADRDA Work Group under the auspices of Department of Health and Human Services Task Force on Alzheimer's Disease. Neurology 34, 939–944.

McKhann, G.M., Knopman, D.S., Chertkow, H., Hyman, B.T., Jack, Jr., C.R., Kawas, C.H., Klunk, W.E., Koroshetz, W.J., Manly, J.J., Mayeux, R., Mohs, R.C., Morris, J.C., Rossor, M.N., Scheltens, P., Carrillo, M.C., Thies, B., Weintraub, S., Phelps, C.H., 2011. The diagnosis of dementia due to Alzheimer's disease: recommendations from the National Institute on Aging-Alzheimer's Association workgroups on diagnostic guidelines for Alzheimer's disease. Alzheimers Dement. 7, 263–269.

Meek, P.D., McKeithan, K., Schumock, G.T., 1998. Economic considerations in Alzheimer's disease. Pharmacotherapy 18, 68–73.

Oddo, S., Billings, L., Kesslak, J.P., Cribbs, D.H., LaFerla, F.M., 2004. Abeta immunotherapy leads to clearance of early, but not late, hyperphosphorylated tau aggregates via the proteasome. Neuron 43, 321–332.

Plassman, B.L., Havlik, R.J., Steffens, D.C., Helms, M.J., Newman, T.N., Drosdick, D., Phillips, C., Gau, B.A., Welsh-Bohmer, K.A., Burke, J.R., Guralnik, J.M., Breitner, J.C., 2000. Documented head injury in early adulthood and risk of Alzheimer's disease and other dementias. Neurology 55, 1158–1166.

Robakis, N.K., Ramakrishna, N., Wolfe, G., Wisniewski, H.M., 1987. Molecular cloning and characterization of a cDNA encoding the cerebrovascular and the neuritic plaque amyloid peptides. Proc. Natl. Acad. Sci. USA 84, 4190–4194.

Rogaev, E.I., Sherrington, R., Rogaeva, E.A., Levesque, G., Ikeda, M., Liang, Y., Chi, H., Lin, C., Holman, K., Tsuda, T., et al., 1995. Familial Alzheimer's disease in kindreds with missense mutations in a gene on chromosome 1 related to the Alzheimer's disease type 3 gene. Nature 376, 775–778.

Rogaeva, E., Meng, Y., Lee, J.H., Gu, Y., Kawarai, T., Zou, F., Katayama, T., Baldwin, C.T., Cheng, R., Hasegawa, H., Chen, F., Shibata, N., Lunetta, K.L., Pardossi-Piquard, R., Bohm, C., Wakutani, Y., Cupples, L.A., Cuenco, K.T., Green, R.C., Pinessi, L., Rainero, I., Sorbi, S., Bruni, A., Duara, R., Friedland, R.P., Inzelberg, R., Hampe, W., Bujo, H., Song, Y.Q., Andersen, O.M., Willnow, T.E., Graff-Radford, N., Petersen, R.C., Dickson, D., Der, S.D., Fraser, P.E., Schmitt-Ulms, G., Younkin, S., Mayeux, R., Farrer, L.A., St George-Hyslop, P., 2007. The neuronal sortilin-related receptor SORL1 is genetically associated with Alzheimer disease. Nat. Genet. 39, 168–177.

Saunders, A.M., Strittmatter, W.J., Schmechel, D., George-Hyslop, P.H., Pericak-Vance, M.A., Joo, S.H., Rosi, B.L., Gusella, J.F., Crapper-MacLachlan, D.R., Alberts, M.J., et al., 1993. Association of apolipoprotein E allele epsilon 4 with late-onset familial and sporadic Alzheimer's disease. Neurology 43, 1467–1472.

Selkoe, D.J., 2008. Soluble oligomers of the amyloid beta-protein impair synaptic plasticity and behavior. Behav. Brain Res. 192, 106–113.

Sherrington, R., Rogaev, E.I., Liang, Y., Rogaeva, E.A., Levesque, G., Ikeda, M., Chi, H., Lin, C., Li, G., et al., 1995. Cloning of a gene bearing missense mutations in early-onset familial Alzheimer's disease. Nature 375, 754–760.

Soscia, S.J., Kirby, J.E., WashICosky, K.J., Tucker, S.M., Ingelsson, M., Hyman, B., Burton, M.A., Goldstein, L.E., Duong, S., Tanzi, R.E., Moir, R.D., 2010. TheAlzheimer's disease associated Amyloid betaprotein is an antimicrobial peptide. PLoS One 5, e9505.

Sperling, R.A., Aisen, P.S., Beckett, L.A., Bennett, D.A., Craft, S., Fagan, A.M., Iwatsubo, T., Jack, Jr., C.R., Kaye, J., Montine, T.J., Park, D.C., Reiman, E.M., Rowe, C.C., Siemers, E., Stern, Y., Yaffe, K., Carrillo, M.C., Thies, B., Morrison-Bogorad, M., Wagster, M.V., Phelps, C.H., 2011. Toward defining the preclinical stages of Alzheimer's disease: recommendations from the National Institute on Aging-Alzheimer's Association workgroups on diagnostic guidelines for Alzheimer's disease. Alzheimers Dement. 7, 280–292.

Strassnig, M., Ganguli, M., 2005. About a peculiar disease of the cerebral cortex: Alzheimer's original case revisited. Psychiatry 2, 30–33.

Strittmatter, W.J., Saunders, A.M., Schmechel, D., Pericak-Vance, M., Enghild, J., Salvesen, G.S., Roses, A.D., 1993. Apolipoprotein E: high-avidity binding to beta-amyloid and increased frequency of type 4 allele in late-onset familial Alzheimer disease. Proc. Natl. Acad. Sci. USA 90, 1977–1981.

Summers, W.K., Majovski, L.V., Marsh, G.M., Tachiki, K., Kling, A., 1986. Oral tetrahydroaminoacridine in long-term treatment of senile dementia, Alzheimer type. N. Engl. J. Med. 315, 1241–1245.

Tanzi, R.E., Gusella, J.F., Watkins, P.C., Bruns, G.A., St George-Hyslop, P., Van Keuren, M.L., Patterson, D., Pagan, S., Kurnit, D.M., Neve, R.L., 1987. Amyloid beta protein gene: cDNA, mRNA distribution, and genetic linkage near the Alzheimer locus. Science 235, 880–884.

Terry, R.D., Masliah, E., Salmon, D.P., Butters, N., DeTeresa, R., Hill, R., Hansen, L.A., Katzman, R., 1991. Physical basis of cognitive alterations in Alzheimer's disease: synapse loss is the major correlate of cognitive impairment. Ann. Neurol. 30, 572–580.

The Ronald and Nancy Reagan Research Institute of the Alzheimer's Association and the National Institute on Aging Working Group, 1998. Consensus report of the Working Group on: "Molecular and Biochemical Markers of Alzheimer's Disease". Neurobiol. Aging 19, 109–116.

Yaffe, K., Vittinghoff, E., Lindquist, K., Barnes, D., Covinsky, K.E., Neylan, T., Kluse, M., Marmar, C., 2010. Posttraumatic stress disorder and risk of dementia among US veterans. Arch. Gen. Psychiatry 67, 608–613.

Zetterberg, H., Mörtberg, E., Song, L., Chang, L., Provuncher, G.K., Patel, P.P., Ferrell, E., Fournier, D.R., Kan, C.W., Campbell, T.G., Meyer, R., Rivnak, A.J., Pink, B.A., Minnehan, K.A., Piech, T., Rissin, D.M., Duffy, D.C., Rubertsson, S., Wilson, D.H., Blennow, K., 2011. Hypoxia due to cardiac arrest induces a time-dependent increase in serum amyloid β levels in humans. PLoS One 6 (12), e28263.

CHAPTER 2
Clinical Diagnosis of Alzheimer's Disease

Chapter Outline

2.1 CLINICAL DIAGNOSIS OF ALZHEIMER'S DISEASE: HISTORICAL PERSPECTIVE

The evolution of the criteria used for the clinical diagnosis of Alzheimer's disease (AD) is depicted in Table 2.1. A set of simplified clinical criteria for the diagnosis of AD was first introduced more than 30 years ago by the National Institute

Biomarkers in Alzheimer's Disease. http://dx.doi.org/10.1016/B978-0-12-804832-0.00002-X
Copyright © 2016 Elsevier Inc. All rights reserved.

Table 2.1	Historical Perspective of Criteria of Clinical Diagnosis of Alzheimer's Disease		
Working Group	**Description of AD Criteria**	**Comments**	**References**
The National Institute of Neurological and Communicative Disorders and Stroke and the Alzheimer's Disease and Related Disorders Association (NINCDS-ADRDA) 1984	AD is a clinical diagnosis and can be categorized into two sections: probable AD and possible AD. Definitive diagnosis of AD is possible by histopathological confirmation	No statement of neuropathologic changes during disease progression. No information of non-AD dementia, and MCI cases at that time	McKhann et al. (1984)
Consortium to Establish a Registry for Alzheimer's Disease (CERAD) 1991	To develop a practical and standardized neuropathology protocol for the postmortem assessment of dementia and control subjects	This consortium defined the terms "definite AD," "probable AD," "possible AD," and "normal brain" by neuropathologic assessment	Mirra et al. (1991)
The National Institute on Aging and Reagan Institute Working Group 1997	Differentiation of cases in high, intermediate, or low likelihood categories of dementia is due to AD vs non-AD dementia cases such as Lewy body disease, vascular dementia, PSP, etc.	Non-AD dementia cases were likely to be categorized as intermediate or low likelihood that dementia was due to AD. Difficult to diagnose non-AD dementia	Hyman (1997)
Diagnostic and Statistical Manual IV Edition Txt Revision 2000 (DSM-IV-TR) criteria	Presence of both memory disorder and impairment in at least one additional cognitive domain. Such cognitive domains interfere with social function and activities of daily living	Included gradual loss of memory and cognitive disturbances that cause significant impairment. Definitive diagnosis of AD possible only in autopsy	American Psychiatric Association Diagnostic and Statistical Manual of Mental Disorders 2000

Table 2.1	Historical Perspective of Criteria of Clinical Diagnosis of Alzheimer's Disease (cont.)		
Working Group	**Description of AD Criteria**	**Comments**	**References**
The National Institute on Aging and Alzheimer's Association Working Group (NIA-AA) 2011	Total AD progression was divided into three phases: an asymptomatic preclinical phase, symptomatic MCI phase, and dementia phase	AD still diagnosed clinically. Biomarkers are proposed as research tools. Biomarkers cannot firmly diagnose AD, only help to confirm clinical diagnosis. Incorporation of AD biomarkers helps by providing information about pathophysiologic changes related to AD	Jack et al. (2011); McKhann et al. (2011); Sperling et al. (2011)
International Working Group (IWG) 2007, 2010, 2014	IWG criteria for AD: (1) clinical AD phenotype manifested by episodic memory profile; (2) presence of AD pathologic biomarker evidences	IWG-2 criteria recognized the use of biomarkers as integral to the diagnosis of AD	Dubois et al. (2007); Dubois et al. (2010); Dubois et al. (2014)

AD, Alzheimer's disease; MCI, mild cognitive impairment; PSP, progressive supranuclear palsy.

of Neurological and Communicative Disorders and Stroke (NINCDS) and the Alzheimer's Disease and Related Disorders Association (ADRDA) (McKhann et al., 1984). According to these criteria, AD can be categorized into two classes: probable AD and possible AD. A clinical diagnosis of probable AD is supported by low scores on various neuropsychological tests [eg, the mini-mental state examination (MMSE)], progressive worsening of memory, a deficit in two or more cognitive areas of function (eg, language, motor skills, and perception), and impaired activities of daily life, plus exclusion of other possible neurodegenerative diseases. A clinical diagnosis of possible AD is made on the basis of dementia syndrome and absence of other neurologic and psychiatric disorders. According to NINCDS-ADRDA criteria, a definitive AD diagnosis is only possible in probable AD with histopathological evidence noted at autopsy. In the literature, most clinical diagnoses of AD used the NINCDS-ADRDA criteria, and the diagnostic sensitivity (81%) and specificity (70%) of the clinical diagnosis of AD using these criteria was sufficient for clinical research and clinical trial purposes (Knopman et al., 2001). At the time, the concept of sequential stages of AD, including mild cognitive impairment (MCI) had not yet been proposed, and so was not reflected in the NINCDS-ADRDA criteria.

The accumulation of knowledge about AD over the subsequent 3 decades, from both basic research efforts and clinical experience, and particularly what has been learned regarding the clinicopathological mechanisms of AD, led the National Institute on Aging and the Alzheimer's Association (NIA-AA) working group to revise the NINCDS-ADRDA criteria in 2010 (McKhann et al., 2011; Sperling et al., 2011; Jack et al., 2011; Albert et al., 2011). The NIA-AA criteria differ from the NINCDS-ADRDA criteria (1984) in two major ways: preclinical AD and MCI AD phases were included as earlier stages of the disease, and biomarkers of AD pathology were incorporated to confirm clinical diagnoses. The NIA-AA set forth guidelines for diagnosing dementia due to AD and dementia due to MCI, use of pathologic markers of AD at autopsy, and use of AD biomarkers to confirm a clinical diagnosis. Of note, the NIA-AA criteria state that use of AD biomarkers to confirm a clinical diagnosis is only applicable in the research setting, and that these biomarkers are not yet suitable for clinical use. As will be described in following chapters, AD biomarkers need to be validated in order to move from the research and development laboratory into clinical settings.

In parallel with the efforts of the NIA-AA, an International Working Group (IWG) was established to incorporate the outcomes of research and clinical experience using the NINCDS-ADRDA criteria for AD into an updated set of criteria. In 2014, the IGW introduced a set of revised criteria for the clinical diagnosis of AD (Dubois et al., 2007, 2010, 2014). According to these criteria, AD is a clinicobiological entity with a specific clinical phenotype that can be confirmed on the basis of in vivo AD pathology (Dubois et al., 2007, 2010, 2014). IGW-2 defines AD on the basis of three features: (1) clinical features, (2) AD pathological biomarkers, and (3) exclusion criteria. Clinical features are episodic memory change and progressive memory decline over the past 6 months as reported by the patient or an informant (eg, caregiver). AD pathological biomarkers are (1) decreased $A\beta_{1-42}$ and increased total tau and phosphorylated tau in cerebrospinal fluid (CSF), (2) increased retention of $A\beta$ tracers in the brain by amyloid positron emission tomography, and (3) presence of one of the following genetic markers of early-onset AD: presinilin1 (PSEN1), presinilin2 (PSEN2), or amyloid precursor protein (APP). Exclusion criteria consist of sudden-onset non-AD dementia, major depression, cardiovascular disease, other metabolic diseases, extrapyramidal symptoms, and seizures, among others. These simplified AD criteria reflect both the clinical AD phenotype (for both typical and atypical AD), plus a pathological AD biomarker consistent with the presence of AD pathology. Such pathological biomarkers are: CSF biomarkers (decreased $A\beta_{1-42}$, increase tau and phosphorylated tau), increased binding of amyloid positron emission tomography tracer, and decreased volumetric magnetic resonance imaging. According to IGW-2, the criteria for mixed AD cases consist of clinical and biomarker evidence of AD plus clinical and biomarker evidence of non-AD dementias. IGW-2 defines preclinical AD as the absence of the AD clinical phenotype plus in vivo evidence of AD pathology determined by AD biomarkers as listed previously.

Incorporation of AD biomarkers into the NIA-AA (2011) criteria helps provide information about the disease pathology that can support a clinical diagnosis. The IWG criteria are readily applicable to clinical trials and confirmation of a clinical diagnosis when biomarker data are available. Both NIA-AA and IWG criteria were developed on the basis of evidence that pathophysiological changes of the disease start before the appearance of clinical symptoms. As a result, discovery and validation of reliable AD biomarkers will allow for earlier diagnosis of AD, prior to symptom presentation and in MCI stages of the disease, when therapies may have a greater effect. Until the clinical utility of these biomarkers is proven, however, clinical diagnosis of AD using a series of clinical, psychometric, and brain scanning tests remains a mainstay of patient workup and therapeutic decision making.

2.2 NEUROPSYCHOLOGICAL ASSESSMENT TESTS TO DIAGNOSE ALZHEIMER'S DISEASE

The first cognitive test to quantify cognitive and functional decline in adults was established in 1968 (Blessed et al., 1968). The clinical manifestation of AD includes memory impairment, personality changes, language impairment, orientation-related impairment with respect to time and space, progressive functional impairment, and loss of interest in normal activities including activities of daily living. Many cognitive assessment scales have been developed over the last several decades for evaluation of cognitive function and dementia. An ideal clinical AD test scale should have following features: (1) should be quick and easily performed by nurses and other health care professionals, (2) should use multifaceted testing mechanisms that assess all AD-related clinical abnormalities including memory impairment, impaired activities of daily living, and abnormal social behavior, (3) can quantify the severity of the disease, (4) should be validated by the "gold standard" diagnostic markers of AD at autopsy, (5) should have good test-retest reliability, and (6) should be sensitive enough to monitor disease progression and measure therapeutic efficacy over time. Most of the clinical AD test scales are based on attention, calculation, comprehension, construction, naming, orientation, recall, registration, repetition, spelling, and writing. A systematic review of cognitive scales used between 1981 and 2008 found a total of 68 relevant scales for clinical diagnosis of AD (Robert et al., 2010). Several important neuropsychological assessment tests are reviewed in Table 2.2. These tests are useful tools for dementia screening in population-based and community settings.

2.2.1 Mini-Mental State Examination

Since Folstein et al. (1975) first published the Mini-Mental State Examination (MMSE) in 1975, it has become the most widely used and well-validated tool for evaluation of an individual's cognitive functional state in clinical practice, clinical trials, and research. A modified and standardized MMSE test developed further improved its reliability (Molloy et al., 1991; Molloy and Standish, 1997). The MMSE consists of a brief 30-point questionnaire that takes only 5–10 min

| | | Table 2.2 | Psychometric Tests for Clinical Diagnosis of AD | |

Test	Description of Method	Comments Advantages/ Disadvantages	References
Mini mental score examination (MMSE)	Testing 11 different domains. It covers variety of cognitive domains. MMSE scale is longer (0–30)	Test examines orientation of place and time, registration, recall, long-term memory, constructional ability, language, and ability to understand and follow commands	Folstein et al. (1975); Molloy and Standish (1997)
Clock drawing test (CDT)	Quickly (~2 min) adding numbers on a predawn clock. CDT scale is shorter (0–10)	Easy and can be used in primary care setting. Poor performance for MCI cases	Nishiwaki et al. (2004)
Alzheimer's Disease Assessment Scale (ADAS)	ADAS has two subscales: ADAS-cognitive and ADAS-noncognitive	ASAS-cog is a standard tool for AD therapeutic efficacy tests	Rosen et al. (1984)
Clinical dementia rating (CDR)	Based on six cognitive and behavioral domains. Interrater reliability is moderate to high. CDR scale is small (0–3; 0,0.5, 1, 2, 3)	Time consuming. It can differentiate MCI cases. Limitation for detecting early dementia	Hughes et al. (1982); Berg (1988); Morris (1993)
Global dementia scale (GDS)	Based on seven dementia stages	Once clinical information is available, scaling GDS is easy	Reisberg et al. (1982)
Neuropsycho-logical Test Battery (NTB)	Based on nine cognitive measure components	Produces reliable and sensitive measure of cognitive changes in mild to moderate AD cases	Harrison et al. (2007)

to complete. It examines six different modalities of cognitive ability: orientation in place and time, short-term verbal recall, immediate recall, language ability, simple numerical calculation ability, and the ability to construct a simple figure. The MMSE cut-off score for dementia is 27. When compared with Diagnostic and Statistical Manual of Mental Disorders IV (DSM-IV) Ed., published by the American Psychiatric Association as the gold standard for dementia scaling, MMSE provides moderate sensitivity (79%) but high specificity (95%) (Hancock and Larner, 2011). MMSE is most effective for differentiating between patients with cognitive impairment from cognitively normal cases, but is limited in its ability to differentiate between types of dementia. In addition to reading and writing skills, the test is heavily dependent on hearing capability, thus the test has limited utility for hearing or visually impaired patients.

2.2.2 Clock Drawing Test

After the MMSE, clock drawing test (CDT) is the second most widely used test for grading cognitive states. The test includes questions in a number of areas including attention, calculation, comprehension, construction, naming, orientation, recall, registration, repetition, spelling, and writing. The origin of the CDT is not clear. Evidence suggests that it was first used by the British neurologist/psychiatrist Sir Henry Head. In the original testing method, a predrawn clock is given to the subject, who is asked to draw the clock hands indicating 10 min past 11 o'clock. The grading system is very simple and is scaled to 10 points. The scoring depends on how the clock was drawn with appropriate markings of an analog clock: 12 numbers, correctly positioned hour-hand and minute-hand, etc. CDT is a quick, simple, and cost-effective tool for assessing dementia in a primary care setting. It is useful for identifying moderate and severe AD patients, but has limitations for diagnosing MCI and early stage AD. It is also very useful for testing subjects across education levels and cultures (Ainslie and Murden, 1993). Presently, CDT and MMSE have been evaluated in combination as a way to improve detection of MCI cases (Cacho et al., 2010).

2.2.3 Alzheimer's Disease Assessment Scale

This mental assessment scale is designed to assess both cognitive and noncognitive AD-related symptoms (Rosen et al., 1984). The cognitive subscale Alzheimer's Disease Assessment Scale (ADAS-Cog) is comprised of 11 modalities that evaluate memory, praxis, and language deficiencies. The other subscale (ADAS-non-Cog) measures 10 modalities related to mood and behavioral abnormalities. Today, ADAS-Cog is commonly used in AD therapy clinical trials to evaluate drug efficacy; however, ADAS-Cog lacks the required sensitivity for measuring cognitive changes over time in subjects with relatively high MMSE scores.

2.2.4 Neuropsychological Test Battery

MMSE and ADAS are not sufficiently sensitive to detect early-stage AD. Neuropsychological test battery (NTB) was found to exhibit better test–retest reliability than ADAS-Cog and can be used in evaluating therapeutic efficacy test in clinical trials in mild AD (Harrison et al., 2007). NTB is a measure of nine cognitive components: Wechsler-Memory verbal immediate (scale 0–24), Wechsler-Memory visual immediate (scale 0–18), Wechsler-Memory verbal delayed (scale 0–8), Wechsler-Memory visual delayed (scale 0–6), Wechsler-Memory digital (scale 0–24), Rey Auditory Verbal Learning Test immediate (scale 0–105), Rey Auditory Verbal Learning Test delayed (scale 0–30), Controlled Word Association Test, and Category Fluency Test (Harrison et al., 2007). All nine NTB components assess memory and/or executive function.

2.2.4.1 OTHER NEUROPSYCHOLOGICAL TESTS

Several other dementia assessment scales that can be used in cognitive assessment of patients with suspected AD are: Abbreviated Mental Test Score (Qureshi and Hodkinson, 1974), Mini-Cog (Borson et al., 2000), 6-CIT (Brooke and Bullock, 1999), Test Your Memory (Brown et al., 2009), Montreal Cognitive Assessment (Nasreddine et al., 2005), and the Addenbrookes Cognitive Assessment (Mathuranth et al., 2000).

2.2.5 Clinical Dementia Rating

The Clinical Dementia Rating (CDR) is a global rating device that was first introduced in a prospective study of patients with mild "senile dementia of AD type" (SDAT) in 1982 (Hughes et al., 1982). New and revised CDR scoring rules were later introduced (Berg, 1988; Morris, 1993; Morris et al., 1997). CDR is estimated on the basis of a semistructured interview of the subject and the caregiver (informant) and on the clinical judgment of the clinician. CDR is calculated on the basis of testing six different cognitive and behavioral domains such as memory, orientation, judgment and problem solving, community affairs, home and hobbies performance, and personal care. The CDR is based on a scale of 0–3: no dementia (CDR = 0), questionable dementia (CDR = 0.5), MCI (CDR = 1), moderate cognitive impairment (CDR = 2), and severe cognitive impairment (CDR = 3). Two sets of questions are asked, one for the informant and another for the subject. The set for the informant includes questions about the subject's memory problem, judgment and problem solving ability of the subject, community affairs of the subject, home life and hobbies of the subject, and personal questions related to the subject. The set for subject includes memory-related questions, orientation-related questions, and questions about judgment and problem-solving ability. Interrater reliability of CDR in a multicenter trial was moderate to high, but showed limitations in detecting early dementia (Rockwood et al., 2000). A modified CDR (mCDR) has been introduced to diagnose and stage MCI and was found to reliably distinguish between MCI and subjects with normal cognition (Duara et al., 2010).

2.2.6 Global Deterioration Scale

The Global Deterioration Scale (GDS) was introduced in 1982 to assess primary degenerative dementia and delineate between dementia stages (Reisberg et al., 1982). GDS describes seven stages on the basis of severity of dementia: stage 1: subjectively and cognitively normal; stage 2: age-associated very MCI; stage 3: MCI due to dementia; stage 4: early dementia; stage 5: moderate dementia; stage 6: moderately severe dementia; and stage 7: severe dementia. The GDS provides caregivers an overview of the stages of cognitive deterioration associated with a primary neurodegenerative disease such as AD. Like CDR, the GDS estimates the severity of dementia.

2.2.7 Dementia Severity Rating Scale

Dementia Severity Rating Scale (DSRS) is a multiple choice questionnaire that can be used by caregivers to assess the severity of dementia, from the mildest to the most severe, in subjects with AD or other dementias. This scale can be used in both the community and in primary care units.

Cognitive measures made by various psychometric tests have been adopted by the Food and Drug Administration and the European Medicines Authority as a way to demonstrate drug efficacy in AD clinical trials. Most of the dementia rating scales are able to detect severe AD cases; however, they are less able to accurately detect AD at early stages. Because early stage AD, before widespread neuronal loss and clinical symptoms occur, is the desired stage for therapeutic intervention, there is a need for new multidomain AD scales that more accurately detect early stage disease and can evaluate drug efficacy (Robert et al., 2010).

2.3 NEUROPATHOLOGY AT BRAIN AUTOPSY TO DIAGNOSE ALZHEIMER'S DISEASE

Beyond the measurement of dementia, the "gold standard" diagnostic marker of AD is the presence of pathologic brain lesions of amyloid plaques (mainly deposition of Aβ) and neurofibrillary tangles (mainly the aggregated form of hyperphosphorylated tau) in the brain at autopsy. Neurofibrillary changes in the brains of individuals with AD were first described by Braak and Braak (1991). Their assessment method is also the most commonly used and accepted approach to evaluating neurofibrillary changes. It rates the distribution of neurofibrillary tangles on silver-stained sections from various neocortical regions, the entorhinal cortex, and hippocampus on a scale from 0 to VI (the Braak score). Standardization of the neuropathologic assessment of AD was discussed by the Consortium to Establish a Registry for Alzheimer's Disease (CERAD) (Mirra et al., 1991). This set of criteria is the most widely used for the pathological diagnosis of AD in the United States and abroad. The CERAD criteria are based on the semiquantitative assessment of the density of neuritic plaques on silver-stained sections from various neocortical regions, with the resulting neuritic plaque score adjusted for age.

Most structured guidelines for clinical–pathological correlation of AD were established in the consensus recommendations for the postmortem diagnosis of AD from the NIA and the Reagan Working Group on diagnostic criteria for the neuropathological assessment of AD (Hyman and Trojanowski, 1997). These recommendations take both the CERAD neuritic plaque score and the neurofibrillary Braak score to express the likelihood (low, intermediate, or high) that the lesions account for the dementia symptoms of the patient. Neurofibrillary tangle formation in general is proportional to dementia severity and to some extent proportional to cerebral atrophy (Bouras et al., 1994; Braak and Braak, 1997). However, maximum levels of amyloid plaque deposition correlate with MCI stages (Bouras et al., 1994; Braak and Braak, 1997). Most AD autopsy reports have found that the distribution of amyloid plaques varies widely with disease stage, whereas neurofibrilliary tangles and neuropil threads are distributed in a pattern that correlates with dementia severity, allowing characterization of six different disease stages (Braak score). Antemortem brain atrophy measured by structural magnetic resonance imaging (sMRI) was also found to correlate with the postmortem Braak score at brain autopsy (Vemuri et al., 2008).

There were two reasons to consider revision of the guidelines established in 1997 by the NIA and the Reagan Working Group on diagnostic criteria for the neuropathological assessment of AD.

1. *Need for diagnostic guidelines for preclinical AD*: Guidelines established by the NIA and the Reagan Working Group were based on the assumption that the person must have clinical dementia in life to be diagnosed with AD. We now know that AD has long prodromal stages during which pathophysiological changes in the brain are occurring but clinical symptoms of dementia have not yet emerged. There is a need for preclinical criteria to diagnose the disease before more widespread synaptic loss and irreversible neuronal damage occurs.

2. *Cognitively normal cases with amyloid plaques and tangles*: A growing body of evidence demonstrates that older individuals without cognitive deficits also have amyloid plaques and neurofibrillary in their brains at autopsy (Crystal et al., 1988; Price and Morris, 1999; Schmitt et al., 2000; Price et al., 2009; Caselli et al., 2010).

A panel of neuropathologists were asked by the NIA-AA for revised guidelines for the neuropathological assessment of AD. In 2002, the revised NIA-AA guidelines ranked AD neuropathological changes on the basis of Aβ plaque score (Thal phase 1, 2, 3, 4, 5) (Thal et al., 2002), neurofibrillary tangles per the Braak score (Braak stage I, I, III, IV, V, VI), and neuritic plaques per the CERAD score (C 0, 1, 2, 3) (Hyman et al., 2012). The resulting "ABC score"—"A" for Aβ plaques, "B" for Braak score, and "C" for CERAD score—is categorized from 0 to 3 (0, 1, 2, 3) according to the severity of AD pathology. For example, Aβ plaques can be divided into A0 = no amyloid plaques, A1 = Thal phase 1 or 2, A2 = Thal phase 3, and A3 = Thal phase 4 or 5; neurofibrillary tangle levels can be categorized into B0 = no neurofibrillary tangles, B1 = Braak stage I or II, B2 = Braak stage III or IV, and B3 = Braak stage V or VI; and neuritic plaque levels can be categorized into C0 = no neuritic plaque, C1 = CERAD score 1, C2 = CERAD score 2, and C3 = CERAD score 3 (Hyman et al., 2012). According to the revised NIA-AA criteria, AD neuropathological report should be simply reported as Ax, Bx, Cx, with x = 0, 1, 2, or 3.

2.4 ACCURACY OF CLINICAL DIAGNOSIS OF ALZHEIMER'S DISEASE

The clinical diagnosis of AD is based on neuropsychological tests and exclusion of other age-related dementias. Clinical diagnosis of AD is unreliable, particularly during early stages of the disease. Judged by autopsy confirmation, the overall accuracy of clinical diagnosis of AD versus non-AD is 78% (Knopman et al., 2001). Diagnostic accuracy of early AD is even lower due to the fact that AD shares several symptoms with a variety of neurological disorders. Disease progression and increasing severity of symptoms can support a diagnosis of AD, but definitive diagnosis is only possible at autopsy, with the identification of characteristic AD pathologic brain lesions, amyloid plaques, and neurofibrillary tangles. An autopsy-validated caveat study showed that in the early stages of AD (disease duration ≤4 years from the date of the first documented symptom of dementia the date of clinical diagnosis), clinical diagnosis was correct in 50% of cases, compared with 83% of cases with more than 4 years of disease duration (Khan and Alkon, 2010). In other autopsy-validated study, two clinical diagnoses were made, one early in disease progression, and another much later (Hogervorst et al., 2003). The overall accuracy reported for the first clinical diagnosis was approximately 60%, compared with 81% for the second clinical diagnosis. Lower accuracies of clinical diagnosis with respect to autopsy confirmation have been reported in community- and population-based studies compared to specialized centers (Lim et al., 1999; Petrovitch et al., 2001).

2.5 ALZHEIMER'S DISEASE-RELATED NEUROLOGICAL CONDITIONS

2.5.1 Preclinical Alzheimer's Disease

There is some evidence from genetic cohorts and clinically normal patients that the pathology of AD starts decades before its clinical manifestation (Morris, 2005) (Fig. 2.1). This preclinical state of AD is speculative and the existence of a pre-MCI stage has yet to be proven. Evidence of preclinical AD includes the presence

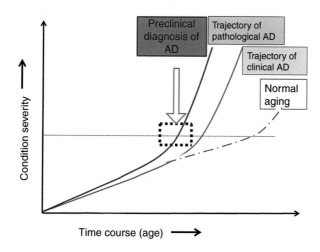

FIGURE 2.1

Hypothetical model of decreasing cognitive function (represented as condition severity) as a function of age for patients with Alzheimer's disease (AD) and normal aging. The model shows the hypothetical deviation of pathological and clinical trajectories of AD from the normal aging process. With normal aging, cognitive ability decreases (normal aging: *broken blue line*). Pathological AD (*red line*) starts before clinical symptoms appear for those individuals who develop AD (*blue line*). Distinguishing preclinical AD from normal aging, prior to the presentation of clinical symptoms of AD, will be possible in the future (*dotted box*).

of genetic risk factors, AD-like brain images, and abnormal CSF biomarkers (low $A\beta_{1-42}$, high tau, and p-tau; see Chapter 5) in cognitively normal individuals, and AD-like neuropathology seen in the normal aged brain. According to the NIA-AA guidelines, evidence of amyloidosis can be measured by in vivo amyloid-PET (see Chapter 3) and by measurement of $A\beta_{1-42}$ in CSF (see Chapter 5), and evidence of neurodegeneration can be assessed by in vivo fluorodeoxyglucose-PET, volumetric MRI (see Chapter 3), and measurement of high tau and p-tau in CSF (see Chapter 5). On the basis of the extent of amyloidosis and neurodegeneration, the NIA-AA workgroup (2011) categorized preclinical AD into following stages: Stage 0 consists of no evidence of amyloidosis (amyloid negative) and no sign of neurodegeneration (neurodegeneration negative); Stage 1 consists of some evidence of amyloidosis (amyloid positive) and no sign of neurodegeneration (neurodegeneration negative); Stage 2 consists of some evidence of amyloidosis (amyloid positive) and signs of neurodegeneration (neurodegeneration positive); Stage 3 consists of evidence of amyloidosis (amyloid positive) and signs of neurodegeneration (neurodegeneration positive), plus subtle cognitive decline that has not reached MCI levels. Those cases with no evidence of amyloidosis (amyloid negative) but signs of neurodegeneration (neurodegeneration positive) are categorized as "suspected non-Alzheimer's pathophysiology (SNAP)" (Jack et al., 2012). Individuals greater than 65 years of age with intact cognitive ability before death and substantial AD lesions at autopsy are classified as having "asymptomatic AD" or "preclinical AD." Identification of individuals in the preclinical phase of AD would provide a critical window of opportunity for therapeutic intervention. However, an autopsy-confirmed study found that these so called "preclinical AD" and MCI cases have similar levels of insoluble $A\beta$ deposition and tau loads, suggesting that cognitive impairment may be caused by other unknown mechanisms (Iacono et al., 2014). Moreover, deposition of amyloid plaques and neurofibrillary tangles are not the principal cause of dementia.

Both NIA-AA and IWG have extended the range of AD from so called "preclinical AD" to the most advanced stages of AD. The NIA-AA guideline (2011) has proposed three distinct phases to describe the progression of AD: preclinical, asymptomatic predementia; symptomatic predementia or MCI; and symptomatic dementia (McKhann et al., 2011; Sperling et al., 2011). According to IWG criteria, a patient with preclinical AD has no sign of dementia, but has one of the following: decreased $A\beta_{1-42}$, together with increased tau or phosphorylated tau-181 in CSF; or increased fibrillary amyloid on PET (Dubois et al., 2014). Although unproven, "asymptomatic preclinical AD" has been recognized by both NIA-AA and IGW-2 as preclinical AD. Both groups hypothesized that the preclinical stage of AD may be the most appropriate time for therapeutic intervention. According to IGW-2, preclinical AD can be further subdivided into two categories: "asymptomatic at risk for AD" and "presymptomatic AD" (Dubois et al., 2014). Asymptomatic at risk for AD have absence of an AD clinical phonotype but evidence of decreased $A\beta_{1-42}$ and increased total tau and phosphorylated tau in CSF, and increased amyloid on PET. Presymptomatic AD cases also have absence of an AD clinical phonotype but have an autosomal dominant mutation of PSEN1,

PSEN2, or APP. At a hypothetical inflection point, the AD pathological trajectory starts to deviate from the normal aging trajectory. At this point, AD pathological changes and abnormalities in brain cell signaling might be occurring without measurable neurodegeneration manifesting as clinical symptoms. The clinical trajectory deviates from normal aging later (Fig. 2.1). Meaningful preclinical diagnosis of AD should be achieved within the early part of the AD pathological trajectory and distinguish between pathological changes associated with normal aging and those associated with AD.

2.5.2   Mild Cognitive Impairment

The term MCI did not exist in 1984, when the NINCDS-ADRDA criteria for AD diagnosis were established. The concept of MCI evolved in the 1990s to define early clinical signs of AD in individuals that did not yet to fulfill all NINCDS-ADRDA criteria for an AD diagnosis. MCI is defined as a prodromal stage of cognitive decline that is greater than expected due to normal aging. It is the earliest clinical manifestation of AD. MCI can be divided into two categories: amnestic MCI (aMCI) and multimodal MCI (mMCI) or nonamnestic MCI. MCI with primarily memory deficits is called as amnestic MCI. Multimodal MCI includes MCI with problems in thinking skills, inability to make sound decisions and judgments, and inability to take the sequential steps needed to perform relatively complex tasks. In general, individuals with aMCI eventually develop AD and those with mMCI develop non-AD dementias. MCI proceeds to AD with an annual rate of 10–12% (Petersen et al., 1999) and an individual with MCI is expected to convert to AD within 5 years (about 5–25% per year) (Petersen et al., 2001). Another study found a slightly higher rate (10–25%) of conversion from MCI to AD (Grand et al., 2011). Patients with MCI who are progressing to AD have c-MCI, and those who continue to stay at the MCI stage have stable MCI (s-MCI). Individuals with c-MCI might be better candidates for inclusion in clinical studies of AD drugs to test the efficacy in slowing progression to moderate and severe AD.

2.5.3   Non-Alzheimer's Dementias

2.5.3.1   *VASCULAR DEMENTIA*

Vascular dementia (VaD) occurs secondary to brain cell death caused by cerebrovascular disease, ischemic or hemorrhagic stroke, and ischemic-hypoxic brain lesions. According to the National Institute of Neurological Disorders and Stroke and the Association Internationale pour la Recherche et l'Enseignement en Neurosciences (NINDS-AIREN) (1993) patients diagnosed with VaD must have dementia with evidence of cardiovascular disease from clinical examination or brain imaging. Differentiation of VaD and AD is difficult. In a study by Barber et al. (2000), no significant difference was found in brain atrophy between patients with VaD and AD. In the United States and Europe, VaD is the second most common cause of age-related dementia after AD (~20% of all age-related dementias). These numbers are even higher in China and Japan (~50% of all age-related dementias). VaD can be classified as multi-infarct dementia, strategic single-infarct dementia, small-vessel disease with dementia, hypoperfusion dementia, hemorrhagic dementia, and dementia caused by a combination of the aforementioned lesions (Román et al., 1993). Subcortical vascular dementia (SVaD) is a type of VaD in which many of the neuropsychological deficits are similar to those seen in AD, making it difficult to distinguish between SVaD and AD by neuropsychological tests alone. With normal aging and infarcts, the small blood vessels in the deep brain regions become distorted, which then causes reduced blood flow. Nerve fibers and neuronal signaling are affected by the reduced blood flow, and cause SVaD clinical symptoms.

2.5.3.2 *LEWY BODY DEMENTIA*

In Lewy body dementia (DLB), the main pathological characteristic is intraneuronal deposits of proteins, mainly α-synuclein. Lewy bodies develop in nerve cells in regions of brain involved in thinking, memory, and movement. DLB is the third most common cause of dementia after AD and VaD, accounting for 10–15% of all age-related dementias. Clinical features of patients with DLB are hallucinations, fluctuating cognitive behavior, and extrapyramidal autonomous dysfunction. The sensitivity of clinical diagnoses of pure DLB confirmed at autopsy is quite low (32.2%), and even lower in the presence of AD, although specificity is quite high (95%) (Nelson et al., 2010). An abnormal dopamine transporter PET scan is normally required for a DLB diagnosis.

2.5.3.3 *FRONTOTEMPORAL DEMENTIA*

Frontotemporal dementia (FTD) is a group of age-related neurodegenerative diseases resulting from the progressive neurodegeneration of the frontal and temporal lobes of the brain. FTD covers a wide range of dementias, including behavioral variant frontotemporal dementia, semantic dementia, and progressive nonfluent aphasia (PNFA). The most common clinical manifestation in patients with FTD is problems with language. FTD represents 5–10% of all age-related dementia diagnoses. People with FTD are younger than those who develop sporadic AD. Most FTD patients are between 45 and 64 years of age.

2.5.3.4 *CREUTZFELDT–JACOB DISEASE*

There are three different kinds of Creutzfeldt–Jacob disease (CJD): sporadic CJD, genetic CJD, and variant CJD. Sporadic CJD is the most common. The pathological hallmark of CJD is deposition of prion proteins (PrP) in the brain.

2.5.3.5 *OTHER NON-AD DEMENTIAS*

Notable non-AD dementias include dementia due to Parkinson's disease, tauopathy, primary progressive aphasia, dementia due to vitamin B12 deficiency, multi-infarct dementia, Huntington's disease, progressive supranuclear palsy (PSP), normal pressure hydrocephalus, seizure disorder, subdural hematoma, multiple sclerosis, and history of significant head trauma followed by persistent neurological deficits.

2.6 CHALLENGES IN DIAGNOSING ALZHEIMER'S DISEASE

AD is a heterogeneous, genetically complex neurological disorder, with neurodegenerative processes that probably begin well before the clinical signs and symptoms appear, and it is hypothesized that AD progresses to the advanced stage through multiple prodromal stages over a period of decades. In addition, AD often develops concomitant with other age-related neurodegenerative diseases such as DLB, VaD, FTD, and tauopathy, all of which are associated with declines in cognitive function, or mild neurocognitive disorders, making a definitive clinical diagnosis of AD much more difficult.

2.6.1 Long Prodromal Stages

AD has a gradual onset with continual deterioration in cognitive ability, progressing through a presymptomatic stage to MCI to moderate AD to severe AD. AD progresses over a period of 2–20 years, with an average of 7 years. The clinical manifestation of the disease is preceded by a long prodromal stage, during which neuropathological lesions form (Fig. 2.1). The first symptoms of decline in cognitive performance appear as early as 12 years, at the MCI stage. At the moderate AD stage, cognitive deficits and depressive symptoms emerge. At the severe AD stage, functional impairment occurs.

2.6.2 Comorbidity With Non-Alzheimer's Disease Dementias

The cognitive impairment and dementia due to AD may overlap with those caused by other progressive irreversible dementias such as ischemic VaD or dementia due to stroke, DLB, FTD, Parkinson's disease (PD), and more than 30 other disorders. Even when the best-trained specialists conduct clinical neuropsychological assessments, they may be unable to diagnose AD when it presents with other dementias, such as DLB, VaD, FTD, and tauopathy. AD associated with DLB pathologies show a similar episodic memory impairment as pure AD (Johnson et al., 2005). An autopsy-confirmed AD cohort study found that ~33% of patients with AD also presented with VaD, DLB, and tauopathy (Khan and Alkon, 2010). In a randomly designed AD peripheral biomarker study of 159 patients (within a total pool of 264 patients), clinically affected AD cases were followed to collects autopsy results. A total of 64 autopsy-confirmed AD patients brain pathology data were examined for AD, VaD, DLB, and PPD pathology (Khan and Alkon, 2010). Among the 64 autopsy-confirmed AD cases, 21 had mixed dementia pathology and 43 cases had pure AD pathology. There was no effect of age or sex with the likelihood of comorbidity. In that study, almost one-third of clinically confirmed AD cases would be rediagnosed with mixed-dementia pathology in comorbid state at autopsy. All of these autopsy-confirmed cases were clinically confirmed AD, making is clear that clinical diagnosis alone may not accurately distinguish AD from non-AD dementias.

2.6.3   Multiple Nongenetic and Genetic Factors

The etiopathogenesis of AD is still unclear. AD is a multifactorial, genetically complex, and heterogeneous disease with two distinct categories, namely, early onset familial AD (FAD) with well-defined genetic causes, and late onset sporadic AD. Age is the dominant risk factor for sporadic AD. Only three genes, APP, PS1, and PS2, have been identified as causal factors in FAD. Similar causal genetic factors have not been identified for sporadic AD, which accounts for more than 95% of AD cases, though epidemiologic and twin studies revealed that genetic factors play an important role in the development of both FAD as well as sporadic AD (see Chapter 4). The incidence of inheritance of sporadic AD is also very high (~58–80%, from different studies). Sporadic AD appears to arise as a consequence of a combination of genetic factors, environmental risk factors, and aging (epigenetic). Several low penetrance genes and risk factor genes for sporadic AD have been identified, such as apolipoprotein E-4 (APOE-4) (Green et al., 2009), sortilin-related receptor (SOLR1) (Rogaeva et al., 2007), and genes included in the AlzGene database (www.alzgene.org) for sporadic AD. Among these, ApoE-4 is best-known risk factor for sporadic AD and has confirmed in multiple studies. Genome-wide association studies using the AlzGene database identified ~32 genes as risk factors for sporadic AD, including SORL1, CLU, PICALM, and CR1 (see Chapter 4 for details). Other strong nongenetic risk factors have been implicated such as head trauma, occupational exposures, pre-existing medical conditions (such as cerebrovascular disease, hypertension, diabetes, dyslipidemia, traumatic brain injury, depression and cancer, and unhealthy lifestyle). Therefore, a critical unsolved problem in AD research is to identify the genetic causes of the sporadic AD form of the disease and their interaction with multifactorial epigenetic contributors.

2.6.4   Anomalous Alzheimer's Disease Neuropathology

The pathologic hallmarks of AD in the brain are the formation of amyloid plaques and neurofibrillary tangles at autopsy, though the definitive diagnosis of AD continues to require clinical dementia in life. However, several studies have reported anomalous AD neuropathology in aged individuals without clinical dementia (reviewed by Xekardaki et al., 2015). In a study of an extremely old population, 49% of subjects who seemed to be cognitively normal just before their death had diffused amyloid plaques at autopsy (Knopman et al., 2003). Another study of the very old population similarly found AD pathology without any display of dementia symptoms (Balasubramanian et al., 2012). An autopsy study from the Baltimore Longitudinal Study of Aging found that subjects with MCI and asymptomatic AD did not differ in terms of amyloid plaque and neurofibrillary tangle load (Iacono et al., 2014). A study led by Crary et al. (2014) found only phosphorylated tangles (p-tau) in a subset of AD patients, with no amyloid plaques in autopsy. Tau-only pathology refers to the condition at autopsy consisting of tau tangles only. Researchers concluded that this kind of pathology may be present in a subset of patients with AD.

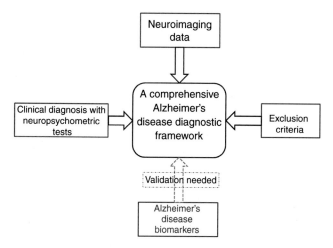

FIGURE 2.2
A simple scheme of a comprehensive Alzheimer's disease (AD) diagnostic framework.
The framework includes four modalities: clinical diagnosis by neuropsychometric tests,
neuroimaging data, exclusion criteria for non-AD dementias, and confirmation with biomarkers.

2.7 COMPREHENSIVE ALZHEIMER'S DISEASE DIAGNOSTIC FRAMEWORK

Comprehensive AD diagnosis and its operationalization into a diagnostic framework has major implications for our understanding of disease pathogenesis and
its application in testing therapeutic interventions. An AD diagnostic framework
with four additive modalities is presented in Fig. 2.2. The four modalities include (1) clinical diagnosis using neuropsychological tests and assessment and
judgment, (2) neuroimaging data, (3) exclusion of other neurodegenerative diseases, and (4) confirmation by biomarkers. Use of exclusion criteria remains an
integral part of the AD diagnostic framework. Many dementia diseases that can
be excluded on the basis of clinical history of dementia, including significant
neurological diseases such as brain tumors, PD, multi-infarct dementia, Huntington's disease, PSP, normal pressure hydrocephalus, seizure disorder, subdural hematoma, multiple sclerosis, or history of significant head trauma followed
by persistent neurological deficits and known structural brain abnormalities.
Other causes of dementia that can be excluded include low thyroid function,
vitamin B12 deficiency, infections, addiction, cancer, and depression.

Incorporation of biomarkers into the AD diagnostic scheme will provide information about the in vivo physiological changes in AD brain. However, a great
deal of validation work lies ahead if current AD biomarkers are to be used in
clinical settings to support a clinical diagnosis of AD. Diagnosis of preclinical
AD presents a particularly challenging hurdle, because neuropsychological tests
alone will not provide information about underlying pathological changes that
have not yet caused symptoms of neurodegeneration. In the future, panels of
well-validated biomarkers may be used to detect preclinical stages of AD.

2.8 CONCLUSIONS

Even with the best-trained specialists using well-validated clinical neuropsychological assessments, AD clinical diagnosis has not reached acceptable levels (>80 sensitivity and >80 specificity: the 1998 Consensus Report of the Working Group [The Ronald and Nancy Reagan Research Institute of the Alzheimer's Association and the National Institute on Aging Working Group (1998) on "Molecular and Biochemical Markers of Alzheimer's Disease"]. According to the NIA-AA criteria, AD can be diagnosed clinically but accuracy, and can be improved by the addition of AD biomarkers. IWG introduced a set of revised and updated criteria for the clinical diagnosis of AD that reconceptualized the disease as a clinicobiological entity with a specific clinical phenotype that could be confirmed in vivo based on pathophysiologic evidence of disease. Some psychometric tests are useful for clinical diagnosis of MCI cases but none can be used for preclinical AD diagnosis. Incorporation of AD biomarkers into the AD diagnostic framework with clinical diagnosis is essential to improve AD diagnostic accuracy as well as detection of preclinical stages of AD. Differential diagnosis of AD, non-AD dementias, and mixed dementias, as well as the diagnosis of preclinical AD are most important challenging aspects of AD diagnosis.

Bibliography

Ainslie, N.K., Murden, R.A., 1993. Effect of education on the clock-drawing dementia screen in non-demented elderly persons. J. Am. Geriatr. Soc. 41, 249–252.

Albert, M.S., DeKosky, S.T., Dickson, D., Dubois, B., Feldman, H.H., Fox, N.C., Gamst, A., Holtzman, D.M., Jagust, W.J., Petersen, R.C., Snyder, P.J., Carrillo, M.C., Thies, B., Phelps, C.H., 2011. The diagnosis of mild cognitive impairment due to Alzheimer's disease: recommendations from the National Institute on Aging-Alzheimer's Association workgroups on diagnostic guidelines for Alzheimer's disease. Alzheimers Dement. 7, 270–279.

Balasubramanian, A.B., Kawas, C.H., Peltz, C.B., Brookmeyer, R., Corrada, M.M., 2012. Alzheimer disease pathology and longitudinal cognitive performance in the oldest-old with no dementia. Neurology 9, 915–921.

Barber, R., Ballard, C., McKeith, I.G., Gholkar, A., O'Brien, J.T., 2000. MRI volumetric study of dementia with Lewy bodies: a comparison with AD and vascular dementia. Neurology 54, 1304–1309.

Berg, L., 1988. Clinical dementia rating (CDR). Psychopharmacol Bull. 24, 637–639.

Blessed, G., Tomlinson, B.E., Roth, M., 1968. The association between quantitative measures of dementia and of senile change in the cerebral grey matter of elderly subjects. Br. J. Psychiatry 114, 797–811.

Borson, S., Scanlan, J., Brush, M., Vitaliano, P., Dokmak, A., 2000. The Mini-Cog: a cognitive vital signs measure for dementia screening in multi-lingual elderly. Int. J. Geriatr. Psychiatry 15, 1021–1027.

Bouras, C., Hof, P.R., Giannakopoulos, P., Michel, J.P., Morrison, J.H., 1994. Regional distribution of neurofibrillary tangles and senile plaques in the cerebral cortex of elderly patients: a quantitative evaluation of a one-year autopsy population from a geriatric hospital. Cereb Cortex 4, 138–150.

Braak, H., Braak, E., 1991. Neuropathological staging of Alzheimer-related changes. Acta Neuropathol. 82, 239–259.

Braak, H., Braak, E., 1997. Frequency of stages of Alzheimer related lesions in different age categories. Neurobiol. Aging 18, 351–357.

Brooke, P., Bullock, R., 1999. Validation of a 6 item cognitive impairment test with a view to primary care usage. Int. J. Geriatr. Psychiatry 14, 936–940.

Brown, J., Pengas, G., Dawson, K., Brown, L.A., Chatworthy, P., 2009. Self administered cognitive-screening test (TYM) for detection of Alzheimer's disease; cross sectional study. BMJ 338, b2030.

Cacho, J., Benito-León, J., García-García, R., Fernández-Calvo, B., Vicente-Villardón, J.L., Mitchell, A.J., 2010. Does the combination of the MMSE and clock drawing test (mini-clock) improve the detection of mild Alzheimer's disease and mild cognitive impairment? J. Alzheimers Dis. 22, 889–896.

Caselli, R.J., Walker, D., Sue, L., Sabbagh, M., Beach, T., 2010. Amyloid load in nondemented brains correlates with APOE e4. Neurosci. Lett. 473, 168–171.

Crary, J.F., Trojanowski, J.Q., Schneider, J.A., Abisambra, J.F., Abner, E.L., Alafuzoff, I., Arnold, S.E., Attems, J., Beach, T.G., Bigio, E.H., Cairns, N.J., Dickson, D.W., Gearing, M., Grinberg, L.T., Hof, P.R., Hyman, B.T., Jellinger, K., Jicha, G.A., Kovacs, G.G., Knopman, D.S., Kofler, J., Kukull, W.A., Mackenzie, I.R., Masliah, E., McKee, A., Montine, T.J., Murray, M.E., Neltner, J.H., Santa-Maria, I., Seeley, W.W., Serrano-Pozo, A., Shelanski, M.L., Stein, T., Takao, M., Thal, D.R., Toledo, J.B., Troncoso, J.C., Vonsattel, J.P., White, 3rd, C.L., Wisniewski, T., Woltjer, R.L., Yamada, M., Nelson, P.T., 2014. Primary age-related tauopathy (PART): a common pathology associated with human aging. Acta Neuropathol. 128, 755–766.

Crystal, H., Dickson, D., Fuld, P., Masur, D., Scott, R., Mehler, M., et al., 1988. Clinico-pathologic studies in dementia: nondemented subjects with pathologically confirmed Alzheimer's disease. Neurology 38, 1682–1687.

Duara, R., Loewenstein, D.A., Greig-Custo, M.T., Raj, A., Barker, W., Potter, E., Schofield, E., Small, B., Schinka, J., Wu, Y., Potter, H., 2010. Diagnosis and staging of mild cognitive impairment, using a modification of the clinical dementia rating scale: the mCDR. Int. J. Geriatr. Psychiatry 25, 282–289.

Dubois, B., Feldman, H.H., Jacova, C., Dekosky, S.T., Barberger-Gateau, P., Cummings, J., Delacourte, A., Galasko, D., Gauthier, S., Jicha, G., Meguro, K., O'brien, J., Pasquier, F., Robert, P., Rossor, M., Salloway, S., Stern, Y., Visser, P.J., Scheltens, P., 2007. Research criteria for the diagnosis of Alzheimer's disease: revising the NINCDS-ADRDA criteria. Lancet Neurol. 6, 734–746.

Dubois, B., Feldman, H.H., Jacova, C., Cummings, J.L., Dekosky, S.T., Barberger-Gateau, P., Delacourte, A., Frisoni, G., Fox, N.C., Galasko, D., Gauthier, S., Hampel, H., Jicha, G.A., Meguro, K., O'Brien, J., Pasquier, F., Robert, P., Rossor, M., Salloway, S., Sarazin, M., de Souza, L.C., Stern, Y., Visser, P.J., Scheltens, P., 2010. Revising the definition of Alzheimer's disease: a new lexicon. Lancet Neurol. 9, 1118–1127.

Dubois, B., Feldman, H.H., Jacova, C., Hampel, H., Molinuevo, J.L., Blennow, K., DeKosky, S.T., Gauthier, S., Selkoe, D., Bateman, R., Cappa, S., Crutch, S., Engelborghs, S., Frisoni, G.B., Fox, N.C., Galasko, D., Habert, M.O., Jicha, G.A., Nordberg, A., Pasquier, F., Rabinovici, G., Robert, P., Rowe, C., Salloway, S., Sarazin, M., Epelbaum, S., de Souza, L.C., Vellas, B., Visser, P.J., Schneider, L., Stern, Y., Scheltens, P., Cummings, J.L., 2014. Advancing research diagnostic criteria for Alzheimer's disease: the IWG-2 criteria. Lancet Neurol. 13, 614–629.

Folstein, M.F., Folstein, S.E., McHugh, P.R., 1975. Minimental state. A practical method for grading the cognitive state of patients for the clinician. J. Psychiatr. Res. 12, 189–198.

Grand, J.H., Caspar, S., MacDonald, S.W., 2011. Clinical features and multidisciplinary approaches to dementia care. J. Multidiscip. Healthc. 4, 125e147.

Green, R.C., Roberts, J.S., Cupples, L.A., Relkin, N.R., Whitehouse, P.J., Brown, T., Eckert, S.L., Butson, M., Sadovnick, A.D., Quaid, K.A., Chen, C., Cook-Deegan, R., Farrer, L.A., REVEAL Study Group, 2009. Disclosure of APOE genotype for risk of Alzheimer's disease. N. Engl. J. Med. 361, 245–254.

Hancock, P., Larner, A.J., 2011. Test Your Memory test: diagnostic utility in a memory clinic population. Int. J. Geriatr. Psychiatry 26, 976–980.

Harrison, J., Minassian, S.L., Jenkins, L., Black, R.S., Koller, M., Grundman, M., 2007. A neuropsychological test battery for use in Alzheimer disease clinical trials. Arch. Neurol. 64, 1323–1329.

Hogervorst, E., Bandelow, S., Combrinck, M., Irani, S.R., Smith, A.D., 2003. The validity and reliability of 6 sets of clinical criteria to classify Alzheimer's disease and vascular dementia in cases confirmed post-mortem: added value of a decision tree approach. Dement. Geriatr. Cogn. Disord. 16, 170–180.

Hughes, C.P., Berg, L., Danziger, W.L., Coben, L.A., Martin, R.L., 1982. A new clinical scale for the staging of dementia. Br. J. Psychiatry 140, 566–572.

Hyman, B.T., 1997. The neuropathological diagnosis of Alzheimer's disease: clinical-pathological studies. Neurobiol. Aging 18 (4 Suppl), S27–S32.

Hyman, B.T., Trojanowski, J.Q., 1997. Consensus recommendations for the postmortem diagnosis of Alzheimer disease from the National Institute on Aging and the Reagan Institute Working Group on diagnostic criteria for the neuropathological assessment of Alzheimer disease. J. Neuropathol. Exp. Neurol. 56, 1095–1097.

Hyman, B.T., Phelps, C.H., Beach, T.G., Bigio, E.H., Cairns, N.J., Carrillo, M.C., Dickson, D.W., Duyckaerts, C., Frosch, M.P., Masliah, E., Mirra, S.S., Nelson, P.T., Schneider, J.A., Thal, D.R., Thies, B., Trojanowski, J.Q., Vinters, H.V., Montine, T.J., 2012. National Institute on Aging—Alzheimer's Association guidelines for the neuropathologic assessment of Alzheimer's disease. Alzheimers Dement. 8, 1–13.

Iacono, D., Resnick, S.M., O'Brien, R., Zonderman, A.B., An, Y., Pletnikova, O., Rudow, G., Crain, B., Troncoso, J.C., 2014. Mild cognitive impairment and asymptomatic Alzheimer disease subjects: equivalent β-amyloid and tau loads with divergent cognitive outcomes. J. Neuropathol. Exp. Neurol. 73, 295–304.

Jack, Jr., C.R., Albert, M.S., Knopman, D.S., McKhann, G.M., Sperling, R.A., Carrillo, M.C., Thies, B., Phelps, C.H., 2011. Introduction to the recommendations from the National Institute on Aging-Alzheimer's Association workgroups on diagnostic guidelines for Alzheimer's disease. Alzheimers Dement. 7, 257–262.

Jack, Jr., C.R., Knopman, D.S., Weigand, S.D., Wiste, H.J., Vemuri, P., Lowe, V., Kantarci, K., Gunter, J.L., Senjem, M.L., Ivnik, R.J., Roberts, R.O., Rocca, W.A., Boeve, B.F., Petersen, R.C., 2012. An operational approach to National Institute on Aging-Alzheimer's Association criteria for preclinical Alzheimer disease. Ann Neurol. 71, 765–775.

Johnson, D.K., Morris, J.C., Galvin, J.E., 2005. Verbal and visuospatial deficits in dementia with Lewy bodies. Neurology 65, 1232–1238.

Khan, T.K., Alkon, D.L., 2010. Early diagnostic accuracy and pathophysiologic relevance of an autopsy-confirmed Alzheimer's disease peripheral biomarker. Neurobiol. Aging 31, 889–900.

Knopman, D.S., DeKosky, S.T., Cummings, J.L., Chui, H., Corey-Bloom, J., Relkin, N., Small, G.W., Miller, B., Stevens, J.C., 2001. Practice parameter: diagnosis of dementia (an evidence-based review). Report of the Quality Standards Subcommittee of the American Academy of Neurology. Neurology 56, 1143–1153.

Knopman, D.S., Parisi, J.E., Salviati, A., Floriach-Robert, M., Boeve, B.F., Ivnik, R.J., Smith, G.E., Dickson, D.W., Johnson, K.A., Petersen, L.E., McDonald, W.C., Braak, H., Petersen, R.C., 2003. Neuropathology of cognitively normal elderly. J. Neuropathol. Exp. Neurol. 62, 1087–1095.

Lim, A., Tsuang, D., Kukull, W., Nochlin, D., Leverenz, J., McCormick, W., Bowen, J., Teri, L., Thompson, J., Peskind, E.R., Raskind, M., Larson, E.B., 1999. Clinico-neuropathological correlation of Alzheimer's disease in a community-based case series. J. Am. Geriatr. Soc. 47, 564–569.

Mathuranth, P., Nestor, P., Berrios, G., Rakowicz, W., Hodges, J.R., 2000. A brief cognitive test battery to differentiate Alzheimer's disease and frontotemporal dementia. Neurology 55, 1613–1620.

McKhann, G., Drachman, D., Folstein, M., Katzman, R., Price, D., Stadlan, E.M., 1984. Clinical diagnosis of Alzheimer's disease: report of the NINCDS-ADRDA Work Group under the auspices of Department of Health and Human Services Task Force on Alzheimer's Disease. Neurology 34, 939–944.

McKhann, G.M., Knopman, D.S., Chertkow, H., Hyman, B.T., Jack, Jr., C.R., Kawas, C.H., Klunk, W.E., Koroshetz, W.J., Manly, J.J., Mayeux, R., Mohs, R.C., Morris, J.C., Rossor, M.N., Scheltens, P., Carrillo, M.C., Thies, B., Weintraub, S., Phelps, C.H., 2011. The diagnosis of dementia due to Alzheimer's disease: recommendations from the National Institute on Aging-Alzheimer's Association workgroups on diagnostic guidelines for Alzheimer's disease. Alzheimers Dement. 7, 263–269.

Mirra, S.S., Heyman, A., McKeel, D., Sumi, S.M., Crain, B.J., Brownlee, L.M., Vogel, F.S., Hughes, J.P., van Belle, G., Berg, L., 1991. The Consortium to Establish a Registry for Alzheimer's Disease (CERAD). Part II. Standardization of the neuropathologic assessment of Alzheimer's disease. Neurology 41, 479–486.

Molloy, D.W., Standish, T.I., 1997. A guide to the standardized Mini-Mental State Examination. Int. Psychogeriatr. 9 (Suppl 1), 87–94.

Molloy, D.W., Alemayehu, E., Roberts, R., 1991. Reliability of a Standardized Mini-Mental State Examination compared with the traditional Mini-Mental State Examination. Am. J. Psychiatry 148, 102–105.

Morris, J.C., 1993. The clinical dementia rating (CDR): current vision and scoring rules. Neurology 43, 2412–2414.

Morris, J.C., 2005. Ealy-stage and preclinical Alzheimer's disease. Alzheimer Dis Assoc Disord 19, 163–165.

Morris, J.C., Ernesto, C., Schafer, K., Coats, M., Leon, S., Sano, M., Thal, L.J., Woodbury, P., 1997. Clinical dementia rating training and reliability in multicenter studies: the Alzheimer's Disease Cooperative Study experience. Neurology 48, 1508–1510.

Nasreddine, Z.S., Phillips, N.A., Bédirian, V., Charbonneau, S., Whitehead, V., Collin, I., Cummings, J.L., Chertkow, H., 2005. The Montreal Cognitive Assessment (MoCA): a brief screening tool for mild cognitive impairment. J. Am. Geriatr. Soc. 53, 695–699.

Nelson, P.T., Jicha, G.A., Kryscio, R.J., Abner, E.L., Schmitt, F.A., Cooper, G., Xu, L.O., Smith, C.D., Markesbery, W.R., 2010. Low sensitivity in clinical diagnoses of dementia with Lewy bodies. J. Neurol. 257, 359–366.

Nishiwaki, Y., Breeze, E., Smeeth, L., Bulpitt, C.J., Peters, R., Fletcher, A.E., 2004. Validity of the Clock-Drawing Test as a screening tool for cognitive impairment in the elderly. Am. J. Epidemiol. 160, 797–807.

Petersen, R.C., Smith, G.E., Waring, S.C., Ivnik, R.J., Tangalos, E.G., Kokmen, E., 1999. Mild cognitive impairment: clinical characterization and outcome. Arch. Neurol. 56, 303–308.

Petersen, R.C., Doody, R., Kurz, A., Mohs, R.C., Morris, J.C., Rabins, P.V., Ritchie, K., Rossor, M., Thal, L., Winblad, B., 2001. Current concepts in mild cognitive impairment. Arch. Neurol. 58, 1985–1992.

Petrovitch, H., White, L.R., Ross, G.W., Steinhorn, S.C., Li, C.Y., Masaki, K.H., Davis, D.G., Nelson, J., Hardman, J., Curb, J.D., Blanchette, P.L., Launer, L.J., Yano, K., Markesbery, W.R., 2001. Accuracy of clinical criteria for AD in the Honolulu-Asia Aging Study, a population-based study. Neurology 57, 226–234.

Price, J.L., Morris, J.C., 1999. Tangles and plaques in nondemented aging and "preclinical" Alzheimer's disease. Ann. Neurol. 45, 358–368.

Price, J.L., McKeel, Jr., D.W., Buckles, V.D., Roe, C.M., Xiong, C., Grundman, M., et al., 2009. Neuropathology of nondemented aging: presumptive evidence for preclinical Alzheimer disease. Neurobiol. Aging 30, 1026–1036.

Qureshi, K., Hodkinson, M., 1974. Evaluation of a 10 question mental test of the institutionalized elderly. Age Ageing 3, 152–157.

Reisberg, B., Ferris, S.H., de Leon, M.J., Crook, T., 1982. The Global Deterioration Scale for assessment of primary degenerative dementia. Am. J. Psychiatry 139, 1136–1139.

Robert, P., Ferris, S., Gauthier, S., Ihl, R., Winblad, B., Tennigkeit, F., 2010. Review of Alzheimer's disease scales: is there a need for a new multi-domain scale for therapy evaluation in medical practice? Alzheimers Res. Ther. 2, 24.

Rockwood, K., Strang, D., MacKnight, C., Downer, R., Morris, J.C., 2000. Interrater reliability of the Clinical Dementia Rating in a multicenter trial. J. Am. Geriatr. Soc. 48, 558–559.

Rogaeva, E., Meng, Y., Lee, J.H., Gu, Y., Kawarai, T., Zou, F., Katayama, T., Baldwin, C.T., Cheng, R., Hasegawa, H., Chen, F., Shibata, N., Lunetta, K.L., Pardossi-Piquard, R., Bohm, C., Wakutani, Y., Cupples, L.A., Cuenco, K.T., Green, R.C., Pinessi, L., Rainero, I., Sorbi, S., Bruni, A., Duara, R., Friedland, R.P., Inzelberg, R., Hampe, W., Bujo, H., Song, Y.Q., Andersen, O.M., Willnow, T.E., Graff-Radford, N., Petersen, R.C., Dickson, D., Der, S.D., Fraser, P.E., Schmitt-Ulms, G., Younkin, S., Mayeux, R., Farrer, L.A., St George-Hyslop, P., 2007. The neuronal sortilin-related receptor SORL1 is genetically associated with Alzheimer disease. Nat. Genet. 39, 168–177.

Román, G.C., Tatemichi, T.K., Erkinjuntti, T., Cummings, J.L., Masdeu, J.C., Garcia, J.H., Amaducci, L., Orgogozo, J.M., Brun, A., Hofman, A., et al., 1993. Vascular dementia: diagnostic criteria for research studies. Report of the NINDS-AIREN International Workshop. Neurology 43, 250–260.

Rosen, W.G., Mohs, R.C., Davis, K.L., 1984. A new rating scale for Alzheimer's disease. Am. J. Psychiatry 141, 1356–1364.

Schmitt, F.A., Davis, D.G., Wekstein, D.R., Smith, C.D., Ashford, J.W., Markesbery, W.R., 2000. Preclinical AD revisited: neuropathology of cognitively normal older adults. Neurology 55, 370–376.

Sperling, R.A., Aisen, P.S., Beckett, L.A., Bennett, D.A., Craft, S., Fagan, A.M., Iwatsubo, T., Jack, Jr., C.R., Kaye, J., Montine, T.J., Park, D.C., Reiman, E.M., Rowe, C.C., Siemers, E., Stern, Y., Yaffe, K., Carrillo, M.C., Thies, B., Morrison-Bogorad, M., Wagster, M.V., Phelps, C.H., 2011. Toward defining the preclinical stages of Alzheimer's disease: recommendations from the National Institute on Aging-Alzheimer's Association workgroups on diagnostic guidelines for Alzheimer's disease. Alzheimers Dement. 7, 280–292.

Thal, D.R., Rüb, U., Orantes, M., Braak, H., 2002. Phases of A beta-deposition in the human brain and its relevance for the development of AD. Neurology 58, 1791–1800.

Vemuri, P., Whitwell, J.L., Kantarci, K., Josephs, K.A., Parisi, J.E., Shiung, M.S., Knopman, D.S., Boeve, B.F., Petersen, R.C., Dickson, D.W., Jack, Jr, C.R., 2008. Antemortem MRI based STructural Abnormality iNDex (STAND)-scores correlate with postmortem Braak neurofibrillary tangle stage. Neuroimage 42, 559–567.

Xekardaki, A., Kövari, E., Gold, G., Papadimitropoulou, A., Giacobini, E., Herrmann, F., Giannakopoulos, P., Bouras, C., 2015. Neuropathological changes in aging brain. Adv. Exp. Med. Biol. 821, 11–17.

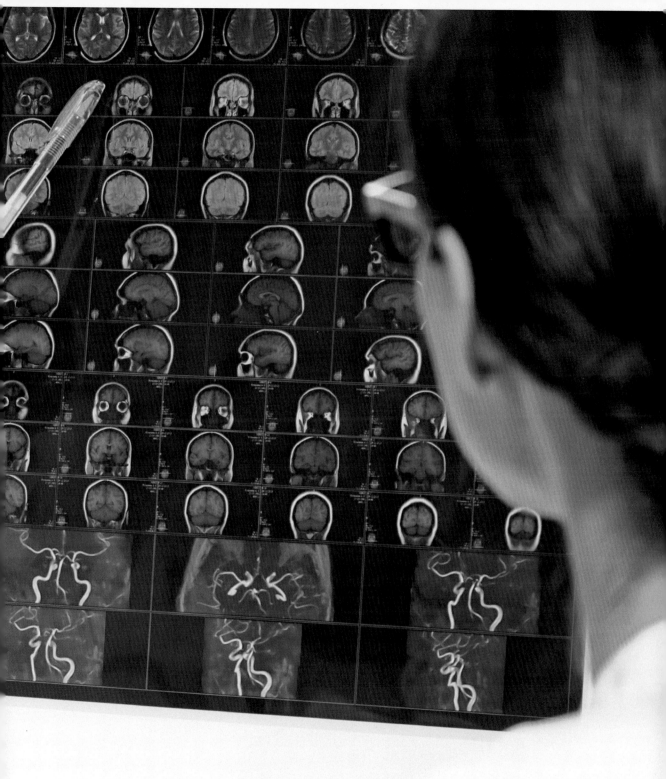

CHAPTER 3

Neuroimaging Biomarkers in Alzheimer's Disease

Chapter Outline

(Continued)

Biomarkers in Alzheimer's Disease. http://dx.doi.org/10.1016/B978-0-12-804832-0.00003-1
Copyright © 2016 Elsevier Inc. All rights reserved.

3.1 NEUROIMAGING IN ALZHEIMER'S DISEASE

Neuroimaging biomarkers are the most widely researched Alzheimer's disease (AD) biomarkers along with cerebrospinal fluid (CSF) AD biomarkers (see Chapter 5 for a discussion of CSF biomarkers). The field of neuroimaging has made revolutionary advances over the last several decades with the development and implementation of new imaging techniques, new scanner and detector technologies, and computing and bioinformatics. Several different neuroimaging modalities are being applied to the detection of AD biomarkers, including single-photon emission computed tomography (SPECT), positron emission tomography (PET), computed tomography (CT), magnetic resonance imaging (MRI), and magnetic resonance spectroscopy (MRS). Although neuroimaging is expensive and technically challenging, it is also noninvasive or semiinvasive in nature and can be performed in vivo in real time and longitudinally, making neuroimaging biomarkers among the most attractive ones for evaluating the progression of neurological disease, including AD. Several neuropathological abnormalities characteristic of AD can be detected by modern neuroimaging methods, including atrophy in specific brain regions or the whole brain (shrinkage), brain Aβ accumulation (amyloid plaques), hyperphosporylated tau (p-tau) deposition, neuronal damage (loss of neurons), abnormal cerebral blood flow, reduced levels of brain metabolites (indicating reduced activity of brain), abnormal neural activity, and regional inflammation of the brain. Recently, neuroimaging has proven useful for detecting abnormal neuronal network connectivity believed to underlie neurological dysfunction in a number of disorders. Neuroimaging may be particularly promising in detecting biomarkers related to preclinical AD, which may allow identification of individuals who are at high risk and who might benefit from early therapeutic intervention prior to widespread neuronal loss.

Depending on the biophysical nature of the AD biomarker, neuroimaging methods can be categorized into three broad groups: (1) structural neuroimaging, (2) functional neuroimaging, and (3) molecular neuroimaging.

3.1.1 Structural Neuroimaging

Brain atrophy (a decrease in brain area due to loss of gray and white matter) is common phenomenon in AD. Brain atrophy and selective loss of white matter (anisotropy) are attributes of progressive neurodegeneration in AD. Both cross-sectional and longitudinal cohort studies have revealed an association between structural changes in the AD brain and cognitive loss. In early stages of AD, shrinkage of specific brain regions such as the hippocampus is commonly observed and well correlated with memory impairment. Structural changes in different brain areas can be monitored by MRI and CT. In general, structural imaging can be used to quantify whole-brain atrophy and region-specific cerebral atrophy, allows a visual rating of hippocampal loss, and provides a volumetric measure of the whole brain. One structural neuroimaging study found an average whole-brain atrophy rate of 2.4 ± 1.1% per year in patients with clinically confirmed AD compared to 0.5 ± 0.4% per year for age-matched healthy control cases (Chan et al., 2001). In patients with AD, the hippocampal atrophy rate is higher than that of the whole brain. The average hippocampal atrophy rate is reported to be approximately 4–6% per year in patients with AD compared to approximately 1–2% per year for age-matched healthy control cases (Jack et al., 1998, 2000). A systematic metaanalysis that included the majority of studies related to hippocampal structural changes due to AD reported consistent rates of atrophy: an annual decline of approximately 4.5% in hippocampal volume for AD patients compared to a 1% decline for age-matched cognitively normal cases (Barnes et al., 2009).

3.1.2 Functional Neuroimaging

A growing body of literature suggests that neuronal networks and synaptic function are affected at the earliest stages of AD, prior to any clinical symptoms. The loss of synaptic integrity is one of major events in AD pathoprogression that can be probed by functional imaging. The main functional neuroimaging modalities are PET, functional MRI (fMRI), and SPECT. Functional monitoring of region-specific brain metabolism by PET and fMRI can be focused to dysfunctional brain areas. Arterial spin-labeled (ASL) MRI has also been used to monitor blood perfusion rate in different regions of the brain, providing an indirect measure of synaptic function and neuronal health. In general, functional imaging in AD can detect spatiotemporal patterns of cerebral hypometabolism, lower neuronal activity, and altered patterns of activation in different brain areas, each of which has been explored as a quantitative biomarker of AD.

3.1.3 Molecular Neuroimaging

Cellular or molecular changes in the AD brain can be detected by neuroimaging. Molecular imaging provides quantitative measure of specific brain metabolites such as N-acetyl aspartate, choline, and lipids. N-acetyl aspartate levels are a measure of neuronal mitochondria activity (Moffett et al., 2007) and are reduced in the AD brain. Changes in brain metabolites due to underlying AD pathology are imaged by nuclear MRS.

The neurodegenerative processes underlying AD begin several years before the clinical signs and symptoms are recognized, and it is believed that AD progresses to its advanced, symptomatic stages through multiple prodromal stages over a period of approximately two decades (for details, see Fig. 5.2). The three prodromal phases of AD are preclinical asymptomatic AD, symptomatic predementia or mild cognitive impairment (MCI) due to AD, and dementia due to AD

(Sperling et al., 2011). In addition, AD may develop with other neurological disorders of old age, including age-related decline in cognitive function or mild neurocognitive disorder, making antemortem definitive diagnosis of AD very difficult. Molecular neuroimaging can increase the accuracy of the differential diagnosis of dementia. Although early treatment of AD can slow disease progression, the ability to diagnose AD in its earliest stages (preclinical stage) is currently limited. PET, fMRI, and SPECT biomarkers can be measured noninvasively to diagnose preclinical AD. Neuroimaging techniques can also aid in identifying asymptomatic individuals who are at high risk of AD. The clinical need for earlier diagnosis and risk stratification of AD has accelerated the search for AD neuroimaging biomarkers. Ideally, these biomarkers should also be able to differentiate AD from non-AD dementias [frontotemporal dementia (FTP), Lewy body dementia (LBD), vascular dementia (VaD), Creutzfeldt–Jacob disease (CJD), corticobasal degeneration (CBD), progressive supranuclear palsy (PSP), and tauopathy, etc.], assess risk of AD when used in combination with other known risk factors, track patients through the prodromal stages of AD, accelerate the discovery and screening of potential therapeutic agents, guide therapeutic decision-making, and monitor therapeutic efficacy.

Due to substantial investment by governments, the pharmaceutical industry, and private donors, neuroimaging biomarkers of preclinical AD is a highly active area of research. In cross-sectional studies, neuroimaging biomarkers have been proven to be excellent tools to support clinical diagnosis of AD. New recommendations from the Alzheimer's Association and the National Institute on Aging (NIA) recommend that neuroimaging and cerebrospinal fluid (CSF) biomarkers should be incorporated into the diagnosis of AD (McKhann et al., 2011). Several MRI and PET neuroimaging biomarkers of AD have been well validated in multicenter studies and are included in the Alzheimer's Association and NIA diagnostic criteria: (1) brain atrophy, (2) increased [^{11}C]-Pittsburgh Compound B (^{11}C-PIB) binding to amyloid plaques, (3) increased ^{18}F-florbetapir (Amyvid)-PET binding to amyloid plaques, and (4) decreased [^{18}F]-fluoro-2-deoxy-D-glucose (^{18}FDG) uptake (Table 3.1). The International Working Group (IGW:

Table 3.1	Advanced Neuroimaging Biomarkers of Alzheimer's Disease	
Neuroimaging Biomarkers		**Abnormality in Alzheimer's Disease**
Neuronal imaging	MRI*	Medial temporal atrophy
	fMRI	Disrupted default-mode neural network
	^{11}C-PIB PET*	Increased amyloid plaques
	^{18}F-florbetapir (Amyvid) PET**	Increased amyloid burden
	^{18}FDG-PET*	Decreased glucose uptake

MRI, magnetic resonance imaging; fMRI, functional MRI; PET, positron emission tomography; ^{11}C-PIB, [^{11}C]-Pittsburgh compound B; ^{18}FDG, [^{18}F]-fluoro-2-deoxy-D-glucose.
*Biomarker is included in 2011 AD criteria (National Institute on Aging and Alzheimer's Association). International Working Group (IGW: 2007, 2010, 2014) has also recommended to include these AD biomarkers to support diagnosis of AD.
**FDA approval for clinical use in April 2012.

Dubois et al., 2007, 2010, 2014) has also recommended that these AD biomarkers be used to support a diagnosis of AD. In this chapter, we provide a critical discussion of the diagnostic performance of neuroimaging biomarkers of AD, their ability to differentiate between AD and common non-AD dementias, their use in longitudinal assessment of prolonged AD prodromal stages, their predictive capacity for risk assessment, and their use for monitoring therapeutic efficacy, as reported in the literature by wide range of research groups.

3.2 COMMON NEUROIMAGING ALZHEIMER'S DISEASE BIOMARKERS AND MODALITIES

Neuroimaging of the brain enables the measurement of various structural and functional biomarkers of AD, including atrophy, deposition of Aβ, accumulation of hyperphosphorylation of tau, changes in metabolism, inflammation, blood flow, and perfusion, and regional neuronal network activity. It detects the pathological condition of the brain and holds great promise for diagnosing dementias, including AD. One of the exciting applications of noninvasive neuroimaging techniques is the ability to quantitatively assess discrete AD-specific alterations in brain anatomical structures and physiological functions. Region-specific Aβ deposition and synaptic dysfunction can be imaged at preclinical (presymptomatic) AD stages and tracked over time as patients move from preclinical disease to MCI to definitive AD. CSF biomarkers, genetic biomarkers, and other biomarkers are unable to quantify regional Aβ deposition. The best-studied neuroimaging biomarkers of AD are detected using a number of imaging modalities, including MRI, fMRI, MRS, PET, SPECT, and electroencephalography (EEG) (Table 3.2).

3.3 MAGNETIC RESONANCE IMAGING

MRI has a high degree of imaging flexibility, high tissue contrast, and no need for ionizing radiation, making it an ideal modality for brain imaging. Brain atrophy is one of the most popular MRI biomarkers used in the diagnosis of AD (Table 3.1). MRI-measured rates of regional atrophy in the hippocampus and entorhinal cortex, as well as global atrophy, have been used as outcome measures in longitudinal studies and clinical trials. MRI used to detect AD biomarkers falls into three major categories: (1) structural MRI (sMRI), (2) functional MRI (fMRI), and (3) magnetic resonance spectroscopy (MRS).

3.3.1 Structural MRI

Because AD is a neurodegenerative disease, noninvasive imaging modalities such as sMRI can be applied in vivo to assess changes in the physical structure of the AD brain. Structural MRI (sMRI) provides useful information about the anatomy of the brain and has been used extensively to examine the changes in the brain associated with normal aging. In the past three decades, advances in scanner technology, image acquisition protocols, experimental design, and analysis methods using bioinformatics tools have moved sMRI from a mere

Table 3.2 **Neuroimaging Biomarkers of Alzheimer's Disease**

Neuroimaging Marker	Imaging Component	Biomarkers	Remarks	References
MRI	Various areas of the brain	■ Brain volume ■ Brain atrophy	■ Useful for longitudinal studies, but low specificity for AD vs non-AD dementias ■ No radiation exposure ■ Claustrophobic reaction ■ No pacemaker	Jack et al. (1999) Visser et al. (1999) Fox et al. (1999a) Fennema-Notestine et al. (2009) Fox and Kennedy (2009) Frisoni et al. (2010) Vemuri and Jack (2010)
fMRI	Blood flow in areas of the brain related to memory processing	■ Paramagnetic properties of oxy-hemoglobin/deoxy-hemoglobin in blood flow	■ No radiation exposure ■ Claustrophobic reaction ■ No pacemaker	Machulda et al. (2003) Pihlajamäki et al. (2009) Vemuri et al. (2012)
DTI	Abnormal diffusion of water molecules	■ Overall mean diffusivity of water molecule ■ Measure fractional anisotropy	■ Claustrophobic reaction ■ No pacemaker	Chua et al. (2009) Zhang et al. (2009) Sexton et al. (2011) Oishi et al. (2011) Nedelska et al. (2015)
ASL	Abnormal cerebral blood flow	■ Employs magnetically labeled water as tracer of blood flow	■ Claustrophobic reaction ■ No pacemaker	Johnson et al. (2005) Dai et al. (2009) Alsop et al. (2010) Wang et al. (2013)

Technique	Description	Features	Limitations	References
MRS	Abnormality in brain metabolite	■ Chemical shift difference of the nuclei (^{1}H, ^{13}C, and ^{31}P) in a given small volume of brain interested region of brain ■ Typical brain metabolites are choline, creatine (Cr), N-acetylaspartate (NAA), and myoinositol (ml)	■ Differential ratios of metabolites ■ Signal quantification presents a major problem ■ Claustrophobic reaction ■ No pacemaker	Bates et al. (1996) Kantarci et al. (2004; Forlenza et al. (2005) Zhu et al. (2006) Mandal (2007)
PET	In vivo, radiotracer binding/uptake by specific brain targets	■ Aβ using ^{11}C-PIB ■ Glucose uptake using ^{18}FDG ■ PHFs-tau binding radio-labeled ligands	■ High cost ■ Radiation exposure ■ Unlikely to be useful for population screening or longitudinal monitoring ■ High association with AD clinical severity and neurodegeneration	Klunk et al. (2004) Scheinin et al. (2009) Jagust et al. (2009) Okamura et al. (2014a)
SPECT	Brain perfusion as an indicator of brain metabolism	■ Blood flow measure using ^{99m}Tc-HMPAO	■ Low resolution compared with MRI ■ Radiotracers have longer half-lives than PET tracers ■ Radiation exposure ■ Unlikely to be useful for population screening or longitudinal monitoring	Dougall et al. (2004) Bonte et al. (2004)
EEG	Analysis of AD EEG signal	■ EEG spectrum is affected by AD	■ Synchrony measures of optimized EEG frequency bands	Gallego-Jutglà et al. (2012) Neto et al. (2015)

ASL–MRI, arterial spin-labeled magnetic resonance imaging; DTI, diffusion tensor imaging; EEG, electroencephalography; fMRI, functional MRI; MRS, magnetic resonance spectroscopy; PET, positron emission tomography; PHFs, paired helical filaments; SPECT, single-photon emission computed tomography; Aβ, beta-amyloid protein; ^{11}C-PIB, [^{11}C]-Pittsburgh Compound B; ^{18}FDG, [^{18}F]-fluoro-2-deoxy-D-glucose; ^{99m}Tc, metastable nuclear isomer of technetium-99; HMPAO, hexamethylpropylene amine oxime.

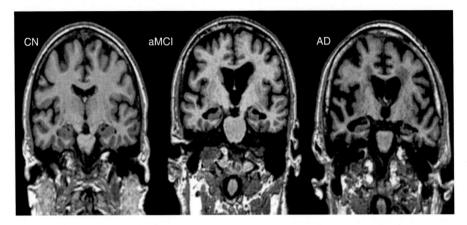

FIGURE 3.1

Typical brain atrophy images recorded by structural MRI. sMRI biomarkers of AD include a significant increase in whole-brain atrophy and atrophy in different parts of the brain. Whole-brain atrophy progressively increases from a cognitively normal (*CN*) older individual to amnestic mild cognitive impairment (*aMCI*) to Alzheimer's disease (*AD*). *(Adapted from Vemuri, P., Jack, C.R., 2010. Alzheimer's Res. Ther. 2, 23; with copyright permission from BioMed Central).*

brain imaging technique to a method for the quantitative measurement of AD biomarkers (Jack et al., 1999; Visser et al., 1999; Fox et al., 1999a,b; Fennema-Notestine et al., 2009; Fox and Kennedy, 2009; Frisoni et al., 2010; Vemuri and Jack, 2010). sMRI is the most widely used neuroimaging technique for characterization of dementia in normal aging, MCI, AD, and non-AD dementias.

Brain atrophy is a common feature of both familial AD as well as sporadic AD (Apostolova et al., 2011). Brain atrophy measured by sMRI correlates with cognitive impairment and tau deposition in AD and is the most widely used neuroimaging biomarker (Fig. 3.1, Tables 3.2 and 3.3). High-resolution sMRI can assess atrophy of critical brain areas such as the parahippocampal gyrus, hippocampus, amygdala, posterior association cortex, and subcortical region (Fox and Freeborough, 1997; Shi et al., 2009; Fennema-Notestine et al., 2009; Frisoni et al., 2010; Vemuri and Jack, 2010; Dickerson and Wolk, 2012). In addition to providing a visual rating of hippocampal atrophy and manual volumetry of the hippocampus, several techniques for quantitative assessment of brain volume and atrophy in MRI images have been introduced, such as automated whole-brain volumetry, quantitative region of interest-based volumetry, the quantitative voxel-based technique, tensor-based morphometric technique, and global atrophy quantification technique. Measurement of hippocampal volume from sMRI images is done by different laboratories using different protocols, leading to heterogeneous results. To address this problem, the European Alzheimer's disease Consortium and Alzheimer's Disease Neuroimaging Initiative (ADNI) introduced a harmonized protocol for hippocampal segmentation in sMRI images (Bocchetta et al., 2015), and advanced automated image analysis software is now available (NeuroQuant; CorTechs Labs Inc., San Diego, CA, USA).

Table 3.3 Differential Diagnosis of Alzheimer's Disease (AD) and Non-AD Dementia Patients Using MRI

AD vs Non-AD Dementia	MRI Techniques	Remarks	References
AD vs FTD Clinically confirmed AD (*n* = 103) FTD (*n* = 17) Control (*n* = 73) Semantic dementia (*n* = 13) Progressive non-fluent aphasia	*Structural MRI:* hippocampal atrophy	■ No significant difference in hippocampal atrophy between FTD and AD	van de Pol et al. (2006)
AD vs non-AD dementia Autopsy confirmed AD (*n* = 11) LBD (*n* = 23) VaD (*n* = 12)	Medial temporal lobe atrophy (MTA) SN: 91% SP: 94%	At autopsy, Braak stage was significant predictor of MTA but not plaques	Burton et al. (2009)
AD vs LBD and VaD Clinically confirmed AD (*n* = 25) LBD (*n* = 27) VaD (*n* = 24) Control (*n* = 26)	*Structural MRI hippocaÜal atrophy:* LBD had significantly larger temporal lobe, hippocampus, and amygdala volumes than those with AD SN, SP, and ACU not determined	■ Significant difference in brain atrophy between LBD and AD; ■ No significant difference in brain atrophy between VaD and AD	Barber et al. (2000)
AD vs LBD Autopsy-confirmed AD (*n* = 30) LBD (*n* = 20) Mixed LBD/AD (*n* = 22) Control (*n* = 15)	*Atrophy rate:* LBD cases were characterized by lower global and regional rates of atrophy, similar to control SN, SP, and ACU not determined	■ Significant difference in brain atrophy between LBD and AD	Nedelska et al. (2015)

AD, Alzheimer's disease; FTD, frontotemporal dementia; MRI, magnetic resonance imaging; SN, sensitivity; SP, specificity; ACU, accuracy; LBD, dementia with Lewy body; VaD, vascular dementia.

There are several potential applications of sMRI detection of AD biomarkers, including early diagnosis of AD, distinguishing AD from MCI (Vemuri and Jack, 2010), evaluation of disease progression (Fox and Freeborough, 1997; Shi et al., 2009), differentiation of AD and non-AD dementias (Jack et al., 2003; McKeith et al., 2005; Barber et al., 1999), prediction of progression from MCI to AD (Duara et al., 1999; Petersen, 2007), screening of high-risk individuals, and measurement of drug efficacy (Krishnan et al., 2003; Fox et al., 2005; Jack et al., 2008; Fox and Kennedy, 2009; Frisoni and Delacourte, 2009; Yuan et al., 2009). sMRI used in longitudinal studies found that the rate of hippocampal atrophy in AD cases is much higher (4–6% per year) compared to age-matched controls (1–2% per year) (Jack et al., 2000; Haller et al., 1997). By comparison, the rate of global atrophy measured by semiautomated brain boundary shift integral was lower (for AD $2.4 \pm 1.1\%$ vs age-matched controls $0.5 \pm 0.4\%$) (Chan et al., 2001). Atrophy measured by sMRI is one of the five AD biomarkers incorporated into the 2011 AD diagnostic criteria (McKhann et al., 2011; Sperling et al., 2011) and the Dubois criteria (Dubois et al., 2007, 2010, 2014). Memory impairment in early AD occurs predominantly in the medial temporal lobe area, hippocampus, and dentate gyrus. Variability of atrophy measured in entorhinal cortex is higher compared to that in the hippocampus. Brain atrophy determined by sMRI correlate with CSF biomarkers and levels of cognitive impairment and the combination was found to provide better discrimination between AD and age-matched normal control cases than either marker alone (de Leon et al., 2006; Vemuri et al., 2009; Walhovd et al., 2010).

Gray matter (GM) atrophy in AD is a reflection of change of brain morphometry and is related to loss of neurons, synapses, and dendritic structures. White matter (WM) changes are related to loss of structural integrity of the brain such as demyelination and dying axonal processes due to AD. Loss of both GM and WM is indicative of the structural damage caused by AD. Areas affected by WM loss due to AD pathology are the temporal, parietal, and frontal lobes as well as the corpus callosum (Double et al., 1996; Teipel et al., 2002; Stoub et al., 2006; Chaim et al., 2007). Compared to late-onset AD, early-onset AD cases have greater cingulum WM atrophy (Migliaccio et al., 2012). WM damage measured using sMRI and sophisticated analysis methods such as voxel-based morphometric analysis can distinguish early-onset AD from late-onset AD (Canu et al., 2012; Migliaccio et al., 2012). Most studies of sMRI to quantify medial temporal atrophy reported reasonably good diagnostic sensitivity for detecting AD compared with control cases. However, the sensitivity and specificity of brain atrophy by MRI are very low for non-AD dementia cases, such as VaD and LBD (McKeith et al., 2005). MRI-based measurements of whole-brain atrophy showed a modest correlation with CSF biomarker levels in patients with AD (Sluimer et al., 2010), but a stronger correlation with clinical progression of AD, measured by changes in the Mini Mental Score Examination (MMSE) score.

Decrease in whole-brain volume (brain atrophy) as an AD biomarker measured by sMRI has been validated in multicenter studies and has been used for study of subject selection and as an outcome measure in several clinical trials. The European Medical Agency (EMA) has endorsed the use of hippocampal volumetric sMRI as a biomarker for patient selection in MCI clinical trials [Committee for Medicinal Products for Human Use (CHMP). Qualification opinion of low hippocampal volume (atrophy) by MRI for use in clinical trials for regulatory purpose in predementia stage of Alzheimer's disease. EMA/CHMP/SAWP/809208/2011. November 17, 2011].

More advances sMRI techniques such as diffusion tensor imaging (DTI) and diffusion weighted imaging (DWI) are now being investigated for monitoring structural changes in the AD brain. More research and validation studies are required for all advanced sMRI techniques before they can be used in larger cohort studies.

3.3.2 Functional MRI

sMRI provides structural information, whereas functional MRI (fMRI) provides both structural and functional information (Machulda et al., 2003; Pihlajamäki et al., 2009). fMRI makes it possible to image the activity of the brain during cognitive, sensory, and motor-related tasks. fMRI was initially developed in early 1990s by Dr Kenneth Kwong in the NMR Center at the Massachusetts General Hospital and Dr Seiji Ogawa of Bell Laboratories and his colleagues at the University of Minnesota. In principle, brain fMRI measures neuronal activity by imaging the paramagnetic properties of oxyhemoglobin/deoxyhemoglobin in blood flowing through the brain. The blood oxygen level dependent (BOLD) signal detected by fMRI is the measure of brain activity and fMRI images reflect the tiny metabolic changes associated with neuronal activity in discrete parts of the brain. Fluctuation in the BOLD signal can provide fine detail about brain activity and network deficiencies in the AD brain with relatively high spatial and medium-grade temporal resolution. fMRI has several potential advantages, such as detecting subtle preclinical changes in neuronal networks and detecting changes in neural activity during therapeutic intervention. fMRI also has the potential to image subtle changes in brain connectivity. Abnormal fMRI signal activity at the very early stages of AD may indicate impaired connectivity and synaptic dysfunction, but not neuronal loss (Braskie et al., 2012). Combined with neuropsychologic and cognitive tests, brain fMRI can identify preclinical structural changes in the posteromedial cortical, frontotemporal, and parietal lobes and functional changes in neuronal activity associated with AD and with therapeutic intervention. One fMRI study found that patients treated with memantine had higher activity measured by a default mode network (DMN) paradigm compared to those who were not treated (placebo) (Lorenzi et al., 2011). Future prospective studies are needed to evaluate the capacity of MRI for differential diagnosis of AD and non-AD dementias and distinguishing early AD from normal age-related cognitive impairment. One challenge encountered by investigators is that fMRI protocols are difficult to conduct in cognitively impaired subjects, and patient-related artifacts have been found in longitudinal studies.

3.3.3 Magnetic Resonance Spectroscopy

MRS is a neuroimaging technique that produces reliable and reproducible quantitative in vivo data on brain neurochemistry. Although both MRI and MRS are

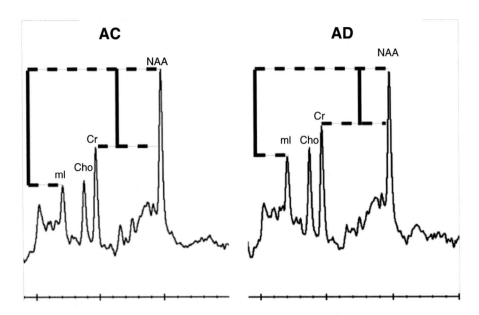

FIGURE 3.2

Example of 1H-MRS of age-matched control (*AC*) and Alzheimer's disease (*AD*) brains scanned from the posterior cingulate voxel with an echo time of 30 mS. Brain metabolites [choline (Cho), creatine (*Cr*), N-acetylaspartate (*NAA*), and myoinositol (*mI*)] concentrations were measured based on the individual peak area. The *NAA/Cr* ratio is lower for *AD* cases compared to *AC* cases, whereas the *mI/Cr* ratio is higher. *(Adapted from Kantarci, K., Petersen, R.C., Boeve, B.F., Knopman, D.S., Tang-Wai, D.F., O'Brien, P.C., Weigand, S.D., Edland, S.D., Smith, G.E., Ivnik, R.J., Ferman, T.J., Tangalos, E.G., Jack, C.R., 2004. Neurology 63, 1393–1398; with copyright permission from the Wolters Kluwer Health, Inc.).*

based on same basic physical principle, MRS consists of NMR (nuclear magnetic resonance) peaks that represent different brain metabolites. MRI produces images based on proton signals from water content in the brain tissue, whereas MRS reports the chemical shift in the nuclei (^{1}H, ^{13}C, and ^{31}P) in a given region of interest of the brain. In MRS, a small volume of tissue (a voxel) is selectively excited in a magnetic field and the free induction decay (FID) is recorded to produce an MR spectrum (for a more detailed description, see Mandal, 2007). The area under each NMR peak corresponds to the concentration of that particular metabolite. Differences in the peak area of metabolites in the brain region of interest provides a measure of the level of neurochemical processing, which is reflective of the pathophysiologic state of brain.

A variety of brain metabolites can be measured in a single session. Two major categories of metabolites are considered to be potential biomarkers of AD: energy metabolites and lipid metabolites. In MRS imaging of the AD brain, the typical energy metabolites measured include choline (Cho), creatine (Cr), N-acetylaspartate (NAA), and myoinositol (mI) (Fig. 3.2). Lipid metabolites are mostly composed of signals from phosphoethanolamine (PE) and phosphocholine in terms of ^{31}P-MRS. Among these specific metabolites, NAA is a neuronal marker seen only in nervous system tissues and is a measure of neuronal mitochondrial health (Moffett et al., 2007). Choline is an indicator of membrane integrity, creatine is thought to be a marker of the energetic status of neurons, and myo-inositol levels reflect the activated glial response in the brain. In general, NAA in the hippocampus is reduced in AD and in MCI cases that convert to AD (Chao et al., 2005). For a quantitative measurement of these metabolites, levels are normalized to an internal standard of creatine concentration (Bates et al., 1996; Zhu et al., 2006). The NAA/Cr ratio was found to be lower in patients with AD, FTD, and VaD compared to age-matched control cases, whereas mI/Cr was higher in patients with AD and FTD compared to age-matched control cases (Kantarci et al., 2004). This study showed that MRS can distinguish AD patients from age-matched control cases with good sensitivity and specificity. Abnormalities of membrane phospholipid metabolism in the AD brain are reflected in the levels of ^{31}P-MRS (Forlenza et al., 2005).

MRS is less sensitive than MRI, since concentrations of MRS-detectable nuclei (^{1}H, ^{13}C, and ^{31}P) in brain metabolites are magnitudes lower than highly abundant MRI-detectable hydrogen (hydrogen from water) in brain tissues. In some brain regions, such as the cerebellum and temporal lobes, metabolites affected by AD are difficult to assess by MRS due to magnetic field heterogeneities. Absolute concentration measurements of brain metabolites are required from MRS peaks to differentiate AD from non-AD cases. In practice, there are major difficulties in quantifying MRS metabolite peaks. Single-center studies of MRS are producing encouraging results. Large-scale clinical trials validating MRS AD biomarkers have yet to be reported.

There are several disadvantages of MRI biomarkers of AD. Loss of brain atrophy can be caused by neurodegenerative diseases other than AD; advanced age may cause wide-spread brain material loss even in the absence of pathology. MRI is expensive compared with other AD biomarker tests and some MRI protocols require highly experienced personnel and advanced bioinformatics specialists for rigorous data analysis. The presence of metal implants in older individuals (eg, pacemakers) and the risk of a claustrophobic reaction are serious limitations to the utility of MRI-based biomarkers in older individuals with dementia. Head motion artifacts are often seen with DTI, fMRI and resting-state fMRI.

3.3.4 Advanced Magnetic Resonance Based Neuroimaging Modalities

The relatively poor specificity and sensitivity of conventional MRI biomarkers of AD has spurred further advances in MRI technology and bioinformatics. Several new MR modalities have been introduced recently that are able to detect subtle changes in brain tissues at the microscopic level. Measurable microscopic-level tissue damage indicative of neurodegeneration in early AD includes axonal integrity loss, demyelination, and break down of microtubule assemblies. These AD biomarkers can be detected by DTI, ASL, and task-free fMRI. All of these technologies are being developed with the goal of distinguishing changes in the functional activity of the AD brain in preclinical stages. Other advanced MRI methods that have been used to detect AD biomarkers include fiber tracking imaging (Taoka et al., 2006) and magnetization transfer imaging (Ridha et al., 2007).

3.3.4.1 DIFFUSION TENSOR IMAGING

DTI is an extension of sMRI that measures the microstructural integrity of brain regions by scanning the diffusion of water molecules. DTI was developed in the 1990s (Moseley et al., 1990; Beaulieu and Allen, 1994; Pierpaoli et al., 1996; Pierpaoli and Basser, 1996). In the literature, DTI is sometimes referred to as diffusion-weighted imaging (DWI). The ideal diffusion of a water molecule is controlled by thermal motion. Therefore, its probability distribution would be isotropic in the ideal case. Water-molecule diffusion in brain white matter is hindered by tightly packed axon bundles and nerve fibers; thus, the probability distribution of the diffusion of water molecules should be anisotropic. In AD, brain white matter integrity is compromised compared to age-matched nondemented controls. Therefore, diffusion of water molecules measured by sMRI should be different in the AD brain compared to controls. In DTI, the diffusion of water molecules is quantified in terms of a tensor model of a 3×3 symmetric matrix. DTI generally measures two parameters from an sMRI scan: (1) overall mean diffusivity (MD) of the water molecules and (2) fractional anisotropy (FA) (Fig. 3.3).

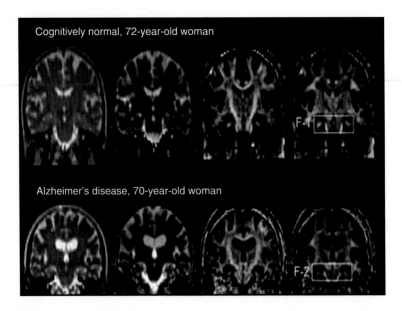

FIGURE 3.3

Typical diffusion tensor imaging (DTI) of the brain of an Alzheimer's disease patient compared to a cognitively normal individual. (a) Conventional MRI images of a 72-year-old cognitively normal woman and a 70-year-old Alzheimer's disease patient. (b) Mean diffusivity (MD), (c) fractional anisotropy (FA) map, and (d) color coded FA for each MRI scan. A significant difference is observed in the cingulum hippocampal area of brain (labeled F-1 on the cognitively normal image and F-2 on the Alzheimer's disease image). *(Adapted from Oishi, K., Mielke, M.M., Albert, M., Lyketsos, C.G., Mori, S., 2011. J. Alzheimers Dis. 26 (Suppl 3), 287–296; with permission from the IOS Press B.V.).*

DTI can quantitatively measure (using MD and FA) the microscopic characteristics of brain tissues in vivo. It can detect abnormalities in brain tissue caused by stroke, brain tumors, multiple sclerosis, and AD. In general, all DTI studies found higher MD and lower FA values for AD and MCI patients compared to age-matched controls (Sexton et al., 2011).

The most important application of DTI is distinguishing MCI cases from age-matched control cases (Chua et al., 2009). DTI was strongly correlated with MMSE and clinical dementia rating (CDR). One of the most important qualities of DTI is that it can differentiate between various brain regions of interest in AD patients and age-matched control cases, including the frontal and temporal lobes, genu, occipital, posterior cingulum, corpus callosum, superior fasciculus, and uncinated fasciculus. It can classify stages of AD. A metaanalysis of 41 DTI studies consisting of 617 patients with AD, 494 with MCI, and 915 age-matched control cases found consistent differences in FA and MD values in most of the specific brain regions studied except for the parietal white matter and internal capsule (Sexton et al., 2011). MD was consistently higher in various brain regions in AD compared to MCI and controls.

There are two drawbacks to DTI techniques: the macroscopic averaging effect and lower specificity due to changes not related to AD pathology. The macroscopic averaging effect comes from the inherent millimeter range resolution of MRI. The ideal diffusion process is measured in the micrometer range with a 0–100 millisecond diffusion time. Therefore, in an MRI scan, the diffusional properties are an average of macroscopic reorganization. Diffusion of water in the brain is an indirect measure of neuronal anatomy, which may be affected by non-AD related factors. Therefore, DTI may not be highly specific to AD pathology. However, a study of DTI in individuals with FTD ($n = 18$) or AD ($n = 18$) and age-matched controls ($n = 19$) found region-specific differences in patterns of FA (Zhang et al., 2009). For example, FTD had reduced FA in the frontal and temporal regions (anterior corpus callosum, bilateral anterior, cingulum tracts, and uncinate tracts) compared to age-matched controls, whereas AD was associated with reduced FA in parietal, temporal and frontal regions (anterior and posterior, cingulum tracts, bilateral descending cingulum tracts) compared to age-matched controls (Zhang et al., 2009). The study also found more damaged white matter in FTD cases. Elevated FA in parahippocampal regions and loss of white mater integrity in parietooccipital regions measured by DTI was shown to distinguish LBD patients from AD patients and age-matched control cases (Nedelska et al., 2015). The study consisted of 120 subjects ($n = 30$ AD, $n = 30$ LBD) and age-matched control cases ($n = 60$) and confirmed that the difference in DTI results for LBD and AD patients was independent of Aβ deposition.

3.3.4.2 *ARTERIAL SPIN LABELING MRI*

In general ^{15}O H2O radiotracer in PET measures cerebral blood flow. Injection of the positron emitter ^{15}O—H_2O into the peripheral vascular system leads to diffusion of ^{15}O—H_2O into brain tissue, which is detected by PET in different areas of brain. Regions with higher positron signals represent a higher

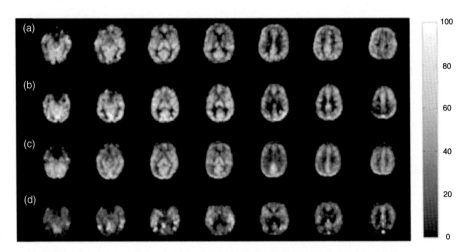

FIGURE 3.4

Cerebral blood flow (CBF) images by arterial spin labeled (ASL) magnetic resonance imaging (MRI). An adult normal control brain (a) has the highest amount of CBF compared to brains from early MCI (b), late MCI (c) and Alzheimer's disease (d) patients. CBF measured by ASL-MRI correlate with prodromal stages of AD. *[Adapted from Wang, Z., Das, S.R., Xie, S.X., Arnold, S.E., Detre, J.A., Wolk, D.A., Alzheimer's Disease Neuroimaging Initiative, 2013. Neuroimage Clin. 2, 630–636; with Creative Commons Attribution License (CC BY)].*

amount of cerebral blood flow (CBF), which is an indirect measure of higher neuronal activity. Magnetically labeled blood water (arterial spin labeling) in the rat brain followed by MRI was first introduced in 1992 (Detre et al., 1992). In principle, ASL-MRI employs magnetically labeled blood water as tracer by inducing a 180-degree radiofrequency pulse in the brain slice of interest (Detre et al., 2009, 2012). Blood perfusion into brain tissues causes a local change in tissue magnetization that can be measured by a standard MRI sequence. The magnetically labeled MRI scan images are compared with images without labeling. Three different arterial spin labeling (ASL) approaches have been reported on the basis of the type of magnetic labeling: single-pulse ASL (PASL), continuous ASL (CASL), and pseudo-continuous pulse (several short pulse) ASL (pCSAL) (Detre et al., 2012). ASL-MRI has great potential as a method to pinpoint vascular factors in neurodegenerative diseases (Detre et al., 2012). Typical CBF images from ASL-MRI have been presented by Wang et al. (2013) (Fig. 3.4), and show a clear difference between AD, MCI, and control groups.

ASL-MRI measures of CBF are essentially equivalent to measures of brain glucose metabolism by [18]FDG-PET. However, there are several advantages of ASL-MRI over [18]FDG-PET. ASL-MRI is not invasive and does not employ any radiotracer or contrast reagents. ASL-MRI is less expensive than [18]FDG-PET and can be performed at the same time as a volumetric MRI scan. ASL-MRI can be quantified in terms of physiologic quantity that is independent of MRI quantification and can be very useful for drug efficacy tests (Wolk and Detre, 2012).

CBF measured by ASL-MRI was found to be reduced in AD patients compared to age-matched control cases in most brain regions, including the posterior cingulate, precuneus, inferior parietal, and lateral prefrontal cortex (Alsop et al., 2000, 2010; Johnson et al., 2005; Dai et al., 2009). Hypoperfusion in AD brain regions measured by ASL-MRI was highly correlated with hypometabolism by ^{18}FDG-PET (Chen et al., 2011; Musiek et al., 2012). One of the unique characteristics of ASL-MRI is that it can distinguish between MCI and AD, with hyperperfusion of the left hippocampus, right amygdala, and ventral striatum in MCI cases but not AD cases (Dai et al., 2009; Wang et al., 2013). Several studies have investigated the capacity of ASL-MRI for differential diagnosis of AD and non-AD dementias. ASL-MRI-based CBF images showed differential region-specific hypoperfusion/hyperperfusion in AD patients versus FTD patients (Du et al., 2006; Hu et al., 2010). These studies found significantly reduced CBF in the frontal cortex and insula region of FTD brains compared to age-matched control cases. By contrast, hypoperfusion was observed in the precuneus and lateral parietal cortices in AD cases. FTD cases showed higher CBF in the precuneus/posterior cingulate and AD brains had higher CBF in anterior cingulate and dorsolateral prefrontal cortex regions. The utility of ASL-MRI for distinguishing AD from VaD has not been explored extensively. One study with a small sample size [14 AD and 8 subcortical ischemic vascular dementia (SiVaD)] found lower CBF for SiVaD cases compared to AD patients (Schuff et al., 2009).

3.3.4.3 *TASK FREE FMRI*

In task-based fMRI, subjects are given specific tasks and fMRI is performed to measure the response of the connectivity networks. In task-free fMRI, the intrinsic neuronal connectivity networks are measured by default mode network (DMN) without any tasks given to the subjects. Use of fMRI for measuring functional connectivity of the resting brain (ie, without any predetermined task) was introduced by several groups (Biswal et al., 1995; Greicius et al., 2003, 2009; Damoiseaux et al., 2006). In the resting state, fMRI measures the spontaneous or intrinsic brain activity in terms of low-frequency fluctuations in the BOLD signal. Some researchers refer to task-free fMRI as intrinsic fMRI. Studies have found certain regions of the cortex that consistently show greater activity during the resting state than during cognitive tasks. Task-free fMRI is easier to perform and less expensive than memory task-based fMRI.

The underlying biology of task-free fMRI is not as clear as that of other MRI paradigms. There are some disadvantages of task-free fMRI. There is always a chance of contamination of resting state fMRI signals with respiratory cycle or cardiac oscillation and small movements. There are no regulatory guidelines from the FDA or EMA regarding task-free MRI biomarkers of AD.

3.4 POSITRON EMISSION TOMOGRAPHY

PET imaging provides information about the pathophysiological and biochemical conditions of the brain. Unlike MRI, PET uses radiolabeled tracers that either bind target proteins or are taken up by target tissues; the tracer emissions are detected and the tracer emission patterns are reconstructed as tomographic images of protein levels or brain metabolic activity. Successful PET neuroimaging is dependent on the radiotracer used. An ideal PET radiotracer for imaging the brain has high biding affinity and selectivity toward the specific analyte, high blood–brain barrier permeability, rapid clearance from the brain to reduce nonspecific binding, moderate lipophilicity, and low metabolism. PET imaging provides a way to study the interaction of Aβ and tau and their influence on neurodegeneration and the metabolic state of the AD brain.

3.4.1   Amyloid Imaging by PET

As amyloid plaque deposition is a hallmark of AD brain pathology at autopsy, PET imaging of the brain to detect Aβ aggregates was considered as a promising ante-mortem AD diagnostic approach (Fig. 3.5). Aβ imaging with PET was also expected to have higher specificity than tau imaging with PET for AD because tau deposition is present in non-AD neurodegenerative diseases such as FTD, PSP, CBD, and senile dementia of neurofibrillary type. Amyloid accumulation also occurs prior to tau pathology in the progression of AD; therefore, Aβ imaging by PET may provide information about pre-symptomatic AD. The Food and Drug Administration (FDA) approved imaging of Aβ for clinical use in April 2012, and according to both FDA and the European Medical Agency (EMA), Aβ-PET can be used as a baseline measure for selecting patients with prede-mentia or mild to moderate AD for inclusion in clinical trials (EMA/CHMP/SAWP/893622/2011 and EMA/CHMP/SAWP/892998/2011).

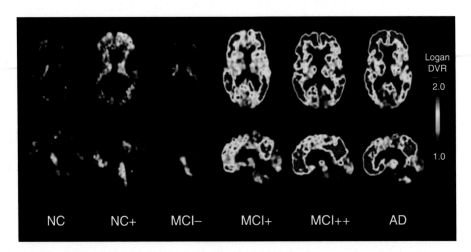

FIGURE 3.5

Typical [11C]-Pittsburgh compound B (11C-PIB) positron emission tomography (PET) images of cognitively normal controls (*NC*), mild cognitive impairment (*MCI*) cases, and an Alzheimer's disease (*AD*) patient. In principle, ^{11}C-PIB accumulation represents the amount of Aβ deposition in the brain. In general Aβ deposition is lower for NC than for MCI cases and AD patients. NC-cases show lower Aβ deposition compared to *NC+* cases. Those cases are called normal but PIB positive. Similarly *MCI+* shows higher Aβ accumulation in the brain compared to *MCI−*. *MCI++* has higher Aβ deposition compared to *MCI+* and *MCI−*. The *bar* indicates the intensity of color. The AD brain has the highest amount of accumulation of Aβ in the brain. (*Adapted from Mathis, C.A., Lopresti, B.J., Klunk, W.E., 2007. Nucl. Med. Biol. 34, 809–822; with permission from the Elsevier Limited*).

Uptake of ^{11}C-PIB in the brain was first developed as a potential PET biomarker of AD in 2004 (Klunk et al., 2004), and extensive studies have since been conducted to validate it. In a study of twins, cognitively impaired subjects (monozygotic and dizygotic individuals combined) showed AD-like patterns of ^{11}C-PIB uptake (Scheinin et al., 2011). Unfortunately, a study by researchers from the Turku PET Center in Turku, Finland found that the rate of ^{11}C-PIB uptake did not correlate with either brain atrophy or cognitive impairment in a group of patients with AD (Scheinin et al., 2009). Furthermore, in another multicenter comparative study conducted by ADNI (supported by the NIH, pharmaceutical companies, and nonprofit funding), there was no relationship between CSF biomarkers (Aβ_{1-42}, t-tau, and p-tau-181), PET neuroimaging of amyloid plaques, and cognitive impairment as measured by MMSE score. Moreover, the brains of aged patients without clinical dementia were found to have a considerable number of amyloid plaques, which increased the false positive rate of ^{11}C-PIB PET neuroimaging (Fig. 3.5). A recent study found that 10 out of 63 patients with probable AD (clinically confirmed by NINCDS-ADRDA criteria, not autopsy) were ^{11}C-PIB PET negative (Shimada et al., 2011). These studies led to the investigation of alternative PET radiotracers with higher accuracy for detecting Aβ accumulation in AD.

The FDA and EMA have approved several ^{18}F-labeled amyloid-binding agents for use as PET radiotracers for Aβ detection: florbetapir (Amyvid, Eli Lilly and Company), flutemetamol (Vizamyl, GE Healthcare), and florbetaben (Neuraceq, Piramal Imaging). Studies with ^{18}F-florbetapir also showed good correlation with amyloid load in AD patients at autopsy (Choi et al., 2012). Recently, an ADNI comparative study found that ^{18}F-florbetapir showed greater specificity than CSF Aβ_{1-42}, although overall diagnostic accuracies were the same (Mattsson et al., 2014). Other well studied Aβ radiotracers include ^{11}C-BF227 and ^{18}F-NAV4694. The positron-emitting isotope ^{18}F (half-life of 109.8 min) has longer half-life than ^{11}C (half-life of 20.4 min), providing a longer window for conducting an imaging study. This is significant, as the cyclotron facility for ^{18}F PET radioisotope production may not be in close proximity to the PET imaging center. Aβ-PET does have some disadvantages: the method is costly and involves injection of a radiotracer, and some control subjects have shown a positive Aβ signal.

3.4.2 Glucose Metabolism Measurement by PET

The human brain consumes approximately 20% of the body's total energy requirement. Glucose is the sole source of energy for the brain due to higher blood–brain barrier permeability. Proteins and fatty acids are bound to albumin, and cannot cross the blood–brain barrier. Using ^{18}FDG-PET neuroimaging, it was found that glucose metabolism was impaired in the brains of AD patients (Ishii et al., 1997). ^{18}FDG-PET is the most extensively researched PET biomarker in AD (Tables 3.2 and 3.4), and the first among neuroimaging biomarkers to receive regulatory approval as an AD biomarker. Patients with AD have a characteristically lower amount of ^{18}FDG-PET activity in affected areas of the brain

Table 3.4   Differential Diagnosis of Alzheimer's Disease (AD) and Non-AD Dementia Patients Using PET Imaging

AD vs Non-AD Dementia	PET Biomarker	Remarks	References
AD vs FTD Clinically confirmed AD (n = 62) FTD (n = 45) Control (n = 25)	*18FDG:* SN: 73% SP: 98% ACU: not determined *11C-PIB:* SN: 89% SP: 83% ACU: not determined	▪ Threshold value was estimated from control cases ▪ 11C-PIB had higher sensitivity (89% vs 73%) while FDG had higher specificity (83% vs 98%)	Rabinovici et al. (2011)
AD vs non-AD dementia Autopsy confirmed AD (n = 20) AD mixed with other dementia (n = 4) Normal control (n = 9) FTD (n = 1); LBD (n = 3) VaD (n = 1); unknown dementia (n = 6)	*18FDG:* SN: 84% SP: 74% ACU: 80%	▪ Addition of 18FDG ▪ Data improved clinical diagnosis	Jagust et al. (2007)
AD vs FTD Autopsy confirmed AD (n = 31) FTD (n = 14)	*18FDG;* SN: 97% SP: 86% ACU: 93%	▪ Addition of 18FDG ▪ Data improved clinical diagnosis	Foster et al. (2007)
AD vs LBD Autopsy confirmed AD (n = 10); LBD (n = 11) Clinically confirmed AD (n = 40); LBD (n = 13)	*18FDG* SN: 90%, SP: 82% ACU: 86%	▪ Addition of 18FDG ▪ Data improved clinical diagnosis	Minoshima et al. (2001)

AD vs LBD and FTD AD ($n = 97$) LBD ($n = 7$) FTD ($n = 6$)	*18FDG* SN: 93% SP: 71% (for LBD) SP: 65% (for FTD)	■ Regional brain metabolism was sensitive indicator of AD and non-AD dementia	Silverman et al. (2001)
AD vs non-AD dementia AD ($n = 199$) Control ($n = 110$) MCI ($n = 114$) FTD ($n = 98$) LBD ($n = 27$)	*18FDG:* *AD vs FTD* SN = 90% SP = 65% *AD vs LBD* SN = 99% SP = 71% *AD vs control* SN = 99% SP = 98%	■ This multicenter study validated differential diagnosis of AD vs non-AD dementias	Mosconi et al. (2008)

AD, Alzheimer's disease; PET, positron emission tomography; FTD, frontotemporal dementia; 18FDG, [18F]-fluoro-2-deoxy-D-glucose; SN, sensitivity; SP, specificity; ACU, accuracy; 11C-PIB, [11C]-Pittsburgh compound B; LBD, dementia with Lewy body; VaD, vascular dementia; MCI, mild cognitive impairment.

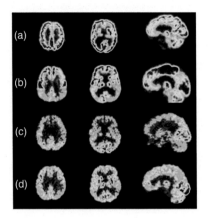

FIGURE 3.6

Typical [18F]-fluoro-2-deoxy-ᴅ-glucose (18FDG) positron emission tomography (PET) images of a normal brain and an Alzheimer's disease (AD) brain. Images show brain regions that are affected by AD have lower glucose metabolism. Severely low levels of glucose metabolism are depicted in cortical region (*yellow* and *blue* color). (a) Cognitively normal case showed higher glucose metabolism. (b) Cognitive symptoms of AD showed lower glucose metabolism compare to normal cases. (c) Frontotemporal dementia (FTD) also shows lower glucose metabolism. (d) Another example of AD case younger than (b). The specific regions affected by AD are the medial temporal lobes and the posteriomedial parietal, lateral parietal, and lateral temporal lobes. *(Adapted from Jagust, W., Reed, B., Mungas, D., Ellis, W., Decarli, C., 2007. Neurology 69, 871–877; with permission from the Wolters Kluwer Health, Inc.).*

compared to age-matched controls (Fig. 3.6). [18]FDG-PET results are strongly correlated with disease progression. In a comparative study of CSF biomarkers and neuroimaging biomarkers, ^{11}C-PIB PET correlated well with CSF biomarkers but not with cognitive impairment, whereas [18]FDG-PET was more strongly associated with MMSE score but not with CSF biomarkers (Jagust et al., 2009). A region-specific hypometabolism signature by [18]FDG-PET may be characteristic of specific non-AD dementia, thereby providing a means for differential diagnosis (Table 3.4). [18]FDG-PET has been validated by autopsy studies and in multicenter cohort studies. A comparison of ^{11}C-PIB-PET and [18]FDG-PET neuroimaging of a normal brain (control) and a brain from a patient with AD is presented in Fig. 3.7. The arrows indicate areas of typical hypometabolism by [18]FDG-PET) and amyloid deposition by ^{11}C-PIB PET (Cohen and Klunk, 2014). The areas with higher amyloid deposition have a lower glucose metabolism signal.

EMA and FDA both indicate that [18]FDG-PET may be useful for differential AD diagnosis, and [18]FDG-PET is included as one of the AD biomarkers in the 2011 AD diagnostic criteria (McKhann et al., 2011; Albert et al., 2011; Sperling et al., 2011). Nevertheless, there are several disadvantages of using [18]FDG-PET for the diagnosis of AD. As mentioned previously, PET is a very expensive and invasive method requiring infusion of a radioactive agent. It is important to point out that some ^{18}F-labeled PET ligands can accumulate in the bone and interfere with PET imaging results. Signals from ^{18}F-labeled PET are affected by inflammation, ischemia, and other conditions (Duara et al., 1987; Shipley et al., 2013). An extensive review by

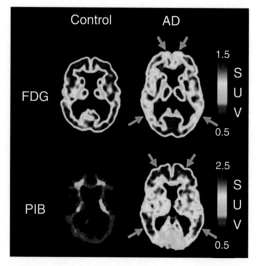

FIGURE 3.7
Comparison of [11C]-Pittsburgh compound B (11C-PIB) positron emission tomography (PET) images and [18F]-fluoro-2-deoxy-ᴅ-glucose (18FDG) PET images of a normal brain (*Control*) and an *AD* brain. *Arrows* indicate areas of typical hypometabolism by ^{18}FDG-PET) and amyloid deposition by ^{11}C-PIB PET. *(Adapted from Cohen, D.A., Klunk, W.E., 2014. Neurobiol. Dis. 72, 117–122; with permission from the Elsevier Limited).*

Cochrance Collaboration concluded that there is no sufficient evidence to support the routine use of ^{18}FDG-PET for determining MCI cases (Smailagic et al., 2015).

3.4.3   Brain Inflammation Imaging by PET

Neuroinflammation caused by activated microglia and astrocytosis has been identified as one of the early events in AD pathophysiology (Serrano-Pozo et al., 2011). PET imaging compounds such as ^{11}C-PK11195 have been developed to measure brain inflammation levels and may be useful in the early diagnosis of AD or MCI (Kropholler et al., 2007). ^{11}C-PK11195 is limited by nonspecific binding, which led to the search for improved active microglia imaging agents (Chauveau et al., 2008). ^{11}C-PBR28 (Owen et al., 2010) and ^{11}C-DED (^{11}C-deuterium-ʟ-deprenyl) (Carter et al., 2012) may be promising new PET biomarkers for imaging of inflammation in AD. A higher ^{11}C-PBR28 signal in the AD brain can distinguish between AD and FTD (Masdeu et al., 2012) (Fig. 3.8).

3.4.4   Tau Imaging by PET

In addition to deposition of amyloid plaques, hyperphosphorylation of tau in AD leads to accumulation of insoluble paired helical filaments (PHF) that form neurofibrillary tangles (NFT), one of the "gold standards" of AD diagnosis at autopsy. Noninvasive detection of tau deposition by PET imaging would be useful for detecting and tracking various neurodegenerative diseases, including AD. Some studies have found better correlation of disease severity with NFT than

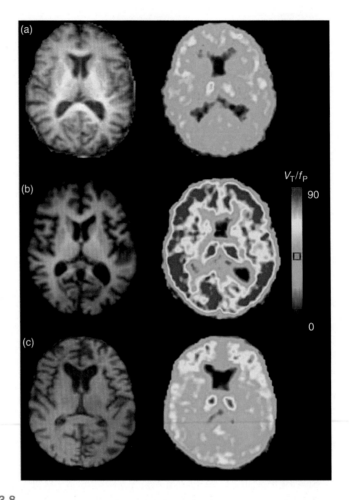

FIGURE 3.8

Typical brain inflammation images related to microglial activation recorded by positron emission tomography (PET). Microglial activation imaged by ^{11}C-PBR28-PET for: (a) age-matched control (AC), (b) Alzheimer's disease (AD), and (c) frontotemporal dementia (FTD). AD patients showed higher ^{11}C-PBR28 signal compared to AC and FTD cases. V_t/f_p, distribution volume normalized by plasma-free fraction of radio ligand. *(Adapted from Masdeu, J.C., Kreisl, W.C., Berman, K.F., 2012. Curr. Opin. Neurol. 25, 410–420; with permission from the Wolters Kluwer Health, Inc.).*

with amyloid plaques in postmortem AD brains (Bierer et al., 1995; Arriagada et al., 1992). The ideal characteristics of a tau PET tracer include high binding affinity for PHF-tau, high selectivity for PHF-tau, and high blood–brain barrier permeability. Tau imaging by PET was first reported using a radiofluorinated derivative of 2-(1-[6-(dimethylamino)-2-naphthyl]ethylidene)malononitrile (DDNP) (^{18}F-DDNP), which showed higher retention times in the brains of AD and MCI patients compared to those of healthy control cases (Small et al., 2006). ^{11}C-phenyl/pyridinyl-butadienyl-benzothiazoles/benzothiazoliums (^{11}C-PBB3) retention was also found in AD and non-AD tauopathy

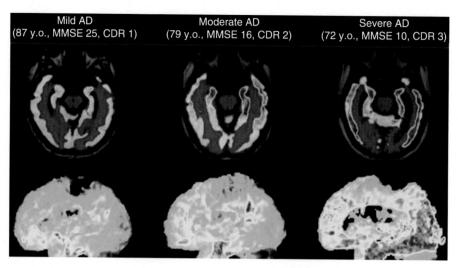

| Mild AD | Moderate AD | Severe AD |
| (87 y.o., MMSE 25, CDR 1) | (79 y.o., MMSE 16, CDR 2) | (72 y.o., MMSE 10, CDR 3) |

FIGURE 3.9

Representative 18F-THK-5117 (a quinolone derivative) positron emission tomography (PET) images of Alzheimer's disease (AD) brains at different stages of disease. The degree of retention of 18F-THK-5117 increases with increasing severity of AD. *(Adapted from Okamura, N., Harada, R., Furumoto, S., Arai, H., Yanai, K., Kudo, Y., 2014. Curr. Neurol. Neurosci. Rep. 14, 500; with permission from the Springer).*

cases (Maruyama et al., 2013). ^{18}F-T807 (fluorine-18 labeled 7-(6-fluoropyridin-3-yl)-5H-pyrido[4,3-*b*]indole; prepared by Siemens Molecular Imaging Biomarker Research [Culver City, CA]) was the first tau-PET radiotracer to be introduced in human trials (Chien et al., 2013). This study found very low nonspecific binding with white matter and demonstrated significant tracer association with areas of PHF-tau in AD brain. Tau-binding novel quinolone derivatives (^{18}F-THK-523, ^{18}F-THK-5105, and ^{18}F-THK-5117) detected by PET were similarly retained in AD brains (Fodero-Tavoletti et al., 2011; Harada et al., 2013; Okamura et al., 2013). Among these, ^{18}F-THK-5117 was found to be superior in terms of signal-to-background ratio and the ability to distinguish between mild, moderate, and severe AD cases (Fig. 3.9) (Okamura et al., 2014b). Brain areas that most often show tau tracer uptake in AD are the medial and lateral temporal cortex, whereas Aβ imaging PET shows signals in the AD neocortex. Recent studies have shown that tau imaging with PET detects tau pathology in brain areas of AD and MCI cases; however, it is less able to distinguish between AD and tau-related non-AD dementias such as FTD, CBD, and PSP. The association of tau-PET with clinical severity of AD and neurodegeneration is stronger than for Aβ-PET.

3.5 SINGLE-PHOTON EMISSION COMPUTED TOMOGRAPHY

Cerebral blood flow (CBF) can be measured by SPECT imaging using either intravenously injected ^{99m}Tc-HMPAO (Hexamethylpropylene amine oxime) or inhaled Xe-133, a gamma ray emitter. Uptake of ^{99m}Tc-HMPAO by brain tissue is propor-

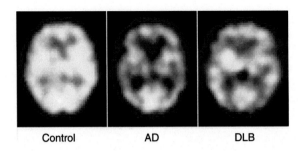

Control   AD   DLB

FIGURE 3.10
Example of single-photon emission computed tomography (SPECT) imaging. SPECT
scan of a *control* case (a), a late onset Alzheimer's disease (*AD*) (b), and an early onset AD (c).
Typical lower perfusion was shown in medial temporal lobes in late onset AD cases. Typical
lower perfusion was shown in parietal region of early onset AD cases. *(Adapted from Colloby, S.J.,
Fenwick, J.D., Williams, E.D., Paling, S.M., Lobotesis, K., Ballard, C., McKeith, I., O'Brien, J.T., 2002. Eur. J.
Nucl. Med. Mol. Imaging 29, 615–622; with permission from the Springer Science and Bus Media B.V.).*

tional to the rate of blood flow in the brain, which is tightly coupled to local brain
metabolism; therefore, differences in blood flow in various areas of the brain correlate with differences in brain metabolism in those areas (Fig. 3.10). In patients
with AD, brain metabolism is impaired. In general, most of the SPECT images of
the temporal and parietal cortex of the AD brain have a pattern of hypoperfusion
that is diagnostic of AD with reasonable sensitivity and specificity. Sensitivity and
specificity of SPECT for diagnosis of AD are in the range of 70–80% in autopsy-
confirmed cohorts (Bonte et al., 1997; Jobst et al., 1998). Though some studies
have shown that SPECT has higher sensitivity for diagnosing advanced AD (Bonte
et al., 2004), differences in regional CBF detected by SPECT can distinguish between AD and non-AD dementias (Matsuda, 2007; Dougall et al., 2004). Both
SPECT and ^{18}FDG-PET neuroimaging provide information about the metabolic
state of the brain, and have comparable diagnostic sensitivity and specificity and
provide independent complementary information for AD. There are several limitations of SPECT, including lower spatial resolution compared with PET images
and a limited number of available single-photon emitting radionuclides. Of the
two modalities, however, SPECT is more widely available and less expensive than
PET, and also uses an isotope (^{99m}Tc) with a longer half-life and less complicated
imaging protocols. Newer, less expensive ^{18}F-compounds for PET imaging are expected to bring down the cost of PET to the level of SPECT.

3.6 ELECTROENCEPHALOGRAPHY
AND MAGNETOENCEPHALOGRAPHY
IN ALZHEIMER'S DISEASE

EEG records the electrophysiological activities of neurons in the brain, whereas
MEG records magnetic fields produced by electrical currents generated by neurons in the brain. In terms of AD biomarkers, EEG measures the electrophysiological changes and MEG measures biomagnetic changes due to AD pathology.
Compared to AD neuroimaging biomarkers for modalities such as MRI and PET,

neurophysiological features affected by AD pathology have not received enough attention as potential biomarkers. Both noninvasive EEG and MEG can directly measure neuronal activity in terms of frequency of oscillatory activities. The spatial resolution of both EEG and MEG are same as MRI techniques (on the order of millimeters), but the temporal resolution for both EEG and MEG (milliseconds) are significantly higher than for MRI, PET, and SPECT techniques.

EEG signals observed in AD brains compared to age-matched controls include: (1) higher theta activity, lower alpha activity, and lower beta activity for mild AD (Babiloni et al., 2004) and (2) higher delta power for severe AD (Dierks et al., 1993). Quantitative EEG analysis can be helpful for detecting general cognitive decline (Table 3.5). EEG signals from AD cases show three important features: slowing of the EEG spectra, reduced complexity of the EEG signals, and perturbations in EEG synchrony (Gallego-Jutglà et al., 2012; Neto et al., 2015). However, these features need to be more clearly defined and validated with extensive analysis in order to distinguish AD patients from non-AD cases. Several longitudinal and cross-sectional studies of EEG used in MCI have found significant accuracy to predict the conversion of MCI to AD (Jelic et al., 2000; Huang et al., 2000; Grunwald et al., 2002).

Several potential applications of MEG biomarkers of AD and their associated studies have been summarized in Table 3.5. Some MEG AD biomarkers showed a strong correlation with brain atrophy (Fernández et al., 2003; Maestú et al., 2003). A 2-year follow-up study found that MEG can predict conversion of cognitively normal elderly person to MCI (Maestú et al., 2006). This study underlines the potential for the use of MEG for preclinical detection of AD.

Advantages of EEG and MEG over MRI are: greater temporal resolution, no interference of metal implants in the body, no interference of hemodynamic response, lower cost, and less complicated protocols. Advantages of MEG over EEG are: no need for a reference electrode; unlike electrical currents, the magnetic field is not affected by biological tissues or distorted by skull; higher spatial resolution, and the use of many more detectors than EEG electrodes. The most important disadvantage of MEG is that the relatively stronger earth magnetic field (10^9fT) interferes with weaker magnetic signals from brain (10^2–10^3fT), resulting in a noisy signal with MEG. Because EEG and MEG are relatively less explored areas as modalities for imaging AD biomarkers, there is a need for large cohort studies, longitudinal follow-up, and standardization in experimental protocols and data analysis in order to evaluating their effectiveness as in vivo noninvasive AD biomarkers.

3.7 DISCRIMINATION BETWEEN ALZHEIMER'S DISEASE AND NON-ALZHEIMER'S DISEASE DEMENTIAS BY NEUROIMAGING BIOMARKERS

Application of MRI techniques to distinguish AD patients from those with non-AD dementias is challenging, and is made more complex in patients who present with AD and comorbid non-AD dementias (Table 3.3). Frisoni et al. have

Table 3.5 Differential Diagnosis of Alzheimer's Disease (AD), Age-Matched Control, and Non-AD Dementia Patients Using Electroencephalography (EEG) and Magnetoencephalography (MEG)

Biomarker Modality/Institution	Study Design and Patient Population	EEG/MEG Biomarker	Comments	References
EEG Haukeland University Hospital, Bergen, Norway	*AD vs VaD* Clinically confirmed AD (n = 77), VaD (n = 77), Control (n = 77)	VaD group showed significantly higher low frequency power than AD (P < 0.0002)	No significant difference between groups in EEG spectral decay from lower to higher frequencies	Neto et al. (2015)
EEG Derriford Hospital, Plymouth, UK	*AD vs Control* Clinically confirmed mild AD (n = 17) Age-matched control (n = 24)	Synchrony measures of optimized EEG frequency bands	Synchrony measures of 5–6Hz range yields best data to distinguish AD vs control cases	Gallego-Jutglà et al. (2012)
EEG Geriatric clinic, Huddinge University Hospital, Huddinge, Sweden	*AD vs MCI vs Control* Mild AD (n = 38); MCI (n = 31), and age-matched control (n = 24)	AD patients showed higher delta and theta global field power (GFP) in EEG spectra compared to control and lower alpha and beta GFP in MCI cases	AD vs age-matched control: accuracy = 84%; AD vs MCI: accuracy = 78%	Huang et al. (2000)
MEG Noran Neurology Clinic, Orr Consulting Psychiatric Recovery, Minneapolis VA Medical Center GRECC and Denver, CO (Radiant Research, University of Colorado, Denver)	*AD vs Control* Clinically confirmed mild AD (n = 117) Age-matched control (n = 123) *Follow-up of 10 months*	Centroid frequency calculated from distribution of relative spectral power and functional connectivity; centroid frequency for AD: 6.78 ± 0.25 Hz; for age-matched control: 8.24 ± 0.2 Hz; Functional connectivity for AD = 0.66 ± 0.001; for age-matched control: 0.59 ± 0.0007.	Centroid frequency was significantly lower for AD cases and functional connectivity was higher for AD cases compared to age-matched controls	Verdoorn et al. (2011)
MEG The Hospital Universitario San Carlos de Madrid, Spain,	*AD vs Control* Clinically confirmed mild AD (n = 15) Age-matched control (n = 16)	Dipole density in delta and theta bands was higher in AD group compared to age-matched control cases	Accuracy = 87.1% MEG results were correlated with hippocampal atrophy	Fernández et al. (2003)

AD, Alzheimer's disease; EEG, electroencephalography; MCI, mild cognitive impairment; MEG, magnetoencephalography; VaD, vascular dementia.

summarized the application of MRI to the differential diagnosis of non-AD dementias and AD (2010). Previously, attempts have been made by NINDS–AIREN (the National Institute of Neurological Disorders and Stroke (NINDS) and the Association Internationale pour la Recherche et l'Enseignement en Neurosciences (AIREN) International Workshop to diagnose VaD by examining the white matter changes or evidence of infarct on sMRI images (Román et al., 1993). FTD has greater focal frontal or temporal degeneration on sMRI images compared to AD (Neary et al., 1998), but this criterion has only been used as supportive evidence for distinguishing FTD and AD cases. LBD can be differentiated from AD based on evidence of an intact medial temporal lobe structure on sMRI images of AD patients (McKeith et al., 2005). Again, this criterion cannot be used as a definitive diagnosis of LBD and can only be used as supportive condition to distinguish AD versus LBD, as other studies have found substantial overlap of atrophy in sMRI images in AD and LBD cases (Barber et al., 2000). sMRI is most effective in distinguishing between AD and CJD patients, by detecting changes in cortical diffusion measured by diffusion weight imaging and pulvinar sign in the thalamic region that are indicative of CJD but not AD (Collie et al., 2001; Tschampa et al., 2005). There is considerable overlap between brain atrophy of AD and atrophy caused by other neurodegenerative diseases, and it is difficult to distinguish between them using this sMRI biomarker alone. Atrophy in medial temporal lobe detected by sMRI is not significantly different in AD and LBD cases (Barber et al. 2000; McKeith et al., 2005), although an autopsy-confirmed study found that medial temporal lobe atrophy measured by sMRI can distinguish AD from other dementias such as LBD and VaD (91% sensitivity and 94% specificity) (Burton et al., 2009). Generalization of the data from a highly selective set of cohort studies is risky, but overall, sMRI hippocampal atrophy has very modest specificity for AD.

Most published studies have reported that neuroimaging biomarkers are not significantly different in AD and non-AD dementias (VaD, FTP, and LBD), although some PET studies have shown promising results, with several PET biomarkers effective in distinguishing between AD with non-AD dementias (Table 3.4). For example, [18]FDG-PET has been shown to increase the diagnostic accuracy of distinguishing FTD from AD (Foster et al., 2007). In an autopsy-validated study, Minoshima et al showed that [18]FDG-PET can distinguish between AD and LBD with 90% sensitivity, 82% specificity, and 86% accuracy (Minoshima et al., 2001). However, a large multicenter study found lower specificity (71% for LBD and 65% for FTD) but higher sensitivity (93%) (Silverman et al., 2001). With regard to the brain glucose metabolic rate, LBD patients show more prominent hypometabolism in the occipital cortices, and FTD patients show more prominent hypometabolism in the frontal or temporal cortices, compared with AD patients.

Although EEG can distinguish between AD and age-matched control cases, no significant differences in EEG biomarkers have been detected in AD versus non-AD dementias (Table 3.5) (Neto et al., 2015).

3.8 LONGITUDINAL ASSESSMENT OF ALZHEIMER'S DISEASE NEUROIMAGING BIOMARKERS

Meaningful longitudinal neuroimaging studies have been conducted by ADNI, with the aim of improving methods used in clinical trials of AD and related disorders. Longitudinal cohort studies of neuroimaging biomarkers not only increase our knowledge about disease progression but also identify biomarkers with prognostic value that are predictive of AD progression, for example, the rate of synaptic loss, neuronal damage, amyloid load, and the deposition of tau. One of the most important results emerging from longitudinal neuroimaging studies is the prognostic significance of hippocampal atrophy in MCI cases that converted to AD. Hippocampal size was found to be strongly correlated with the rate of conversion of MCI to AD (Jack et al., 1999). When compared with other neuroimaging modalities to predict MCI to AD conversion, MRI had the highest accuracy (67%) (Trzepacz et al., 2014). Longitudinal MRI has been used to measure the rate of brain atrophy that is directly related to neuronal degeneration. The hippocampal atrophy rate for AD cases is 4–6% per year compared to 1–2% per year in age-matched healthy controls as measured in longitudinal MRI studies (Jack et al., 1998, 2000). The whole-brain atrophy rate was less than that of the hippocampus (Table 3.6). Chan et al. (2001) estimated annual whole-brain atrophy for AD cases to be 2.4 ± 1.1% versus 0.5 ±0.4% for age-matched healthy control cases. Whole-brain atrophy rate in AD was found to be strongly correlated with cognitive decline (Fox et al., 1999a, 2005; Jack et al., 2004). A longitudinal study found that the age- and sex-adjusted whole-brain atrophy rate was not correlated with the

Table 3.6	Longitudinal Atrophy Rate in Alzheimer's Disease (AD), Age-Matched Control, and Non-AD Dementia Patients Using MRI		
Study Design	**Atrophy Rate**	**Comments**	**References**
Metaanalysis of 9 studies (*n* = 595) Control (*n* = 212)	AD = 4.66% Control = 1.41%	Hippocampal atrophy rate	Barnes et al. (2009)
Follow-up study 3 years AD (*n* = 28), MCI (*n* = 43) Control (*n* = 58)	AD = 3.5%, MCI (converter) = 3.69% MCI (stable) = 2.55% Control = 1.73%	Hippocampal atrophy rate	Jack et al. (2000)
Average scan interval = 12.8 months AD (*n* = 54), FTD (*n* = 30) Control (*n* = 27)	AD = 2.37 ± 1.03% FTD = 3.15 ± 2.08% Control = 0.47 ± 0.40%	Whole-brain atrophy rate	Chan et al. (2001)

AD, Alzheimer's disease; FTD, frontotemporal dementia; MCI, mild cognitive impairment; MRI, magnetic resonance imaging.

CSF biomarkers $A\beta_{1-42}$ and total tau, but was modestly correlated with the CSF biomarker p-tau (Sluimer et al., 2010).

3.9 EARLY DIAGNOSIS OF ALZHEIMER'S DISEASE USING NEUROIMAGING BIOMARKERS

Therapeutic interventions for AD are likely to have the greatest effect if initiated in the early, preclinical stages of the disease, before synaptic loss and neuronal death occur. Preventive therapy for AD is only possible if AD can be detected in its earliest stages (preclinical), before the widespread loss of dendritic spines and synapses. Ideally, diagnosis of AD should occur even before the MCI stage when the first symptoms of memory decline are manifest. Today, researchers are investigating the ability of various biomarkers to predict which patients have a higher likelihood of developing AD in the future, even as they strive to identify and validate definitive diagnostic biomarkers of preclinical disease. Fox et al. (1999b) first reported evidence of cerebral atrophy in preclinical AD measured by serial MRI scans. Several studies have suggested that patients with preclinical AD are those who have higher risk of developing AD in future, such as cognitively normal elderly people with greater $A\beta$ plaque burden as measured by $A\beta$-PET (Mintun et al., 2006; Villemagne et al., 2008; Johnson et al., 2012). The NIA-AA working group defined preclinical AD as the first stage of a prodromal AD course consisting of three stages. In the first stage, a patient is positive for amyloid plaques by PET imaging or has low CSF $A\beta_{1-42}$, with no sign of neurodegeneration by sMRI. In the second stage, the patient has evidence of elevated CSF tau, neuronal injury, and amyloid plaques on imaging. In the third stage, the patient begins to experience subtle cognitive deficits that are less severe than those seen in MCI (Sperling et al., 2011). According to the IWG-2 criteria, a patient with preclinical AD has no clinical signs or symptoms but has one of the following: (1) decreased $A\beta_{1-42}$, together with increased tau or p-tau in CSF; or (2) increased fibrillary amyloid on PET (Dubois et al., 2014). Both working groups agree that neuroimaging biomarkers may provide valuable information when combined with CSF biomarkers for identifying the preclinical stages of AD.

3.10 CROSS-CORRELATION BETWEEN NEUROIMAGING WITH OTHER AD BIOMARKERS

Cross-correlation of neuroimaging biomarkers with CSF biomarkers is presented in Chapter 5 and with peripheral fluid given in Chapter 6. The relationship between genetic variants of AD and neuroimaging biomarkers has been studied extensively (Chauhan et al., 2015; Biffi et al., 2010; Potkin et al., 2009; Shen et al., 2010; Furney et al., 2011; Ramanan et al., 2014, 2015) and is summarized in Table 3.7. The most important AD-related genes with strong associations with neuroimaging biomarkers are APOE4, TOMM40, BIN1, CD33, CR1, PICALM, IL1RAP, and BCHE. Most studies have found a strong correlation between the APOE4 allele and smaller hippocampal volume measured by neuroimaging (Table 3.7), although some studies could not verify this association (Reiman et al., 1998; Ferencz et al., 2013; Khan et al., 2014).

Table 3.7 Correlation of Neuroimaging Biomarkers With Alzheimer's Disease Genetic Variants

Study Design/Patient Population	Neuroimaging Biomarker/Genomic Approach	Association of Neuroimaging Biomarkers and Alzheimer's Disease Genetic Markers	References
10 population-bases big cohorts* Total subject (n = 8175)	sMRI markers and GWAS	APOE was associated with smaller hippocampal volume. CD33 was associated with smaller intracranial volume	Chauhan et al. (2015)
ADNI AD (n = 168) MCI (n = 357) Age-matched control (n = 215)	sMRI markers: volume of hippocampus, amygdala, white matter lesion and thickness of cortex, parahippocampal gyrus, and temporal pole cortex correlation with GWAS results	APOE was associated with all sMRI markers except white matter lesion volume. CR1, PICALM and BIN1 were associated multiple sMRI biomarkers	Biffi et al. (2010)
ADNI AD (n = 229) Age-matched control (n = 194)	Region specific MRI volumetry and GWAS	TOMM40 was associated with lower MRI volumetry in AD cases	Potkin et al. (2009)
ADNI AD (n = 175) MCI (n = 354) Age-matched control (n = 204)	MRI voxel-based morphometry and GWAS	APOE and TOMM40 genes were strongly associated with multiple brain regions from MRI voxel-based brain morphometry	Shen et al. (2010)
ADNI and AddNeuroMed Consortium. AD (n = 94), 96 MCI (n = 96), Healthy control (n = 91)	Atrophy in regions associated with neurodegeneration measured by MRI and GWAS	PICALM was found to be the most significant gene associated with entorhinal cortical thickness	Furney et al. (2011)

ADNI AD (n = 71) Healthy control (n = 179) Early-MCI (n = 190) Late-MCI (n = 115)	Aβ-florbetapir PET and GWAS	APOE and BCHE (rs509208) associated with cerebral amyloid deposition	Ramanan et al. (2014)
ADNI (n = 495), Indiana Memory and Aging Study (n = 25), Rush Memory and Aging Project (n = 178), and Religious Orders Study (n = 178)	Aβ-florbetapir PET and GWAS	APOE and IL1RAP (rs12053868-G) associated with cerebral amyloid deposition	Ramanan et al. (2015)

AD, Alzheimer's disease; ADNI, Alzheimer's Disease Neuroimaging Initiative; BCHE, butyrylcholinesterase; BIN1, bridging integrator 1; CD33, CD33 antigen; CR1, complement cell-surface receptor; GWAS, genome-wide association study; MCI, mild cognitive impairment; s-MRI, structural magnetic resonance; IL1RAP, Interleukin-1 receptor accessory protein PET, positron emission tomography; PICALM, phosphatidylinositol binding clathrin assembly protein; TOMM40, translocase of outer mitochondrial membrane 40 homolog.

*Aging Gene-Environment Susceptibility-Reykjavik Study, Atherosclerosis Risk in Communities Study, Austrian Stroke Prevention Study, Cardiovascular Health Study, Framingham Heart Study (FHS), Rotterdam Study, Erasmus Rucphen Family study, Religious Order Study & Rush Memory and Aging Project, Tasmanian Study of Cognition and Gait, and the 3C Dijon study.

3.11 PERFORMANCE OF NEUROIMAGING BIOMARKERS IN ASSESSING DRUG EFFICACY IN ALZHEIMER'S DISEASE CLINICAL TRIALS

Neuroimaging biomarkers in AD clinical trials can be used (1) for selection of patient population (as exclusion and inclusion criteria) and (2) as clinical endpoints (surrogate biomarkers of efficacy). Neuroimaging may play a particularly important role as markers of AD therapeutic efficacy in early stages of AD, as conventional clinical outcome measures may be unable to accurately capture treatment effects prior to symptomatic disease. Neuroimaging biomarkers indicating therapeutic efficacy in AD include increased brain volume and decreased brain atrophy, which would be assessed in parallel with cognitive functional outcomes. Volumetric magnetic resonance imaging (vMRI) of the whole brain, hippocampal area, and ventricular enlargement; retention of ^{11}C-PIB by amyloid plaques detected by PET (^{11}C-PIB PET); assessment of brain glucose metabolism by ^{18}FDG-PET; and cerebral blood flow measured by ^{99m}Tc-HMPAO SPECT have all been tested as potential AD biomarkers to assess the efficacy of AD therapies in clinical trials (Table 3.8).

vMRI has been used extensively as an imaging endpoint in clinical trials in mild to moderate AD (Cash et al., 2014). As a surrogate endpoint, vMRI and SPECT imaging results have in some cases indicated responses to treatment that were opposite to clinical results (Table 3.8). For example, a phase 3 trial of tramiprosate (ALZHEMED) showed positive vMRI biomarker results with no significant clinical improvement (Saumier et al., 2009). Supplementation with docosahexaenoic acid (DHA) compared with placebo did not slow the rate of cognitive and functional decline or brain atrophy determined by vMRI in patients with mild to moderate AD (Quinn et al., 2010). A 1-year randomized controlled clinical trial of memantine in AD showed that there were no statistically significant differences between memantine and placebo in total brain or hippocampal atrophy rates in patients with probable AD who treated for 1 year (Wilkinson et al., 2012). A multicenter, randomized clinical trial of rosiglitazone found that treatment was associated with an early increase in whole-brain glucose metabolism by ^{18}FDG-PET imaging but without clinical evidence of slowing the progression of AD (Tzimopoulou et al., 2010). Lassere (2008) has proposed a qualitative scheme for evaluation of AD biomarkers as surrogate endpoints of drug efficacy, on the basis of the character and performance of the biomarker in the context of specific targets, study design, statistical strength, and conflicting results. According to this scheme, neuroimaging biomarkers have not yet reached a level of accuracy to be considered as accurate surrogate endpoints for AD clinical trials.

3.12 COST OF NEUROIMAGING BIOMARKERS FOR ALZHEIMER'S DISEASE

Neuroimaging to detect AD biomarkers is expensive, and often requires specific protocol modifications for imaging patients with AD or other dementias. Neuroimaging centers must purchase and maintain expensive imaging instruments,

Table 3.8 Use of Neuroimaging Biomarkers of Alzheimer's Disease in Assessing Drug Efficacy in Clinical Trials

Modality	Cognitive Effect	Biomarkers	Remarks	References
Phase III trial: Aβ-immunotherapy with bapineuzumab	No significant improvement in cognitive function	▪ Whole-brain volume does not change for both APOE ε4 carriers and noncarriers ▪ Slight decrease in Aβ plaques by ¹¹C-PIB-PET in APOE ε4 group	▪ No clinical improvement with treatment	Salloway et al. (2014)
Phase III trial: Aβ-immunotherapy with solanezumab (a humanized monoclonal antibody that binds Aβ)	No significant improvement in cognitive or functional ability	▪ Whole-brain volume does not change ▪ Hippocampal area does not change	▪ No significant clinical improvement with treatment ▪ CSF biomarker results were opposite to the trial results	Doody et al. (2014)
Docosahexaenoic acid supplementation	MRI methods of the ADNI study were used to generate brain volumes	▪ Whole-brain volume does not change ▪ Hippocampal area does not change ▪ Ventricular area does not change	▪ Supplementation did not slow the rate of cognitive and functional decline in patients with mild to moderate AD	Quinn et al. (2010)
Phase III trial of tramiprosate (ALZHEMED™)	Cognitive improvement was lower than anticipated	▪ Hippocampal volume change measured by MRI (vMRI)	▪ No significant clinical improvement vMRI results were opposite to the clinical results	Saumier et al. (2009)
Phase II trial of TAI (tau aggregation inhibitor)	No clinical decline for 24 weeks treatment	▪ Changes in blood flow using ⁹⁹ᵐTc-HMPAO	▪ Brain perfusion as an indicator of brain metabolism ▪ SPECT imaging showed response to treatment	Wischik and Staff (2009)

(Continued)

Table 3.8 Use of Neuroimaging Biomarkers of Alzheimer's Disease in Assessing Drug Efficacy in Clinical Trials *(cont.)*

Modality	Cognitive Effect	Biomarkers	Remarks	References
Rosiglitazone	Feasibility of using ^{18}FDG-PET as part of a multicenter therapeutics trial	Early increase in whole-brain glucose metabolism	No biological or clinical evidence for slowing progression of AD	Tzimopoulou et al. (2010)
Memantine	Brain atrophy with serial sMRI	▪ No significant change in whole-brain volume ▪ No change in hippocampal area	▪ Significant correlation between rate of atrophy and decline in cognitive and behavioral outcomes	Wilkinson et al. (2012)
Memantine	Resting state default mode network (DMN) activity measure by fMRI	▪ Treatment found higher neuronal activity	▪ Demonstrated some beneficial effects of memantine on the DMN in AD	Lorenzi et al. (2011)

MRI, magnetic resonance imaging; fMRI, functional MRI; sMRI, structural MRI; vMRI, volumetric MRI; PET, positron emission tomography; SPECT, single-photon emission computed tomography; Aβ, beta-amyloid protein; ^{11}C-PIB, [^{11}C]-Pittsburgh Compound B; ^{99m}Tc, metastable nuclear isomer of technetium-99; HMPAO, hexamethylpropylene amine oxime.

have a CLIA (Clinical Laboratory Improvement Amendments) compliant facility, specialized and trained staff, and the capacity for bioinformatic analyses of imaging data. The cost of an average brain MRI scan has been estimated to be $1181 (estimated by OKCOPAY); others have placed the cost as $1179 ± 722.92 (http://www.nerdwallet.com/blog/health/2014/05/09/how-much-does-an-mri-cost/) in metropolitan areas of the United States. The cost of brain PET scan is even higher ($7700 per scan, National PET Scan Brain Procedure; www.new-choicehealth.com). A normal SPECT scan costs roughly the same as an MRI ($1100; www.amenclinics.com). Regardless of the modality used, neuroimaging for detection of AD biomarkers adds extra cost into the procedural charges for sophisticated bioinformatics-based analysis of AD biomarkers.

3.13 LIMITATIONS OF ALZHEIMER'S DISEASE NEUROIMAGING BIOMARKERS

3.13.1 Sophisticated and Expensive Technology

The main limitation to using neuroimaging of AD biomarkers modality is the technical sophistication required for the scans. Only very specialized centers that meet all infrastructure and regulatory compliance requirements and have highly technically trained expert teams of neuroscientists, radiologists, and bioinformatics specialists can perform this type of imaging. In addition, the imaging equipment and its maintenance are expensive, and for PET radiotracers with short half-lives (eg, ^{11}C-PIB with a half-life of ~20 min), ready access to a cyclotron is also required. For these reasons, neuroimaging of AD biomarkers is more costly and geographically limited compared to other testing approaches. The cost of PET imaging technology prohibits its use as a routine biomarker test, including screening patients and to evaluate drug efficacy in clinical trials.

3.13.2 Radioactivity Exposure

Both PET and SPECT neuroimaging techniques require the use of radioactive tracers, which raises issues regarding radiation exposure safety. However, a study designed specifically to monitor the health effects of administration of radionuclides for medical imaging found that the dose administered is significantly below the threshold dose for deterministic effects due to radiation exposure to the fetus (Takalkar et al., 2011). The average dose limit of ^{18}FDG activity is 185–370 MBq (5–10 mCi).

3.13.3 Nonspecific PET Tracer Binding

Large-scale clinical trial data on Aβ-PET has shown no association between the amount of Aβ-PET signal in AD brains and disease severity (Salloway et al., 2014; Sperling et al., 2012). The most widely studied PET radiotracers used to detect AD biomarkers are ^{11}C-PIB for amyloid plaques and ^{18}FDG for glucose uptake. Yet Dr. William E. Klunk (one of the discoverers of ^{11}C-PIB PET) and his coinvestigators detected amyloid plaques in 22% of healthy age-matched control cases by ^{11}C-PIB-PET (Pike et al., 2007). Moreover, soluble Aβ, which is the form toxic to neurons, cannot be detected by ^{11}C-PIB-PET. It has been reported that elevated soluble Aβ, but not amyloid plaque formation, is the main cause of memory impairment in AD transgenic mice (Koistinaho et al., 2001). Rowe et al. (2010) found that ^{11}C-PIB binding to Aβ increases from <10% in patients under 70 years age to 40% in those aged 80 years, suggesting some nonspecific binding activity that may obscure test results. In addition, 22% of healthy age-matched controls (without any cognitive impairment) were considered to be AD positive based on their biomarker value with ^{11}C-PIB PET (Rowe et al., 2010). While [18]FDG-PET imaging might be able to distinguish between FTD and AD, ^{18}F compounds (flutemetamol, flornetapir) have a high affinity to brain white matter that may increase non-specific binding. For some ^{18}F-labeled PET ligands, defluorination can cause bone accumulation of ^{18}F that may influence in imaging results. [18]FDG uptake in the brain is also affected by normal aging and the level of local brain inflammation, independent of AD-related pathology (Foster et al., 2007; Shipley et al., 2013).

3.13.4 Variability Between Imaging Centers and Standardization of Neuroimaging Biomarkers

One of the goals of WW-ADNI (World Wide Alzheimer's Disease Neuroimaging Initiative (WW-ADNI) is to standardize the methods used for conducting neuroimaging scans. For example, most studies have found that MRI measurement of the loss of hippocampal volume over time is a biomarker of AD and can be used to track disease progression. Yet standardization of assessing volume changes by MRI is needed for its wider clinical use. Without proper standardized methods for measuring the change in hippocampal volume, MRI will continue to yield inconsistent results across different clinics and laboratories. To overcome this challenge, a global harmonized protocol was established for manual segmentation of hippocampal MRI images of AD brains (Frisoni and Jack, 2011; Jack et al., 2011)

3.14 CONCLUSIONS

Alzheimer's disease (AD) biomarkers that can diagnose the disease in its preclinical stages and can be used to assess treatment efficacy are urgently needed. PET and CSF biomarkers are the best predictors of AD pathoprogression. MRI has the potential to identify cortical regions of abnormality known to be affected in AD nearly a decade before clinical symptoms of dementia emerge, providing important preclinical evidence of neurodegeneration. New guidelines for diagnosis of AD set by a joint NIA-AA panel of lead scientists recommend the assessment of: (1) dementia due to AD, (2) dementia due to MCI, (3) pathology for AD autopsy, and the need for (4) biomarker development for preclinical AD. An NIA-AA working group (2011) has named hippocampal atrophy, decreased [18]FDG-PET, and increased Aβ-PET as three of five AD biomarkers recommended for use in AD fundamental and clinical research. The EMA has gone a step further in this respect by approving hippocampal volume measurement by sMRI for patient enrichment in clinical trials of prodromal AD/MCI cases. The

International Working Group-2 (IGW-2) for AD biomarkers also proposed the use of volumetric MRI and ^{18}FDG-PET as AD biomarkers for disease monitoring. MRI has been used extensively as a surrogate endpoint of therapeutic efficacy and safety in AD clinical trials, but there are some limitations.

Despite decades of expensive research on neuroimaging biomarkers for AD, the conclusion remains that they are costly and have yet to be standardized in a clinical setting. A combination of neuroimaging and CSF biomarkers may be more accurate for diagnosing dementia due to AD. Neuroimaging biomarkers of AD include hippocampal and entorhinal cortex volume change measured by sMRI, ^{18}FDG-PET glucose uptake, and regional blood flow within the temporoparietal cortex by SPECT. Prospective studies are needed to evaluate the capacity of neuroimaging biomarkers for differential detection of AD, non-AD dementias, and normal aging and for early diagnosis of AD. Aβ-PET imaging cannot be used as surrogate marker for drug efficacy testing because changes of amyloid plaque burden do not necessarily correspond to disease severity. CSF biomarkers (Chapter 5) provide comparable, in some cases even better, sensitivity and specificity than either Aβ-PET or tau-PET neuroimaging. Neuroimaging of AD biomarkers requires more test-retest verification, standardization across different laboratories, and more specifically, concerted multimodal approaches for detecting different aspects of a complex disease such as AD.

Bibliography

Albert, M.S., DeKosky, S.T., Dickson, D., Dubois, B., Feldman, H.H., Fox, N.C., et al., 2011. The diagnosis of mild cognitive impairment due to Alzheimer's disease: recommendations from the National Institute on Aging-Alzheimer's Association workgroups on diagnostic guidelines for Alzheimer's disease. Alzheimers Dement. 7, 270–279.

Alsop, D.C., Detre, J.A., Grossman, M., 2000. Assessment of cerebral blood flow in Alzheimer's disease by spin-labeled magnetic resonance imaging. Ann. Neurol. 47, 93–100.

Alsop, D.C., Dai, W., Grossman, M., Detre, J.A., 2010. Arterial spin labeling blood flow MRI: its role in the early characterization of Alzheimer's disease. J. Alzheimers Dis. 20, 871–880.

Apostolova, L.G., Hwang, K.S., Medina, L.D., Green, A.E., Braskie, M.N., Dutton, R.A., Lai, J., Geschwind, D.H., Cummings, J.L., Thompson, P.M., Ringman, J.M., 2011. Cortical and hippocampal atrophy in patients with autosomal dominant familial Alzheimer's disease. Dement. Geriatr. Cogn. Disord. 32, 118–125.

Arriagada, P.V., Growdon, J.H., Hedley-Whyte, E.T., Hyman, B.T., 1992. Neurofibrillary tangles but not senile plaques parallel duration and severity of Alzheimer's disease. Neurology 42, 631–639.

Babiloni, C., Binetti, G., Cassetta, E., Cerboneschi, D., Dal Forno, G., Del Percio, C., Ferreri, F., Ferri, R., Lanuzza, B., Miniussi, C., Moretti, D.V., Nobili, F., Pascual-Marqui, R.D., Rodriguez, G., Romani, G.L., Salinari, S., Tecchio, F., Vitali, P., Zanetti, O., Zappasodi, F., Rossini, P.M., 2004. Mapping distributed sources of cortical rhythms in mild Alzheimer's disease. A multicentric EEG study. Neuroimage 22, 57–67.

Barber, R., Gholkar, A., Scheltens, P., Ballard, C., McKeith, I.G., O'Brien, J.T., 1999. Medial temporal lobe atrophy on MRI in dementia with Lewy bodies: a comparison with Alzheimer's disease, vascular dementia and normal ageing. Neurology 52, 1153–1158.

Barber, R., Ballard, C., McKeith, I.G., Gholkar, A., O'Brien, J.T., 2000. MRI volumetric study of dementia with Lewy bodies: a comparison with AD and vascular dementia. Neurology 54, 1304–1309.

Barnes, J., Bartlett, J.W., van de Pol, L.A., Loy, C.T., Scahill, R.I., Frost, C., Thompson, P., Fox, N.C., 2009. A meta-analysis of hippocampal atrophy rates in Alzheimer's disease. Neurobiol. Aging 30, 1711–1723.

Bates, T.E., Strangward, M., Keelan, J., Davey, G.P., Munro, P.M., Clark, J.B., 1996. Inhibition of N-acetylaspartate production: implications for 1H MRS studies in vivo. Neuroreport 7, 1397–1400.

Beaulieu, C., Allen, P.S., 1994. Determinants of anisotropic water diffusion in nerves. Magn. Reson. Med. 31, 394–400.

Bierer, L.M., Hof, P.R., Purohit, D.P., Carlin, L., Schmeidler, J., Davis, K.L., Perl, D.P., 1995. Neocortical neurofibrillary tangles correlate with dementia severity in Alzheimer's disease. Arch. Neurol. 52, 81–88.

Biffi, A., Anderson, C.D., Desikan, R.S., Sabuncu, M., Cortellini, L., Schmansky, N., Salat, D., Rosand, J., Alzheimer's Disease Neuroimaging Initiative (ADNI), 2010. Genetic variation and neuroimaging measures in Alzheimer disease. Arch. Neurol. 67, 677–785.

Biswal, B., Yetkin, F.Z., Haughton, V.M., Hyde, J.S., 1995. Functional connectivity in the motor cortex of resting human brain using echo-planar MRI. Magn. Reson. Med. 34, 537–541.

Bocchetta, M., Boccardi, M., Ganzola, R., Apostolova, L.G., Preboske, G., Wolf, D., Ferrari, C., Pasqualetti, P., Robitaille, N., Duchesne, S., Jack, Jr., C.R., Frisoni, G.B., EADC-ADNI Working Group on The Harmonized Protocol for Manual Hippocampal Segmentation and for the Alzheimer's Disease Neuroimaging Initiative, 2015. Harmonized benchmark labels of the hippocampus on magnetic resonance: the EADC-ADNI project. Alzheimers Dement. 11, 151–160.

Bonte, F.J., Weiner, M.F., Bigio, E.H., White, 3rd., C.L., 1997. Brain blood flow in the dementias: SPECT with histopathologic correlation in 54 patients. Radiology 202, 793–797.

Bonte, F.J., Harris, T.S., Roney, C.A., Hynan, L.S., 2004. Differential diagnosis between Alzheimer's and frontotemporal disease by the posterior cingulate sign. J. Nuclear Med. 45, 771–774.

Braskie, M.N., Medina, L.D., Rodriguez-Agudelo, Y., Geschwind, D.H., Macias-Islas, M.A., Cummings, J.L., Bookheimer, S.Y., Ringman, J.M., 2012. Increased fMRI signal with age in familial Alzheimer's disease mutation carriers. Neurobiol. Aging 33, 424.e11–424.e21.

Burton, E.J., Barber, R., Mukaetova-Ladinska, E.B., Robson, J., Perry, R.H., Jaros, E., Kalaria, R.N., O'Brien, J.T., 2009. Medial temporal lobe atrophy on MRI differentiates Alzheimer's disease from dementia with Lewy bodies and vascular cognitive impairment: a prospective study with pathological verification of diagnosis. Brain 132, 195–203.

Canu, E., Frisoni, G.B., Agosta, F., Pievani, M., Bonetti, M., Filippi, M., 2012. Early and late onset Alzheimer's disease patients have distinct patterns of white matter damage. Neurobiol. Aging 33, 1023–1033.

Carter, S.F., Schöll, M., Almkvist, O., Wall, A., Engler, H., Långström, B., Nordberg, A., 2012. Evidence for astrocytosis in prodromal Alzheimer disease provided by 11C-deuterium-L-deprenyl: a multitracer PET paradigm combining 11C-Pittsburgh compound B and 18F-FDG. J. Nucl. Med. 53, 37–46.

Cash, D.M., Rohrer, J.D., Ryan, N.S., Ourselin, S., Fox, N.C., 2014. Imaging endpoints for clinical trials in Alzheimer's disease. Alzheimers Res. Ther. 6, 87.

Chaim, T.M., Duran, F.L., Uchida, R.R., Perico, C.A., de Castro, C.C., Busatto, G.F., 2007. Volumetric reduction of the corpus callosum in Alzheimer's disease in vivo as assessed with voxel-based morphometry. Psychiatry Res. 154, 59–68.

Chan, D., Fox, N.C., Jenkins, R., Scahill, R.I., Crum, W.R., Rossor, M.N., 2001. Rates of global and regional cerebral atrophy in AD and frontotemporal dementia. Neurology 57, 1756–1763.

Chao, L.L., Schuff, N., Kramer, J.H., Du, A.T., Capizzano, A.A., O'Neill, J., Wolkowitz, O.M., Jagust, W.J., Chui, H.C., Miller, B.L., Yaffe, K., Weiner, M.W., 2005. Reduced medial temporal lobe N-acetylaspartate in cognitively impaired but nondemented patients. Neurology 64, 282–289.

Chauhan, G., Adams, H.H., Bis, J.C., Weinstein, G., Yu, L., Töglhofer, A.M., Smith, A.V., van der Lee, S.J., Gottesman, R.F., Thomson, R., Wang, J., Yang, Q., Niessen, W.J., Lopez, O.L., Becker,

J.T., Phan, T.G., Beare, R.J., Arfanakis, K., Fleischman, D., Vernooij, M.W., Mazoyer, B., Schmidt, H., Srikanth, V., Knopman, D.S., Jack, Jr., C.R., Amouyel, P., Hofman, A., DeCarli, C., Tzourio, C., van Duijn, C.M., Bennett, D.A., Schmidt, R., Longstreth, Jr., W.T., Mosley, T.H., Fornage, M., Launer, L.J., Seshadri, S., Ikram, M.A., Debette, S., 2015. Association of Alzheimer's disease GWAS loci with MRI markers of brain aging. Neurobiol. Aging 36, 1765, e7-1765–e16-1765.

Chauveau, F., Boutin, H., Van Camp, N., Dollé, F., Tavitian, B., 2008. Nuclear imaging of neuro-inflammation: a comprehensive review of [11C]-PK11195 challengers. Eur. J. Nucl. Med. Mol. Imaging 35, 2304–2319.

Chen, Y., Wolk, D.A., Reddin, J.S., Korczykowski, M., Martinez, P.M., Musiek, E.S., Newberg, A.B., Julin, P., Arnold, S.E., Greenberg, J.H., Detre, J.A., 2011. Voxel-level comparison of arterial spin-labeled perfusion MRI and FDG-PET in Alzheimer disease. Neurology 77, 1977–1985.

Chien, D.T., Bahri, S., Szardenings, A.K., Walsh, J.C., Mu, F., Su, M.Y., Shankle, W.R., Elizarov, A., Kolb, H.C., 2013. Early clinical PET imaging results with the novel PHF-tau radioligand [F-18]-T807. J. Alzheimers Dis. 34, 457–468.

Choi, S.R., Schneider, J.A., Bennett, D.A., Beach, T.G., Bedell, B.J., Zehntner, S.P., Krautkramer, M.J., Kung, H.F., Skovronsky, D.M., Hefti, F., Clark, C.M., 2012. Correlation of amyloid PET ligand florbetapir F 18 binding with Abeta aggregation and neuritic plaque deposition in postmortem brain tissue. Alzheimer Dis. Assoc. Disord. 26, 8–16.

Chua, T.C., Wen, W., Chen, X., Kochan, N., Slavin, M.J., Trollor, J.N., Brodaty, H., Sachdev, P.S., 2009. Diffusion tensor imaging of the posterior cingulate is a useful biomarker of mild cognitive impairment. Am. J. Geriatr. Psychiatry 17, 602–613.

Cohen, D.A., Klunk, W.E., 2014. Early detection of Alzheimer's disease using PiB and FDG PET. Neurobiol Dis. 72, 117–122.

Collie, D.A., Sellar, R.J., Zeidler, M., Colchester, A.C., Knight, R., Will, R.G., 2001. MRI of Creutzfeldt–Jakob disease: imaging features and recommended MRI protocol. Clin. Radiol. 56, 726–739.

Dai, W., Lopez, O.L., Carmichael, O.T., Becker, J.T., Kuller, L.H., Gach, H.M., 2009. Mild cognitive impairment and Alzheimer disease: patterns of altered cerebral blood flow at MR imaging. Radiology 250, 856–866.

Damoiseaux, J.S., Rombouts, S.A., Barkhof, F., Scheltens, P., Stam, C.J., Smith, S.M., Beckmann, C.F., 2006. Consistent resting-state networks across healthy subjects. Proc. Natl. Acad. Sci. USA 103, 13848–13853.

de Leon, M.J., DeSanti, S., Zinkowski, R., Mehta, P.D., Pratico, D., Segal, S., Rusinek, H., Li, J., Tsui, W., Saint Louis, L.A., Clark, C.M., Tarshish, C., Li, Y., Lair, L., Javier, E., Rich, K., Lesbre, P., Mosconi, L., Reisberg, B., Sadowski, M., DeBernadis, J.F., Kerkman, D.J., Hampel, H., Wahlund, L.O., Davies, P., 2006. Longitudinal CSF and MRI biomarkers improve the diagnosis of mild cognitive impairment. Neurobiol. Aging 27, 394–401.

Detre, J.A., Leigh, J.S., Williams, D.S., Koretsky, A.P., 1992. Perfusion imaging. Magn. Reson. Med. 23, 37–45.

Detre, J.A., Wang, J., Wang, Z., Rao, H., 2009. Arterial spin-labeled perfusion MRI in basic and clinical neuroscience. Curr. Opin. Neurol. 22, 348–355.

Detre, J.A., Rao, H., Wang, D.J., Chen, Y.F., Wang, Z., 2012. Applications of arterial spin labeled MRI in the brain. J. Magn. Reson. Imaging 35, 1026–1037.

Dickerson, B.C., Wolk, D.A., Alzheimer's Disease Neuroimaging Initiative, 2012. MRI cortical thickness biomarker predicts AD-like CSF and cognitive decline in normal adults. Neurology 78, 84–90.

Dierks, T., Ihl, R., Frölich, L., Maurer, K., 1993. Dementia of the Alzheimer type: effects on the spontaneous EEG described by dipole sources. Psychiatry Res. 50, 151–162.

Doody, R.S., Thomas, R.G., Farlow, M., Iwatsubo, T., Vellas, B., Joffe, S., Kieburtz, K., Raman, R., Sun, X., Aisen, P.S., Siemers, E., Liu-Seifert, H., Mohs, R., Alzheimer's Disease Cooperative Study

Steering Committee; Solanezumab Study Group, 2014. Phase 3 trials of solanezumab for mild-to-moderate Alzheimer's disease. N. Engl. J. Med. 370, 311–321.

Double, K.L., Halliday, G.M., Kril, J.J., Harasty, J.A., Cullen, K., Brooks, W.S., Creasey, H., Broe, G.A., 1996. Topography of brain atrophy during normal aging and Alzheimer's disease. Neurobiol. Aging 17, 513–521.

Dougall, N.J., Bruggink, S., Ebmeier, K.P., 2004. Systematic review of the diagnostic accuracy of 99mTc-HMPAO-SPECT in dementia. Am. J. Geriatr. Psychiatry 12, 554–570.

Du, A.T., Jahng, G.H., Hayasaka, S., Kramer, J.H., Rosen, H.J., Gorno-Tempini, M.L., Rankin, K.P., Miller, B.L., Weiner, M.W., Schuff, N., 2006. Hypoperfusion in frontotemporal dementia and Alzheimer disease by arterial spin labeling MRI. Neurology 67, 1215–1220.

Duara, R., Gross-Glenn, K., Barker, W.W., Chang, J.Y., Apicella, A., Loewenstein, D., Boothe, T., 1987. Behavioral activation and the variability of cerebral glucose metabolic measurements. J. Cereb. Blood Flow Metab. 7, 266–271.

Duara, R., Barker, W., Luis, C.A., 1999. Frontotemporal dementia and Alzheimer's disease: differential diagnosis. Dement. Geriatr. Cogn. Disord. 10 (Suppl1), 37–42.

Dubois, B., Feldman, H.H., Jacova, C., Dekosky, S.T., Barberger-Gateau, P., Cummings, J., Delacourte, A., Galasko, D., Gauthier, S., Jicha, G., Meguro, K., O'brien, J., Pasquier, F., Robert, P., Rossor, M., Salloway, S., Stern, Y., Visser, P.J., Scheltens, P., 2007. Research criteria for the diagnosis of Alzheimer's disease: revising the NINCDS-ADRDA criteria. Lancet Neurol. 6, 734–746.

Dubois, B., Feldman, H.H., Jacova, C., Cummings, J.L., Dekosky, S.T., Barberger-Gateau, P., Delacourte, A., Frisoni, G., Fox, N.C., Galasko, D., Gauthier, S., Hampel, H., Jicha, G.A., Meguro, K., O'Brien, J., Pasquier, F., Robert, P., Rossor, M., Salloway, S., Sarazin, M., de Souza, L.C., Stern, Y., Visser, P.J., Scheltens, P., 2010. Revising the definition of Alzheimer's disease: a new lexicon. Lancet Neurol. 9, 1118–1127.

Dubois, B., Feldman, H.H., Jacova, C., Hampel, H., Molinuevo, J.L., Blennow, K., DeKosky, S.T., Gauthier, S., Selkoe, D., Bateman, R., Cappa, S., Crutch, S., Engelborghs, S., Frisoni, G.B., Fox, N.C., Galasko, D., Habert, M.O., Jicha, G.A., Nordberg, A., Pasquier, F., Rabinovici, G., Robert, P., Rowe, C., Salloway, S., Sarazin, M., Epelbaum, S., de Souza, L.C., Vellas, B., Visser, P.J., Schneider, L., Stern, Y., Scheltens, P., Cummings, J.L., 2014. Advancing research diagnostic criteria for Alzheimer's disease: the IWG-2 criteria. Lancet Neurol. 13, 614–629.

Fennema-Notestine, C., Hagler, Jr., D.J., McEvoy, L.K., Fleisher, A.S., Wu, E.H., Karow, D.S., Dale, A.M., Alzheimer's Disease Neuroimaging Initiative, 2009. Structural MRI biomarkers for preclinical and mild Alzheimer's disease. Hum. Brain Mapp. 30, 3238–3253.

Ferencz, B., Laukka, E.J., Lovden, M., Kalpouzos, G., Keller, L., Graff, C., Wahlund, L.O., Fratiglioni, L., Backman, L., 2013. The influence of APOE and TOMM40 polymorphismson hippocampal volume and episodic memory in old age. Front. Hum. Neurosci. 7, 198.

Fernández, A., Arrazola, J., Maestú, F., Amo, C., Gil-Gregorio, P., Wienbruch, C., Ortiz, T., 2003. Correlations of hippocampal atrophy and focal low-frequency magnetic activity in Alzheimer disease: volumetric MR imaging-magnetoencephalographic study. AJNR Am. J. Neuroradiol. 24, 481–487.

Fodero-Tavoletti, M.T., Okamura, N., Furumoto, S., Mulligan, R.S., Connor, A.R., McLean, C.A., Cao, D., Rigopoulos, A., Cartwright, G.A., O'Keefe, G., Gong, S., Adlard, P.A., Barnham, K.J., Rowe, C.C., Masters, C.L., Kudo, Y., Cappai, R., Yanai, K., Villemagne, V.L., 2011. 18F-THK523: a novel in vivo tau imaging ligand for Alzheimer's disease. Brain 134, 1089–1100.

Forlenza, O.V., Wacker, P., Nunes, P.V., Yacubian, J., Castro, C.C., Otaduy, M.C., Gattaz, W.F., 2005. Reduced phospholipid breakdown in Alzheimer's brains: a ^{31}P spectroscopy study. Psychopharmacology 180, 359–365.

Foster, N.L., Heidebrink, J.L., Clark, C.M., Jagust, W.J., Arnold, S.E., Barbas, N.R., DeCarli, C.S., Turner, R.S., Koeppe, R.A., Higdon, R., Minoshima, S., 2007. FDG-PET improves accuracy in distinguishing frontotemporal dementia and Alzheimer's disease. Brain 130, 2616–2635.

Fox, N.C., Freeborough, P.A., 1997. Brain atrophy progression measured from registered serial MRI: validation and application to Alzheimer's disease. J. Magn. Reson. Imaging 7, 1069–1075.

Fox, N.C., Kennedy, J., 2009. Structural imaging markers for therapeutic trials in Alzheimer's disease. J. Nutr. Health Aging 13, 350–352.

Fox, N.C., Scahill, R.I., Crum, W.R., Rossor, M.N., 1999a. Correlation between rates of brain atrophy and cognitive decline in AD. Neurology 52, 1687–1689.

Fox, N.C., Warrington, E.K., Rossor, M.N., 1999b. Serial magnetic resonance imaging of cerebral atrophy in preclinical Alzheimer's disease. Lancet 353, 2125–12125.

Fox, N.C., Black, R.S., Gilman, S., Rossor, M.N., Griffith, S.G., Jenkins, L., Koller, M., AN1792(QS-21)-201 Study, 2005. Effects of Aβ immunization (AN1792) on MRI measures of cerebral volume in Alzheimer disease. Neurology 64, 1563–1572.

Frisoni, G.B., Delacourte, A., 2009. Neuroimaging outcomes in clinical trials in Alzheimer's disease. J. Nutr. Health Aging 13, 209–212.

Frisoni, G.B., Jack, C.R., 2011. Harmonization of magnetic resonance-based manual hippocampal segmentation: a mandatory step for wide clinical use. Alzheimers Dement. 7, 171–174.

Frisoni, G.B., Fox, N.C., Jack, Jr., C.R., Scheltens, P., Thompson, P.M., 2010. The clinical use of structural MRI in Alzheimer disease. Nat. Rev. Neurol. 6, 67–77.

Furney, S.J., Simmons, A., Breen, G., Pedroso, I., Lunnon, K., Proitsi, P., Hodges, A., Powell, J., Wahlund, L.O., Kloszewska, I., Mecocci, P., Soininen, H., Tsolaki, M., Vellas, B., Spenger, C., Lathrop, M., Shen, L., Kim, S., Saykin, A.J., Weiner, M.W., Lovestone, S., Alzheimer's Disease Neuroimaging Initiative; AddNeuroMed Consortium, 2011. Genome-wide association with MRI atrophy measures as a quantitative trait locus for Alzheimer's disease. Mol. Psychiatry 16, 1130–1138.

Gallego-Jutglà, E., Elgendi, M., Vialatte, F., Solé-Casals, J., Cichocki, A., Latchoumane, C., Jeong, J., Dauwels, J., 2012. Diagnosis of Alzheimer's disease from EEG by means of synchrony measures in optimized frequency bands. Conf. Proc. IEEE Eng. Med. Biol. Soc. 2012, 4266–4270.

Greicius, M.D., Krasnow, B., Reiss, A.L., Menon, V., 2003. Functional connectivity in the resting brain: a network analysis of the default mode hypothesis. Proc. Natl. Acad. Sci. USA 100, 253–258.

Greicius, M.D., Supekar, K., Menon, V., Dougherty, R.F., 2009. Resting-state functional connectivity reflects structural connectivity in the default mode network. Cereb. Cortex 19, 72–78.

Grunwald, M., Busse, F., Hensel, A., Riedel-Heller, S., Kruggel, F., Arendt, T., Wolf, H., Gertz, H.J., 2002. Theta-power differences in patients with mild cognitive impairment under rest condition and during haptic tasks. Alzheimer Dis. Assoc. Disord. 16, 40–48.

Haller, J.W., Banerjee, A., Christensen, G.E., Gado, M., Joshi, S., Miller, M.I., Sheline, Y., Vannier, M.W., Csernansky, J.G., 1997. Three-dimensional hippocampal MR morphometry with high-dimensional transformation of a neuroanatomic atlas. Radiology 202, 504–510.

Harada, R., Okamura, N., Furumoto, S., Tago, T., Maruyama, M., Higuchi, M., Yoshikawa, T., Arai, H., Iwata, R., Kudo, Y., Yanai, K., 2013. Comparison of the binding characteristics of [18F]THK-523 and other amyloid imaging tracers to Alzheimer's disease pathology. Eur. J. Nucl. Med. Mol. Imaging 40, 125–132.

Hu, W.T., Wang, Z., Lee, V.M., Trojanowski, J.Q., Detre, J.A., Grossman, M., 2010. Distinct cerebral perfusion patterns in FTLD and AD. Neurology 75, 881–888.

Huang, C., Wahlund, L., Dierks, T., Julin, P., Winblad, B., Jelic, V., 2000. Discrimination of Alzheimer's disease and mild cognitive impairment by equivalent EEG sources: a cross-sectional and longitudinal study. Clin. Neurophysiol. 111, 1961–1967.

Ishii, K., Sasaki, M., Kitagaki, H., Yamaji, S., Sakamoto, S., Matsuda, K., Mori, E., 1997. Reduction of cerebellar glucose metabolism in advanced Alzheimer's disease. J. Nuclear Med. 38, 925–928.

Jack, Jr., C.R., Petersen, R.C., Xu, Y., O'Brien, P.C., Smith, G.E., Ivnik, R.J., Tangalos, E.G., Kokmen, E., 1998. Rate of medial temporal lobe atrophy in typical aging and Alzheimer's disease. Neurology 51, 993–999.

Jack, Jr., C.R., Petersen, R.C., Xu, Y.C., O'Brien, P.C., Smith, G.E., Ivnik, R.J., Boeve, B.F., Waring, S.C., Tangalos, E.G., Kokmen, E., 1999. Prediction of AD with MRI-based hippocampal volume in mild cognitive impairment. Neurology 52, 1397–1403.

Jack, Jr., C.R., Petersen, R.C., Xu, Y., O'Brien, P.C., Smith, G.E., Ivnik, R.J., Boeve, B.F., Tangalos, E.G., Kokmen, E., 2000. Rates of hippocampal atrophy correlate with change in clinical status in aging and AD. Neurology 55, 484–489.

Jack, Jr., C.R., Slomkowski, M., Gracon, S., Hoover, T.M., Felmlee, J.P., Stewart, K., Xu, Y., Shiung, M., O'Brien, P.C., Cha, R., Knopman, D., Petersen, R.C., 2003. MRI as a biomarker of disease progression in a therapeutic trial of milameline for AD. Neurology 60, 253–260.

Jack, Jr., C.R., Shiung, M.M., Gunter, J.L., O'Brien, P.C., Weigand, S.D., Knopman, D.S., Boeve, B.F., Ivnik, R.J., Smith, G.E., Cha, R.H., Tangalos, E.G., Petersen, R.C., 2004. Comparison of different MRI brain atrophy, rate measures with clinical disease progression in AD. Neurology 62, 591–600.

Jack, Jr., C.R., Petersen, R.C., Grundman, M., Jin, S., Gamst, A., Ward, C.P., Sencakova, D., Doody, R.S., Thal, L.J., Members of the Alzheimer's Disease Cooperative Study (ADCS), 2008. Members of the Alzheimer's Disease Cooperative Study (ADCS). Longitudinal MRI findings from the vitamin E and donepezil treatment study for MCI. Neurobiol. Aging 29, 1285–1295.

Jack, Jr., C.R., Barkhof, F., Bernstein, M.A., Cantillon, M., Cole, P.E., Decarli, C., Dubois, B., Duchesne, S., Fox, N.C., Frisoni, G.B., Hampel, H., Hill, D.L., Johnson, K., Mangin, J.F., Scheltens, P., Schwarz, A.J., Sperling, R., Suhy, J., Thompson, P.M., Weiner, M., Foster, N.L., 2011. Steps to standardization and validation of hippocampal volumetry as a biomarker in clinical trials and diagnostic criterion for Alzheimer's disease. Alzheimers Dement. 7, 474–485.

Jagust, W., Reed, B., Mungas, D., Ellis, W., Decarli, C., 2007. What does fluorodeoxyglucose PET imaging add to a clinical diagnosis of dementia? Neurology 69, 871–877.

Jagust, W.J., Landau, S.M., Shaw, L.M., Trojanowski, J.Q., Koeppe, R.A., Reiman, E.M., Foster, N.L., Petersen, R.C., Weiner, M.W., Price, J.C., Mathis, C.A., 2009. Relationships between biomarkers in aging and dementia. Neurology 73, 1193–1199.

Jelic, V., Johansson, S.E., Almkvist, O., Shigeta, M., Julin, P., Nordberg, A., Winblad, B., Wahlund, L.O., 2000. Quantitative electroencephalography in mild cognitive impairment: longitudinal changes and possible prediction of Alzheimer's disease. Neurobiol. Aging 21, 533–540.

Jobst, K.A., Barnetson, L.P., Shepstone, B.J., 1998. Accurate prediction of histologically confirmed Alzheimer's disease and the differential diagnosis of dementia: the use of NINCDS-ADRDA and DSM-III-R criteria, SPECT, X-ray CT, and Apo E4 in medial temporal lobe dementias. Oxford Project to Investigate Memory and Aging. Int. Psychogeriatr. 10, 271–302.

Johnson, N.A., Jahng, G.H., Weiner, M.W., Miller, B.L., Chui, H.C., Jagust, W.J., Gorno-Tempini, M.L., Schuff, N., 2005. Pattern of cerebral hypoperfusion in Alzheimer disease and mild cognitive impairment measured with arterial spin-labeling MR imaging: initial experience. Radiology 234, 851–859.

Johnson, K.A., Fox, N.C., Sperling, R.A., Klunk, W.E., 2012. Brain imaging in Alzheimer disease. Cold Spring Harb. Perspect. Med. 2, a006213.

Kantarci, K., Petersen, R.C., Boeve, B.F., Knopman, D.S., Tang-Wai, D.F., O'Brien, P.C., Weigand, S.D., Edland, S.D., Smith, G.E., Ivnik, R.J., Ferman, T.J., Tangalos, E.G., Jack, Jr., C.R., 2004. 1H MR spectroscopy in common dementias. Neurology 63, 1393–1398.

Khan, W., Giampietro, V., Ginestet, C., Dell'Acqua, F., Bouls, D., Newhouse, S., Dobson, R., Banaschewski, T., Barker, G.J., Bokde, A.L., Büchel, C., Conrod, P., Flor, H., Frouin, V., Garavan, H., Gowland, P., Heinz, A., Ittermann, B., Lemaître, H., Nees, F., Paus, T., Pausova, Z., Rietschel, M., Smolka, M.N., Ströhle, A., Gallinat, J., Westman, E., Schumann, G., Lovestone, S., Simmons, A., IMAGEN consortium, 2014. No differences in hippocampal volume between carriers and noncarriers of the ApoE ε4 and ε2 alleles in young healthy adolescents. J. Alzheimers Dis. 40, 37–43.

Klunk, W.E., Engler, H., Nordberg, A., Wang, Y., Blomqvist, G., Holt, D.P., Bergström, M., Savitcheva, I., Huang, G.F., Estrada, S., Ausén, B., Debnath, M.L., Barletta, J., Price, J.C., Sandell, J., Lopresti,

B.J., Wall, A., Koivisto, P., Antoni, G., Mathis, C.A., Långström, B., 2004. Imaging brain amyloid in Alzheimer's disease with Pittsburgh Compound-B. Ann. Neurol. 55, 306–319.

Koistinaho, M., Ort, M., Cimadevilla, J.M., Vondrous, R., Cordell, B., Koistinaho, J., Bures, J., Higgins, L.S., 2001. Specific spatial learning deficits become severe with age in beta -amyloid precursor protein transgenic mice that harbor diffuse beta -amyloid deposits but do not form plaques. Proc. Natl. Acad. Sci. USA 98, 14675–14680.

Krishnan, K.R., Charles, H.C., Doraiswamy, P.M., Mintzer, J., Weisler, R., Yu, X., Perdomo, C., Ieni, J.R., Rogers, S., 2003. Randomized, placebo-controlled trial of the effects of donepezil on neuronal markers and hippocampal volumes in Alzheimer's disease. Am. J. Psychiatry 160, 2003–2011.

Kropholler, M.A., Boellaard, R., van Berckel, B.N., Schuitemaker, A., Kloet, R.W., Lubberink, M.J., Jonker, C., Scheltens, P., Lammertsma, A.A., 2007. Evaluation of reference regions for (R)-[(11) C]PK11195 studies in Alzheimer's disease and mild cognitive impairment. J. Cereb. Blood Flow Metab. 27, 1965–1974.

Lassere, M.N., 2008. The Biomarker-Surrogacy Evaluation Schema: a review of the biomarker-surrogate literature and a proposal for a criteria-based, quantitative, multidimensional hierarchical levels of evidence schema for evaluating the status of biomarkers as surrogate endpoints. Stat. Methods Med. Res. 17, 303–340.

Lorenzi, M., Beltramello, A., Mercuri, N.B., Canu, E., Zoccatelli, G., Pizzini, F.B., Alessandrini, F., Cotelli, M., Rosini, S., Costardi, D., Caltagirone, C., Frisoni, G.B., 2011. Effect of memantine on resting state default mode network activity in Alzheimer's disease. Drugs Aging 28, 205–217.

Machulda, M.M., Ward, H.A., Borowski, B., Gunter, J.L., Cha, R.H., O'Brien, P.C., Petersen, R.C., Boeve, B.F., Knopman, D., Tang-Wai, D.F., Ivnik, R.J., Smith, G.E., Tangalos, E.G., Jack, Jr., C.R., 2003. Comparison of memory fMRI response among normal, MCI, and Alzheimer's patients. Neurology 61, 500–506.

Maestú, F., Arrazola, J., Fernández, A., Simos, P.G., Amo, C., Gil-Gregorio, P., Fernandez, S., Papanicolaou, A., Ortiz, T., 2003. Do cognitive patterns of brain magnetic activity correlate with hippocampal atrophy in Alzheimer's disease? J. Neurol. Neurosurg. Psychiatry 74, 208–212.

Maestú, F., Campo, P., Gil-Gregorio, P., Fernández, S., Fernández, A., Ortiz, T., 2006. Medial temporal lobe neuromagnetic hypoactivation and risk for developing cognitive decline in elderly population: a 2-year follow-up study. Neurobiol. Aging 27, 32–37.

Mandal, P.K., 2007. Magnetic resonance spectroscopy (MRS) and its application in Alzheimer's disease. Concept. Magn. Reson. A 30A, 40–64.

Maruyama, M., Shimada, H., Suhara, T., Shinotoh, H., Ji, B., Maeda, J., Zhang, M.R., Trojanowski, J.Q., Lee, V.M., Ono, M., Masamoto, K., Takano, H., Sahara, N., Iwata, N., Okamura, N., Furumoto, S., Kudo, Y., Chang, Q., Saido, T.C., Takashima, A., Lewis, J., Jang, M.K., Aoki, I., Ito, H., Higuchi, M., 2013. Imaging of tau pathology in a tauopathy mouse model and in Alzheimer patients compared to normal controls. Neuron 79, 1094–1108.

Masdeu, J.C., Kreisl, W.C., Berman, K.F., 2012. The neurobiology of Alzheimer disease defined by neuroimaging. Curr. Opin. Neurol. 25, 410–420.

Matsuda, H., 2007. Role of neuroimaging in Alzheimer's disease, with emphasis on brain perfusion SPECT. J. Nucl. Med. 48, 1289–1300.

Mattsson, N., Insel, P.S., Landau, S., Jagust, W., Donohue, M., Shaw, L.M., Trojanowski, J.Q., Zetterberg H6, Blennow, K., Weiner, M., Alzheimer's Disease Neuroimaging Initiative, 2014. Diagnostic accuracy of CSF Ab42 and florbetapir PET for Alzheimer's disease. Ann. Clin. Transl. Neurol. 1, 534–543.

McKeith, I.G., Dickson, D.W., Lowe, J., Emre, M., O'Brien, J.T., Feldman, H., Cummings, J., Duda, J.E., Lippa, C., Perry, E.K., Aarsland, D., Arai, H., Ballard, C.G., Boeve, B., Burn, D.J., Costa, D., Del Ser, T., Dubois, B., Galasko, D., Gauthier, S., Goetz, C.G., Gomez-Tortosa, E., Halliday, G., Hansen, L.A., Hardy, J., Iwatsubo, T., Kalaria, R.N., Kaufer, D., Kenny, R.A., Korczyn, A., Kosaka, K., Lee, V.M., Lees, A., Litvan, I., Londos, E., Lopez, O.L., Minoshima, S., Mizuno, Y., Molina, J.A., Mukaetova-Ladinska, E.B., Pasquier, F., Perry, R.H., Schulz, J.B., Trojanowski, J.Q., Yamada, M.,

Consortium on LBD, 2005. Diagnosis and management of dementia with Lewy bodies: third report of the LBD Consortium. Neurology 65, 1863–1872.

McKhann, G.M., Knopman, D.S., Chertkow, H., Hyman, B.T., Jack, Jr., C.R., Kawas, C.H., Klunk, W.E., Koroshetz, W.J., Manly, J.J., Mayeux, R., Mohs, R.C., Morris, J.C., Rossor, M.N., Scheltens, P., Carrillo, M.C., Thies, B., Weintraub, S., Phelps, C.H., 2011. The diagnosis of dementia due to Alzheimer's disease: recommendations from the National Institute on Aging-Alzheimer's Association workgroups on diagnostic guidelines for Alzheimer's disease. Alzheimers Dement. 7, 263–269.

Migliaccio, R., Agosta, F., Possin, K.L., Rabinovici, G.D., Miller, B.L., Gorno-Tempini ML, 2012. White matter atrophy in Alzheimer's disease variants. Alzheimers Dement. 8 (5 Suppl), S78–S87.

Minoshima, S., Foster, N.L., Sima, A.A., Frey, K.A., Albin, R.L., Kuhl, D.E., 2001. Alzheimer's disease versus dementia with Lewy bodies: cerebral metabolic distinction with autopsy confirmation. Ann. Neurol. 50, 358–365.

Mintun, M.A., Larossa, G.N., Sheline, Y.I., Dence, C.S., Lee, S.Y., Mach, R.H., Klunk, W.E., Mathis, C.A., DeKosky, S.T., Morris, J.C., 2006. [11C]PIB in a nondemented population: potential antecedent marker of Alzheimer disease. Neurology 67, 446–452.

Moffett, J.R., Ross, B., Arun, P., Madhavarao, C.N., Namboodiri, A.M., 2007. N-Acetylaspartate in the CNS: from neurodiagnostics to neurobiology. Prog. Neurobiol. 81, 89–131.

Mosconi, L., Tsui, W.H., Herholz, K., Pupi, A., Drzezga, A., Lucignani, G., Reiman, E.M., Holthoff, V., Kalbe, E., Sorbi, S., Diehl-Schmid, J., Perneczky, R., Clerici, F., Caselli, R., Beuthien-Baumann, B., Kurz, A., Minoshima, S., de Leon, M.J., 2008. Multicenter standardized 18F-FDG PET diagnosis of mild cognitive impairment, Alzheimer's disease, and other dementias. J. Nucl. Med. 49, 390–398.

Moseley, M.E., Cohen, Y., Kucharczyk, J., Mintorovitch, J., Asgari, H.S., Wendland, M.F., Tsuruda, J., Norman, D., 1990. Diffusion-weighted MR imaging of anisotropic water diffusion in cat central nervous system. Radiology 176, 439–445.

Musiek, E.S., Chen, Y., Korczykowski, M., Saboury, B., Martinez, P.M., Reddin, J.S., Alavi, A., Kimberg, D.Y., Wolk, D.A., Julin, P., Newberg, A.B., Arnold, S.E., Detre, J.A., 2012. Direct comparison of fluorodeoxyglucose positron emission tomography and arterial spin labeling magnetic resonance imaging in Alzheimer's disease. Alzheimers Dement. 8, 51–59.

Neary, D., Snowden, J.S., Gustafson, L., Passant, U., Stuss, D., Black, S., Freedman, M., Kertesz, A., Robert, P.H., Albert, M., Boone, K., Miller, B.L., Cummings, J., Benson, D.F., 1998. Frontotemporal lobar degeneration: a consensus on clinical diagnostic criteria. Neurology 51, 1546–1554.

Nedelska, Z., Schwarz, C.G., Boeve, B.F., Lowe, V.J., Reid, R.I., Przybelski, S.A., Lesnick, T.G., Gunter, J.L., Senjem, M.L., Ferman, T.J., Smith, G.E., Geda, Y.E., Knopman, D.S., Petersen, R.C., Jack, Jr., C.R., Kantarci, K., 2015. White matter integrity in dementia with Lewy bodies: a voxel-based analysis of diffusion tensor imaging. Neurobiol. Aging 36, 2010–2017.

Neto, E., Allen, E.A., Aurlien, H., Nordby, H., Eichele, T., 2015. EEG spectral features discriminate between Alzheimer's and vascular dementia. Front Neurol. 6, 25.

Oishi, K., Mielke, M.M., Albert, M., Lyketsos, C.G., Mori, S., 2011. DTI analyses and clinical applications in Alzheimer's disease. J. Alzheimers Dis. 26 (Suppl 3), 287–296.

Okamura, N., Furumoto, S., Harada, R., Tago, T., Yoshikawa, T., Fodero-Tavoletti, M., Mulligan, R.S., Villemagne, V.L., Akatsu, H., Yamamoto, T., Arai, H., Iwata, R., Yanai, K., Kudo, Y., 2013. Novel 18F-labeled arylquinoline derivatives for noninvasive imaging of tau pathology in Alzheimer disease. J. Nucl. Med. 54, 1420–1427.

Okamura, N., Furumoto, S., Fodero-Tavoletti, M.T., Mulligan, R.S., Harada, R., Yates, P., Pejoska, S., Kudo, Y., Masters, C.L., Yanai, K., Rowe, C.C., Villemagne, V.L., 2014a. Non-invasive assessment of Alzheimer's disease neurofibrillary pathology using 18F-THK5105 PET. Brain 137, 1762–1771.

Okamura, N., Harada, R., Furumoto, S., Arai, H., Yanai, K., Kudo, Y., 2014b. Tau PET imaging in Alzheimer's disease. Curr. Neurol. Neurosci. Rep. 14, 500.

Owen, D.R., Howell, O.W., Tang, S.P., Wells, L.A., Bennacef, I., Bergstrom, M., Gunn, R.N., Rabiner, E.A., Wilkins, M.R., Reynolds, R., Matthews, P.M., Parker, C.A., 2010. Two binding sites for [3H] PBR28 in human brain: implications for TSPO PET imaging of neuroinflammation. J. Cereb. Blood Flow Metab. 30, 1608–1618.

Petersen, R.C., 2007. Mild cognitive impairment. Continuum lifelong learning. Neurol. 13, 15–38.

Pierpaoli, C., Basser, P.J., 1996. Toward a quantitative assessment of diffusion anisotropy. Magn. Reson. Med. 36, 893–906.

Pierpaoli, C., Jezzard, P., Basser, P.J., Barnett, A., Di Chiro, G., 1996. Diffusion tensor MR imaging of human brain. Radiology 201, 637–648.

Pihlajamäki, M., O' Keefe, K., Bertram, L., Tanzi, R.E., Dickerson, B.C., Blacker, D., Albert, M.S., Sperling, R.A., 2009. Evidence of altered posteromedial cortical fMRI activity in subjects at risk for Alzheimer disease. Alzheimer Dis. Assoc. Disord. 24, 28–36.

Pike, K.E., Savage, G., Villemagne, V.L., Ng, S., Moss, S.A., Maruff, P., Mathis, C.A., Klunk, W.E., Masters, C.L., Rowe, C.C., 2007. Beta-amyloid imaging and memory in non-demented individuals: evidence for preclinical Alzheimer's disease. Brain 130, 2837–2844.

Potkin, S.G., Guffanti, G., Lakatos, A., Turner, J.A., Kruggel, F., Fallon, J.H., Saykin, A.J., Orro, A., Lupoli, S., Salvi, E., Weiner, M., Macciardi, F., Alzheimer's Disease Neuroimaging Initiative, 2009. Hippocampal atrophy as a quantitative trait in a genome-wide association study identifying novel susceptibility genes for Alzheimer's disease. PLoS One 4, e6501.

Quinn, J.F., Raman, R., Thomas, R.G., Yurko-Mauro, K., Nelson, E.B., Van Dyck, C., Galvin, J.E., Emond, J., Jack, Jr., C.R., Weiner, M., Shinto, L., Aisen, P.S., 2010. Docosahexaenoic acid supplementation and cognitive decline in Alzheimer disease: a randomized trial. JAMA 304, 1903–1911.

Rabinovici, G.D., Rosen, H.J., Alkalay, A., Kornak, J., Furst, A.J., Agarwal, N., Mormino, E.C., O'Neil, J.P., Janabi, M., Karydas, A., Growdon, M.E., Jang, J.Y., Huang, E.J., Dearmond, S.J., Trojanowski, J.Q., Grinberg, L.T., Gorno-Tempini, M.L., Seeley, W.W., Miller, B.L., Jagust, W.J., 2011. Amyloid vs FDG-PET in the differential diagnosis of AD and FTLD. Neurology 77, 2034–2042.

Ramanan, V.K., Risacher, S.L., Nho, K., Kim, S., Swaminathan, S., Shen, L., Foroud, T.M., Hakonarson, H., Huentelman, M.J., Aisen, P.S., Petersen, R.C., Green, R.C., Jack, C.R., Koeppe, R.A., Jagust, W.J., Weiner, M.W., Saykin, A.J., Alzheimer's Disease Neuroimaging Initiative, 2014. APOE and BCHE as modulators of cerebral amyloid deposition: a florbetapir PET genome-wide association study. Mol. Psychiatry 19, 351–357.

Ramanan, V.K., Risacher, S.L., Nho K3, Kim, S., Shen, L., McDonald, B.C., Yoder, K.K., Hutchins GD5, West, J.D., Tallman, E.F., Gao, S., Foroud, T.M., Farlow, M.R., De Jager, P.L., Bennett, D.A., Aisen, P.S., Petersen, R.C., Jack, Jr., C.R., Toga, A.W., Green, R.C., Jagust, W.J., Weiner, M.W., Saykin, A.J., Alzheimer's Disease Neuroimaging Initiative (ADNI), 2015. GWAS of longitudinal amyloid accumulation on 18F-florbetapir PET in Alzheimer's disease implicates microglial activation gene IL1RAP. Brain 138, 3076–3088.

Reiman, E.M., Uecker, A., Caselli, R.J., Lewis, S., Bandy, D., de Leon, M.J., De Santi, S., Convit, A., Osborne, D., Weaver, A., Thibodeau, S.N., 1998. Hippocampal volumes in cognitively normal persons at genetic risk for Alzheimer's disease. Ann. Neurol. 44, 288–291.

Ridha, B.H., Symms, M.R., Tozer, D.J., Stockton, K.C., Frost, C., Siddique, M.M., Lewis, E.B., MacManus, D.G., Boulby, P.A., Barker, G.J., Rossor, M.N., Fox, N.C., Tofts, P.S., 2007. Magnetization transfer ratio in Alzheimer disease: comparison with volumetric measurements. AJNR Am. J. Neuroradiol. 28, 965–970.

Román, G.C., Tatemichi, T.K., Erkinjuntti, T., Cummings, J.L., Masdeu, J.C., Garcia, J.H., Amaducci, L., Orgogozo, J.M., Brun, A., Hofman, A., et al., 1993. Vascular dementia: diagnostic criteria for research studies. Report of the NINDS-AIREN International Workshop. Neurology 43, 250–260.

Rowe, C.C., Ellis, K.A., Rimajova, M., Bourgeat, P., Pike, K.E., Jones, G., Fripp, J., Tochon-Danguy, H., Morandeau, L., O'Keefe, G., Price, R., Raniga, P., Robins, P., Acosta, O., Lenzo, N., Szoeke, C., Salvado, O., Head, R., Martins, R., Masters, C.L., Ames, D., Villemagne, V.L., 2010. Amyloid imaging results from the Australian Imaging, Biomarkers and Lifestyle (AIBL) study of aging. Neurobiol. Aging 31, 1275–1283.

Salloway, S., Sperling, R., Fox, N.C., Blennow, K., Klunk, W., Raskind, M., Sabbagh, M., Honig, L.S., Porsteinsson, A.P., Ferris, S., Reichert, M., Ketter, N., Nejadnik, B., Guenzler, V., Miloslavsky, M., Wang, D., Lu, Y., Lull, J., Tudor, I.C., Liu, E., Grundman, M., Yuen, E., Black, R., Brashear, H.R., Bapineuzumab 301 and 302 Clinical Trial Investigators, 2014. Two phase 3 trials of bapineuzumab in mild-to-moderate Alzheimer's disease. N. Engl. J. Med. 370, 322–333.

Saumier, D., Aisen, P.S., Gauthier, S., Vellas, B., Ferris, S.H., Duong, A., Suhy, J., Oh, J., Lau, W., Garceau, D., Haine, D., Sampalis, J., 2009. Lessons learned in the use of volumetric MRI in therapeutic trials in Alzheimer's disease: the ALZHEMED (Tramiprosate) experience. J. Nutr. Health Aging 13, 370–372.

Scheinin, N.M., Aalto, S., Koikkalainen, J., Lötjönen, J., Karrasch, M., Kemppainen, N., Viitanen, M., Någren, K., Helin, S., Scheinin, M., Rinne, J.O., 2009. Follow-up of [11C]PIB uptake and brain volume in patients with Alzheimer disease and controls. Neurology 73, 1186–1192.

Scheinin, N.M., Aalto, S., Kaprio, J., Koskenvuo, M., Räihä, I., Rokka, J., Hinkka-Yli-Salomäki, S., Rinne, J.O., 2011. Early detection of Alzheimer disease: 11C-PIB PET in twins discordant for cognitive impairment. Neurology 77, 453–460.

Schuff, N., Matsumoto, S., Kmiecik, J., Studholme, C., Du, A., Ezekiel, F., Miller, B.L., Kramer, J.H., Jagust, W.J., Chui, H.C., Weiner, M.W., 2009. Cerebral blood flow in ischemic vascular dementia and Alzheimer's disease, measured by arterial spin-labeling magnetic resonance imaging. Alzheimers Dement. 5, 454–462.

Serrano-Pozo, A., Mielke, M.L., Gómez-Isla, T., Betensky, R.A., Growdon, J.H., Frosch, M.P., Hyman, B.T., 2011. Reactive glia not only associates with plaques but also parallels tangles in Alzheimer's disease. Am. J. Pathol. 179, 1373–1384.

Sexton, C.E., Kalu, U.G., Filippini, N., Mackay, C.E., Ebmeier, K.P., 2011. A meta-analysis of diffusion tensor imaging in mild cognitive impairment and Alzheimer's disease. Neurobiol. Aging 32, 2322.e5–2322.e18.

Shen, L., Kim, S., Risacher, S.L., Nho, K., Swaminathan, S., West, J.D., Foroud, T., Pankratz, N., Moore, J.H., Sloan, C.D., Huentelman, M.J., Craig, D.W., Dechairo, B.M., Potkin, S.G., Jack, Jr., C.R., Weiner, M.W., Saykin, A.J., Alzheimer's Disease Neuroimaging Initiative, 2010. Whole genome association study of brain-wide imaging phenotypes for identifying quantitative trait loci in MCI and AD: a study of the ADNI cohort. Neuroimage 53, 1051–1063.

Shi, F., Liu, B., Zhou, Y., Yu, C., Jiang, T., 2009. Hippocampal volume and asymmetry in mild cognitive impairment and Alzheimer's disease: meta-analyses of MRI studies. Hippocampus 19, 1055–1064.

Shimada, H., Ataka, S., Takeuchi, J., Mori, H., Wada, Y., Watanabe, Y., Miki, T., 2011. Pittsburgh compound B-negative dementia—a possibility of misdiagnosis of patients with non-Alzheimer disease-type dementia as having AD. J. Geriatr. Psychiatry Neurol. 24, 123–126.

Shipley, S.M., Frederick, M.C., Filley, C.M., Kluger, B.M., 2013. Potential for misdiagnosis in community-acquired PET scans for dementia. Neurol. Clin. Pract. 3, 305–312.

Silverman, D.H., Small, G.W., Chang, C.Y., Lu, C.S., Kung De Aburto, M.A., Chen, W., Czernin, J., Rapoport, S.I., Pietrini, P., Alexander, G.E., Schapiro, M.B., Jagust, W.J., Hoffman, J.M., Welsh-Bohmer, K.A., Alavi, A., Clark, C.M., Salmon, E., de Leon, M.J., Mielke, R., Cummings, J.L., Kowell, A.P., Gambhir, S.S., Hoh, C.K., Phelps, M.E., 2001. Positron emission tomography in evaluation of dementia: regional brain metabolism and long-term outcome. JAMA 286, 2120–2127.

Sluimer, J.D., Bouwman, F.H., Vrenken, H., Blankenstein, M.A., Barkhof, F., van der Flier, W.M., Scheltens, P., 2010. Whole-brain atrophy rate and CSF biomarker levels in MCI and AD: a longitudinal study. Neurobiol. Aging 31, 758–764.

Smailagic, N., Vacante, M., Hyde, C., Martin, S., Ukoumunne, O., Sachpekidis, C., 2015. 18F-FDG PET for the early diagnosis of Alzheimer's disease dementia and other dementias in people with mild cognitive impairment (MCI). Cochrane Database Syst. Rev. 1, CD010632.

Small, G.W., Kepe, V., Ercoli, L.M., Siddarth, P., Bookheimer, S.Y., Miller, K.J., Lavretsky, H., Burggren, A.C., Cole, G.M., Vinters, H.V., Thompson, P.M., Huang, S.C., Satyamurthy, N., Phelps, M.E., Barrio, J.R., 2006. PET of brain amyloid and tau in mild cognitive impairment. N. Engl. J. Med. 355, 2652–2663.

Sperling, R.A., Aisen, P.S., Beckett, L.A., Bennett, D.A., Craft, S., Fagan, A.M., Iwatsubo, T., Jack, Jr., C.R., Kaye, J., Montine, T.J., Park, D.C., Reiman, E.M., Rowe, C.C., Siemers, E., Stern, Y., Yaffe, K., Carrillo, M.C., Thies, B., Morrison-Bogorad, M., Wagster, M.V., Phelps, C.H., 2011. Toward defining the preclinical stages of Alzheimer's disease: recommendations from the National Institute on Aging-Alzheimer's Association workgroups on diagnostic guidelines for Alzheimer's disease. Alzheimers Dement. 7, 280–292.

Sperling, R., Salloway, S., Brooks, D.J., Tampieri, D., Barakos, J., Fox, N.C., Raskind, M., Sabbagh, M., Honig, L.S., Porsteinsson, A.P., Lieburg, I., Arrighi, H.M., Morris, K.A., Lu, Y., Liu, E., Gregg, K.M., Brashear, H.R., Kinney, G.G., Black, R., Grundman, M., 2012. Amyloid-related imaging abnormalities in patients with Alzheimer's disease treated with bapineuzumab: a retrospective analysis. Lancet Neurol. 11, 241–249.

Stoub, T.R., de Toledo-Morrell, L., Stebbins, G.T., Leurgans, S., Bennett, D.A., Shah, R.C., 2006. Hippocampal disconnection contributes to memory dysfunction in individuals at risk for Alzheimer's disease. Proc. Natl. Acad. Sci. USA 103, 10041–10045.

Takalkar, A.M., Khandelwal, A., Lokitz, S., Lilien, D.L., Stabin, M.G., 2011. 18F-FDG PET in pregnancy and fetal radiation dose estimates. J. Nucl. Med. 52, 1035–1040.

Taoka, T., Iwasaki, S., Sakamoto, M., Nakagawa, H., Fukusumi, A., Myochin, K., Hirohashi, S., Hoshida, T., Kichikawa, K., 2006. Diffusion anisotropy and diffusivity of white matter tracts within the temporal stem in Alzheimer disease: evaluation of the "tract of interest" by diffusion tensor tractography. AJNR Am. J. Neuroradiol. 27, 1040–1045.

Teipel, S.J., Bayer, W., Alexander, G.E., Zebuhr, Y., Teichberg, D., Kulic, L., Schapiro, M.B., Moller, H.J., Rapoport, S.I., Hampel, H., 2002. Progression of corpus callosum atrophy in Alzheimer disease. Arch. Neurol. 59, 243–248.

Trzepacz, P.T., Yu, P., Sun, J., Schuh, K., Case, M., Witte, M.M., Hochstetler, H., Hake, A., Alzheimer's Disease Neuroimaging Initiative, 2014. Comparison of neuroimaging modalities for the prediction of conversion from mild cognitive impairment to Alzheimer's dementia. Neurobiol. Aging 35, 143–151.

Tschampa, H.J., Kallenberg, K., Urbach, H., Meissner, B., Nicolay, C., Kretzschmar, H.A., Knauth, M., Zerr, I., 2005. MRI in the diagnosis of sporadic Creutzfeldt–Jakob disease: a study on interobserver agreement. Brain 128, 2026–2033.

Tzimopoulou, S., Cunningham, V.J., Nichols, T.E., Searle, G., Bird, N.P., Mistry, P., Dixon, I.J., Hallett, W.A., Whitcher, B., Brown, A.P., Zvartau-Hind, M., Lotay, N., Lai, R.Y., Castiglia, M., Jeter, B., Matthews, J.C., Chen, K., Bandy, D., Reiman, E.M., Gold, M., Rabiner, E.A., Matthews, P.M., 2010. A multi-center randomized proof-of-concept clinical trial applying [^{18}F]FDG-PET for evaluation of metabolic therapy with rosiglitazone XR in mild to moderate Alzheimer's disease. J. Alzheimers Dis. 22, 1241–1256.

van de Pol, L.A., Hensel, A., van der Flier, W.M., Visser, P.J., Pijnenburg, Y.A., Barkhof, F., Gertz, H.J., Scheltens, P., 2006. Hippocampal atrophy on MRI in frontotemporal lobar degeneration and Alzheimer's disease. J. Neurol. Neurosurg. Psychiatry 77, 439–442.

Vemuri, P., Jack, Jr., C.R., 2010. Role of structural MRI in Alzheimer's disease. Alzheimer's Res. Ther. 2, 23.

Vemuri, P., Wiste, H.J., Weigand, S.D., Shaw, L.M., Trojanowski, J.Q., Weiner, M.W., Knopman, D.S., Petersen, R.C., Jack, Jr., C.R., Alzheimer's Disease Neuroimaging Initiative, 2009. MRI and CSF biomarkers in normal, MCI, and AD subjects: diagnostic discrimination and cognitive correlations. Neurology 73, 287–293.

Vemuri, P., Jones, D.T., Jack, Jr., C.R., 2012. Resting state functional MRI in Alzheimer's Disease. Alzheimers Res. Ther. 4, 2.

Verdoorn, T.A., McCarten, J.R., Arciniegas, D.B., Golden, R., Moldauer, L., Georgopoulos, A., Lewis, S., Cassano, M., Hemmy, L., Orr, W., Rojas, D.C., 2011. Evaluation and tracking of Alzheimer's disease severity using resting-state magnetoencephalography. J. Alzheimers Dis. 26 (Suppl 3), 239–255.

Villemagne, V.L., Pike, K.E., Darby, D., Maruff, P., Savage, G., Ng, S., Ackermann, U., Cowie, T.F., Currie, J., Chan, S.G., Jones, G., Tochon-Danguy, H., O'Keefe, G., Masters, C.L., Rowe, C.C., 2008. Abeta deposits in older non-demented individuals with cognitive decline are indicative of preclinical Alzheimer's disease. Neuropsychologia 46, 1688–1697.

Visser, P.J., Scheltens, P., Verhey, F.R., Schmand, B., Launer, L.J., Jolles, J., Jonker, C., 1999. Medial temporal lobe atrophy and memory dysfunction as predictors for dementia in subjects with mild cognitive impairment. J. Neurol. 246, 477–485.

Walhovd, K.B., Fjell, A.M., Brewer, J., McEvoy, L.K., Fennema-Notestine, C., Hagler, Jr., D.J., Jennings, R.G., Karow, D., Dale, A.M., Alzheimer's Disease Neuroimaging Initiative, 2010. Combining MR imaging, positron-emission tomography, and CSF biomarkers in the diagnosis and prognosis of Alzheimer disease. AJNR Am. J. Neuroradiol. 31, 347–354.

Wang, Z., Das, S.R., Xie, S.X., Arnold, S.E., Detre, J.A., Wolk, D.A., Alzheimer's Disease Neuroimaging Initiative, 2013. Arterial spin labeled MRI in prodromal Alzheimer's disease: a multi-site study. Neuroimage Clin. 2, 630–636.

Wilkinson, D., Fox, N.C., Barkhof, F., Phul, R., Lemming, O., Scheltens, P., 2012. Memantine and brain atrophy in Alzheimer's disease: a 1-year randomized controlled trial. J. Alzheimers Dis. 29, 459–469.

Wischik, C., Staff, R., 2009. Challenges in the conduct of disease-modifying trails in AD: practical experience from a phase 2 trial of tau-aggregation inhibitor therapy. J. Nutr. Health Aging 13, 367–369.

Wolk, D.A., Detre, J.A., 2012. Arterial spin labeling MRI: an emerging biomarker for Alzheimer's disease and other neurodegenerative conditions. Curr. Opin. Neurol. 25, 421–428.

Yuan, Y., Gu, Z.X., Wei, W.S., 2009. Fluorodeoxyglucose-positron-emission tomography, single-photon emission tomography, and structural MR imaging for prediction of rapid conversion to Alzheimer disease in patients with mild cognitive impairment: a meta-analysis. AJNR Am. J. Neuroradiol. 30, 404–410.

Zhang, Y., Schuff, N., Du, A.T., Rosen, H.J., Kramer, J.H., Gorno-Tempini, M.L., Miller, B.L., Weiner, M.W., 2009. White matter damage in frontotemporal dementia and Alzheimer's disease measured by diffusion MRI. Brain 132, 2579–2592.

Zhu, X., Schuff, N., Kornak, J., Soher, B., Yaffe, K., Kramer, J.H., Ezekiel, F., Miller, B.L., Jagust, W.J., Weiner, M.W., 2006. Effects of Alzheimer disease on fronto-parietal brain N-acetyl aspartate and myo-inositol using magnetic resonance spectroscopic imaging. Alzheimer Dis. Assoc. Disord. 20, 77–85.

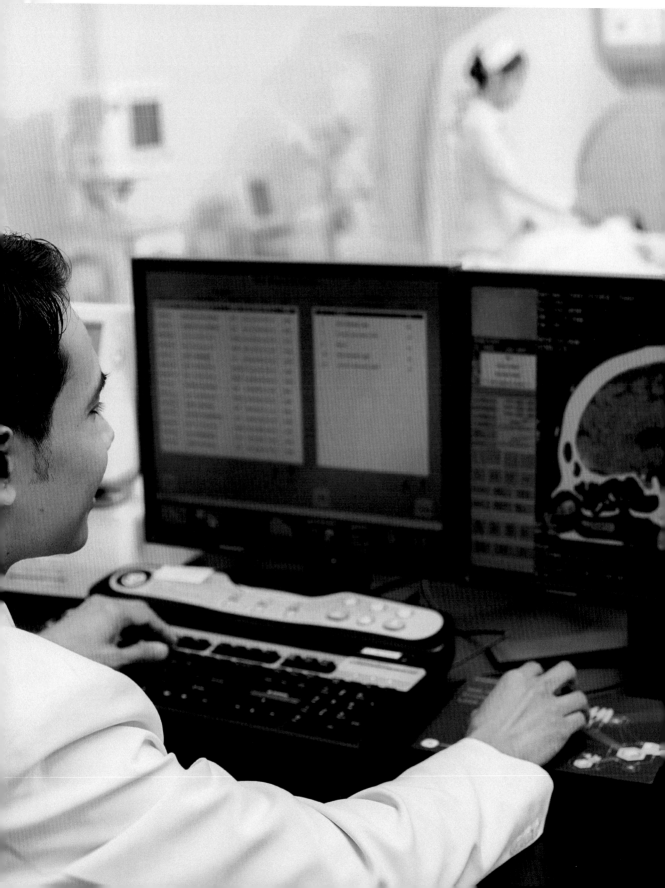

CHAPTER 4

Genetic Biomarkers in Alzheimer's Disease

4.1 BACKGROUND

Alzheimer's disease (AD) represents between 60% and 80% of all dementias among the elderly. One of every 9 individuals 65 years of age or older, and 1 of every 3 individuals 85 years of age or older have been diagnosed with AD (2014 Alzheimer's Disease Facts and Figures). Two distinct types of AD dementia exist: early-onset AD (EOAD), which has a well-defined genetic cause, typically develops before age 65 in those with an autosomal dominant family history of AD, and is very rare (<5% of AD cases); and late-onset AD (LOAD), which is the most common type, is sporadic and heterogeneous in nature, and occurs at late in life (onset ≥65 years of age). EOAD is caused by mutations in specific AD genes, including amyloid precursor protein (APP), presenilin1 (PSEN1),

Biomarkers in Alzheimer's Disease. http://dx.doi.org/10.1016/B978-0-12-804832-0.00004-3
Copyright © 2016 Elsevier Inc. All rights reserved.

Table 4.1	Confirmed Genes Related to Alzheimer's Disease	
Gene/ Chromosomal Location	**AD-Related Pathways**	**References**
APP mutation 21q21	Increases $A\beta_{1-42}/A\beta_{1-40}$ ratio by increasing both β- and γ-secretase activity; associated with EOAD	Tanzi et al. (1987), Goldgaber et al. (1987), Kang et al. (1987), Goate et al. (1991), Tanzi et al. (1992), Kamino et al. (1992)
PSEN1 14q24	Increases $A\beta_{1-42}/A\beta_{1-40}$ ratio by increasing γ-secretase activity; associated with EOAD	Sherrington et al. (1995)
PSEN2 1q31	Increases $A\beta_{1-42}/A\beta_{1-40}$ ratio by increasing γ-secretase activity; associated with EOAD	Rogaev et al. (1995), Levy-Lahad et al. (1995)
APOE 19q13.2	The ε4 allele of APOE. Decreases Aβ clearance; associated with LOAD	Strittmatter et al. (1993), Saunders et al. (1993)

AD, Alzheimer's disease; EOAD, early-onset familial AD; FAD, familial AD; LOAD, late-onset AD; APP, amyloid precursor protein; PSEN1, presenilin1; PSEN2, presenilin2; APOE, apolipoprotein E.

and presenilin2 (PSEN2) (Table 4.1). Mutations in these genes cause abnormal β- and γ-secretase cleavage of APP, which leads to overproduction of toxic Aβ. Excess toxic Aβ causes synaptic dysfunction and neurodegeneration, leading to memory loss and cognitive decline. The amyloid cascade hypothesis is the most widely accepted explanation for AD pathology (Hardy and Selkoe, 2002); consistent with this hypothesis, LOAD is believed to be caused by defective toxic Aβ clearance from the brain, which leads to synaptic loss and neurodegeneration. It follows that genes and signaling pathways involved in Aβ clearance are promising targets for diagnostic tests and therapeutic interventions for LOAD.

Evidence of the heritability of LOAD has been reported in both epidemiological and neuroimaging studies (Table 4.2). Family history of AD is the second strongest risk factor of AD, after old age. The genetic predisposition of developing LOAD (60–80%) has been estimated to be lower than that of schizophrenia (~80%), but much higher than that of diabetes (~40%) or Parkinson's disease (~30%) (Elbein, 1997; Sullivan et al., 2003; Warner and Schapira, 2003; Lambert and Amouyel, 2011). Epidemiological studies of twins with AD found strong evidence supporting the heritability of LOAD (Bergem et al., 1997; Gatz et al., 1997, 2006), while neuroimaging of APOE4 carriers found increased hippocampal atrophy (Potkin et al., 2009) and lower glucose metabolism by [32]F-fluordeoxyglucose positron emission tomography (FDG PET) (Reiman et al., 2005), suggesting a link between APOE4 and LOAD. A large-scale collaborative analysis of neuroimaging and genetic data by the Enhancing Neuro Imaging Genetics through Meta-Analysis (ENIGMA) Consortium, a group including

Table 4.2	Heritability of Late-Onset Alzheimer's Disease (LOAD)		
Study	**Methodology**	**Results**	**References**
Study of twin pairs from the Norwegian Twin Registry	Concordance rate for LOAD: monozygotic = 78% and dizygotic = 39%	Estimated LOAD heritability = 60%, no significant heritability among vascular dementia	Bergem et al. (1997)
Swedish Twin Registry aged 65 years and older	Twin pairs in which one or both individuals had AD	Estimated LOAD heritability = 58–79%	Gatz et al. (2006)
Large family-based group of older adults	Memory test of 277 with AD and 622 unaffected	Cognitive functions strongly influenced by genetics in older persons with and without AD	Wilson et al. (2011)
Enhancing Neuroimaging Genetics through Meta-Analysis (ENIGMA) Consortium	Total of 24,997 subjects	Genome profile associated with hippocampal volume	Thompson et al. (2014)

AD, Alzheimer's disease; LOAD, late-onset AD.

70 institutes worldwide, found a relationship between specific genetic markers, including APOE4, and brain imaging results in AD cases (Thompson et al., 2014).

Epidemiological studies have found that in addition to genetics, age and environmental factors contribute to LOAD risk. Genetics plays an important role in aging (~30%), and aging is one of the main risk factors for LOAD; therefore, the genetics of aging indirectly influences the risk of LOAD. Thus, LOAD is inherited in a non-Mendelian fashion. The exact genes associated with LOAD continue to remain a mystery, yet decades of researches on the genetics of AD suggest that a genetic risk factor or factors exist. Researchers are currently investigating several candidate genes that may confer susceptibility to LOAD. Identification of these genes, how they interact with age and environment, and their role in LOAD pathogenesis will aid researchers and clinicians in diagnosing, treating, and perhaps preventing LOAD. The possibility of early diagnosis and treatment, and even prevention, of AD support the drive to invest in research that will decipher the complex genetic determinants of LOAD and the clinical application of AD biomarkers to diagnosis and risk assessment. This chapter addresses recent advances and continuing challenges to determining the genetic risk factors of LOAD and developing new diagnostic biomarkers of this heterogeneous and complex disease. Detection of a set of gene markers connected within intricate networks, rather than individual genes, might be a more suitable approach to developing a genetic diagnostic test for LOAD.

Table 4.3	Methods for Discovering Alzheimer's Disease (AD)-Related Genes		
Method	**Methodology**	**Comments**	**References**
Positional cloning approach	Linkage mapping method is an unbiased search	No need for prior knowledge of the gene	Sherrington et al. (1995)
Candidate-gene approach	Hypothesis-driven and limited by how much is known of the biology of the disorder being investigated	Tests a predetermined hypothesis about the genetic basis of a disease; conflicting results reported in the literature	Pericak-Vance et al. (1991), Strittmatter et al. (1993)
Genome-wide association approach	Hypothesis-independent; very high-throughput	Higher chance to identify novel candidate genes; higher proportion of false-positive results, expensive, need bigger sample size	Bertram et al. (2008), Harold et al. (2009), Lambert et al. (2009), Seshadri et al. (2010)
Pharma-cogenomic approaches	Pharmacogenomic response read-out by conventional methods, such as microarray	Applicable to personalized medicine for AD treatment in future	Martinelli-Boneschi et al. (2013)

4.2 APPROACHES TO IDENTIFYING GENETIC BIOMARKERS IN ALZHEIMER'S DISEASE

The various genetic approaches that have been used to identify potential AD biomarkers are summarized in Table 4.3.

4.2.1 Positional Cloning Approach

Positional cloning is used to identify human genes based on their chromosomal location. Positional cloning is one of the important approaches to finding the genes underlying genetic disorders with Mendelian inheritance, including familial EOAD. The process of positional cloning is shown in Fig. 4.1. This process is also called "reverse genetics." A systematic scan of the entire genome of the members of an affected family is performed and compared with those of highly specific matched control cases. Genetic linkage analysis, or "linkage mapping" reveals the most likely chromosomal location of a particular gene. Positional cloning is thus an unbiased search of the whole genome without any assumption about the role of a certain gene. One of the best examples of the application of the positional cloning approach is the identification of the gene for Huntington disease in 1983 (Gusella et al., 1983). In 1986, the positional

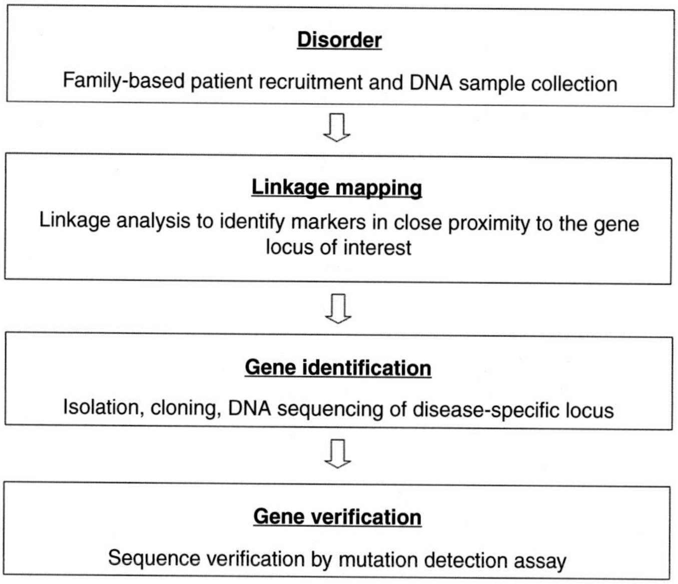

```
┌─────────────────────────────────────────────────────────────┐
│                         Disorder                            │
│    Family-based patient recruitment and DNA sample collection│
└─────────────────────────────────────────────────────────────┘
                              ⇩
┌─────────────────────────────────────────────────────────────┐
│                      Linkage mapping                         │
│    Linkage analysis to identify markers in close proximity   │
│             to the gene locus of interest                    │
└─────────────────────────────────────────────────────────────┘
                              ⇩
┌─────────────────────────────────────────────────────────────┐
│                    Gene identification                       │
│    Isolation, cloning, DNA sequencing of disease-specific    │
│                         locus                                │
└─────────────────────────────────────────────────────────────┘
                              ⇩
┌─────────────────────────────────────────────────────────────┐
│                    Gene verification                         │
│    Sequence verification by mutation detection assay         │
└─────────────────────────────────────────────────────────────┘
```

FIGURE 4.1
Flow diagram of the positional cloning approach.

cloning method was also used to identify the gene causing chronic granulomatous disease in humans (Royer-Pokora et al., 1986). Since then, ~100 genetic disease genes have been identified by positional cloning, including the three EOAD genes, APP, PSEN1, and PSEN2 (Cruts et al., 1995; Tanzi et al., 1996) (Table 4.1). Discovery of these EOAD genes provided the foundation for the amyloid cascade hypothesis of AD. The limitations of this approach are that it is a time-consuming, expensive process, and that molecular characterization of gene variants is needed to understand the link to disease pathogenesis.

4.2.2 Candidate Gene Approach

In this hypothesis-driven method, some degree of prior knowledge about the pathophysiology of the disease and the potential role of the gene of interest (the candidate gene) is necessary (Fig. 4.2). PCR is the most common tool used to genotype a small number of known variants. In the candidate gene approach, gene profiling of affected individuals (AD) and unrelated cases and controls can be performed relatively quickly and inexpensively. Identification of APOE as a risk factor of LOAD is one example of the successful application of the candidate gene approach (Strittmatter et al., 1993). With the prior knowledge of a gene for LOAD located at chromosome 19 from linkage analysis, the candidate gene approach was conducted on 30 AD cases and 91 control cases. The study found overexpression of the APOE4 gene in AD cases. This robust observation was repeated and confirmed by several researchers. The candidate gene approach is particularly useful for identifying genes involved in multifactorial diseases like LOAD. Other important genes discovered using the candidate gene approach are

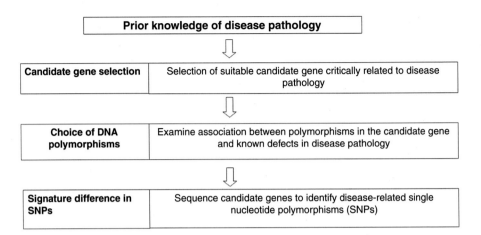

FIGURE 4.2
Flow diagram of candidate gene approach.

SORL1 (Rogaeva et al., 2007) and SORCS1 (Xu et al., 2013). Unlike the positional cloning method, the candidate gene approach does not require genomic testing of a large number of affected and unaffected family members.

4.2.3 Genome-Wide Association Study

The completion of the Human Genome Project in 2003 opened up new avenues to detect sets of genes associated with diseases, by identifying multiple genetic variations across the entire genome. The genome-wide association study (GWAS) approach involves detection of small sequence variations (single nucleotide polymorphisms, or SNPs) in the genome that are present in individuals with a particular disorder but not in control cases (Fig. 4.3). This approach is extraordinarily powerful for identifying biomarkers of complex diseases like LOAD because it involves scanning the whole genome to find sets of SNPs in related or unrelated genes that are specifically associated with the disease. GWAS has been used to identify SNP–trait associations for many diseases, and has led to the discovery of novel susceptibility genes for LOAD (Grupe et al., 2007; Lambert et al., 2009; Harold et al., 2009; Seshadri et al., 2010; Lambert et al., 2013). Although the GWAS approach has identified several promising disease biomarkers for LOAD, inconsistencies between studies have posed a challenge to further development for clinical application (Table 4.4). Thus far, GWAS have identified the association between APOE and LOAD, underscoring the utility of the GWAS method for LOAD gene hunting. However, in a recent GWAS conducted to verify the most promising LOAD susceptibility genes previously identified through GWAS produced conflicting results (Shi et al., 2012). Apart from the APOE locus (rs2075650), none of the other LOAD gene loci demonstrated significant evidence of association. Furthermore, GWAS results for AD LOAD genes identified in a Caucasian population (United States and Western European) were not reproducible in other ethnic groups. For example, there is a lack of association

| **Patient selection** |
| Sample (usually blood) collection from patients and similar set of people without disease |

| **RNA extraction** |
| RNA extraction and cDNA preparation |

| **Microarray** |
| Fragmentation→Hybridization onto microarray chip (sample binding to known gene array)→Ligation (fluorescent probes bind differentially to sample) → Detection (fluorescence intensity measurement) |

| **Data analysis** |
| Analysis and data interpretation |

FIGURE 4.3
Flow diagram for genome-wide association study (GWAS).

between LOAD and CR1, PICALM, and CLU gene polymorphisms in the Polish population (Klimkowicz-Mrowiec et al., 2013). Among the top AD susceptibility genes identified by GWAS, only MS4A6A and CD33 are associated with LOAD in the Han Chinese population (Deng et al., 2012; Tan et al., 2013b), and only PICALM is associated with AD in the Korean population (Chung et al., 2013).

4.2.4 Pharmacogenomics in Alzheimer's Disease

Pharmacogenomics is the study of genetic changes in response to a pharmaceutical agent, and can be used to identify biomarkers of disease, measure treatment response, and identify new drug targets. The role of pharmacogenomic approaches to uncover biomarkers for AD has not been extensively explored. The pharmacogenomics of brain disorders, including AD, has been studied by one group (Cacabelos et al., 2007, 2012; Cacabelos, 2007, 2009). Those studies found that pharmacogenomic responses in AD depend on the interaction of genes involved in drug metabolism and genes associated with AD pathogenesis. For example, carriers of the genotype APOE4/4 CYP2D6-PMs, CYP2D6-UMs (CYP2D6 is located on chromosome 22q13.2, spanning 4.38 kb with nine exons) were the worst responders to conventional AD drug treatments (Cacabelos, 2007; Cacabelos et al., 2012). The results of another pharmacogenomics study were published in an Italian cohort (Martinelli-Boneschi et al., 2013). The study was designed to examine the response of cholinesterase inhibitor treatment in AD in terms of the changes elicited in gene expression. The study found response of SNP rs6720975 that maps in the intronic region

TABLE 4.4		Most Important Genes Related to LOAD and Their Relation to the Disease Pathology	
SNP ID	**Genes**	**Important Function Related to LOAD**	**References**
rs2075650	APOE4	Almost all GWAS found APOE locus associated with LOAD. APOE 2, 3, 4 differentially regulate toxic Aβ clearance	Strittmatter et al. (1993), Castellano et al. (2011)
rs744373	BIN1	BIN1 variant increases risk of AD through tau. Influence calcium homeostasis, inflammation, and apoptosis. BIN1 was associated with smaller hippocampal volume measured by sMRI	Chapuis et al. (2013), Tan et al. (2013a), Chauhan et al. (2015)
rs11136000	CLU	Most GWAS found CLU locus associated with LOAD. Influences Aβ aggregation and clearance, neuroinflammation, lipid metabolism. Similar to APOE, is elevated in brain areas such as hippocampus and entorhinal cortex mostly affected by AD and present in CSF and amyloid plaques	Yu and Tan (2012), McGeer et al. (1992), Nuutinen et al. (2009), Calero et al. (2000)
rs3764650	ABCA7	Highly expressed in CA1 neurons and microglia. Regulates apolipoprotein-mediated phospholipid and cholesterol efflux from cells	Chan et al. (2008)
rs3818361	CR1	Interacts with Aβ and plays important role in peripheral clearance of $Aβ_{1-42}$. Higher level in CSF of AD patients	Crehan et al. (2012), Daborg et al. (2012)
rs3851179	PICALM	Endocytosis, associated with tau pathology and Aβ clearance. Nonselectively expresses in both pre and postsynaptic areas	Ando et al. (2013)
rs610932	MS4A4A, MS4A4E, MS4A6A	Regulation of cellular signaling, including calcium signaling; immunity regulation	Ma et al. (2015), Liang et al. (2001)
rs3865444	CD33	Increase in AD brain and decrease in Aβ uptake by microglia. Strong association with smaller intracranial volume measured by sMRI	Griciuc et al. (2013), Jiang et al. (2014), Chauhan et al. (2015)
rs11767557	EPHA1	EPHA1 is a membrane-bound receptor that plays roles in axonal guidance, synaptic development, plasticity, apoptosis, and inflammation	Kullander and Klein (2002), Martínez and Soriano (2005), Lai and Ip (2009)
rs9349407	CD2AP	Receptor-mediated endocytosis	Lynch et al. (2003)

ABC7, ATP-binding cassette subfamily A member 7; Aβ, beta amyloid; AD, Alzheimer's disease; APOE4, apolipoproteinE4; BIN1, bridging integrator 1; CD33, CD33 antigen; CD2AP, CD2-associated protein; CLU, clusterin; CR1, complement cell-surface receptor; CSF, cerebrospinal fluid; EPHA1, ephrin type-A receptor 1; GWAS, genome-wide association study; LOAD, late-onset Alzheimer's disease; MS4A, membrane-spanning 4-domains subfamily A gene cluster; sMRI, structural magnetic resonance; MS4A6A, membrane-spanning 4-domains subfamily A gene cluster 6A; MS4A4A, membrane-spanning 4-domains subfamily A gene cluster 4A; MS4A4E, membrane-spanning 4-domains subfamily A gene cluster 4E; PICALM, phosphatidylinositol binding clathrin assembly protein.

of a gene PRKCE. Incidentally this gene encodes PKCepsilon (PKCε) that plays important role in AD pathology (Khan et al., 2015). There are several potential types of genes that could be monitored in pharmacogenomic studies in AD, including (a) genes known to be associated with AD, (b) genes that are involved in the metabolism or absorption of the drug. Optimization AD pharmacological treatment by pharmacogenomic procedures would be important in era of personalized medicine, making it possible to differentiate between patients who are likely to respond to a particular treatment and those who are not.

4.3 EARLY-ONSET (FAMILIAL) ALZHEIMER'S DISEASE GENES

Patients affected by EOAD account for fewer than 1–5% of all AD cases. EOAD genes are inherited in a Mendelian fashion, with very little influence from environment or epigenetics. As a result, EOAD genes were discovered very early in AD research, prior to the availability of modern molecular genetic technologies. Glenner and Wong (1984) were the first to predict that production of Aβ is controlled by specific genes. They also predicted the chromosomal location of genes associated with AD on chromosome 21. Variations in three genes—APP, PSEN1, and PSEN2—on chromosome 21 have since been identified as causal factors in EOAD (Table 4.1). The presence of variations of these three genes almost certainly predicts the onset of EOAD.

APP (chromosomal location: 21q21): In 1987, the first mutations in APP were discovered that caused an inherited form of EOAD (Goldgaber et al., 1987; Kang et al., 1987; Tanzi et al., 1987; Robakis et al., 1987). This discovery was confirmed by several other laboratories (Goate et al., 1991; Kamino et al., 1992; Goate, 2006). APP is located on chromosome 21q21.3, covering ~240 kb with 19 exons, and is expressed in numerous tissues. The most common isoforms are APP695, APP751, and APP770. Among them, APP695 is expressed in neuronal cells. Full-length APP is cleaved by three enzyme systems—α-secretase, β-secretase, and γ-secretase—and the resulting cleavage products depend on the sequence of cleavage. Today, a total of 24 mutations, including duplications, have been reported for the APP gene. APP mutations increase in both production of Aβ and the $A\beta_{1\text{-}42}/A\beta_{1\text{-}40}$ ratio.

PSEN1 (chromosomal location: 14q24) and PSEN2 (chromosomal location: 1q31): In 1995, mutations in a second gene, PSEN1, were discovered to cause inherited AD (Sherrington et al., 1995). The third EOAD gene, PSEN2, is a homolog of PSEN1 and was discovered several months after PSEN1 (Levy-Lahad et al., 1995) and confirmed by Rogaev et al. (1995). Today, 185 mutations in PSEN1 and 14 mutations in PSEN2 have been reported to be associated with EOAD. All three EOAD genes were identified by genetic linkage studies and positional cloning. PSEN1 and PSEN2 mutations increase the $A\beta_{1\text{-}42}/A\beta_{1\text{-}40}$ ratio.

Several attempts have been made to identify additional EOAD-associated genes in addition to APP, PSEN1, and PSEN2. A few genes have been identified, including MAPT (microtubule-associated protein tau) (on chromosome 17q) (Rademakers et al., 2003) and transcription activation domain interacting protein gene (PAXIP1) (Rademakers et al., 2005).

4.4 RECENT ADVANCES IN ALZHEIMER'S DISEASE GENETIC BIOMARKER RESEARCH

4.4.1 Susceptibility Genes in LOAD Identified by GWAS

The 10 most important genes associated with LOAD that were identified by different genetic studies, as listed in the Alzgene database (http://www.alzgene.org/), are as follows: APOE_E2/3/4, BIN1, CLU, ABCA7, CR1, PICALM, MS4A (MS4A4A/MS4A4E/MS4A6A), CD33, EPHA1, and CD2AP. The estimated attributable fractional LOAD genetic risk in Caucasians is 20% for APOE, 9% for CLU, 4% for PICALM, and 4% for BIN1 (Lambert and Amouyel, 2011). Sortilin-1 (SORL1) and translocase of outer mitochondrial membrane 40 homolog (TOMM40) are also strongly associated with LOAD. Other LOAD-associated genes are ACE (angiotensin converting enzyme 1) (Kehoe et al., 1999), IL8 (interleukin 8) (Mines et al., 2007), LDLR (low-density lipoprotein receptor) (Kim et al., 2009b), GAB2 (GRB2-associated binding protein 2) (Reiman et al., 2007), and ATXN1 (ataxin 1) (Bertram et al., 2008).

4.4.1.1 *APOLIPOPROTEIN E (APOE)*

Evidence of an association between APOE and AD was first observed in a genetic linkage study (Pericak-Vance et al., 1991). APOE (chromosomal location 19q13) is the only gene that has been found to be consistently associated with LOAD risk across most studies. Like the EOAD genes (APP, PSEN1, and PSEN2), APOE4 was identified by genetic linkage studies and positional cloning and was later validated by genetic association studies. The main roles of APOE are clearance of Aβ and lipid metabolism in AD-related pathology. APOE transports cholesterol from astrocytes to neurons through APOE receptors. There are several alleles of APOE—APOE2, APOE3, APOE4—which differentially regulate Aβ clearance in brain. Variation of APOE at ε-4 (APOE4) and ε-2 (APOE2) increases the risk for LOAD by 50%. A single copy of the APOE4 allele increases the risk of AD two- to threefold, and two copies increase the risk 12-fold (Bertram and Tanzi, 2009; Roses, 1996). Other apolipoproteins include APOA-1 and APOJ (clusterin). Of the variants, APOE4 is less effective in cholesterol transportation and Aβ clearance. When APOE2 and APOE3 form complexes with toxic Aβ, the rate of clearance across the blood–brain barrier is much higher than that achieved by APOE4-Aβ complexes (Deane et al., 2008; Kim et al., 2009a). Consistent with the amyloid cascade hypothesis, a neuroprotective effect of APOE2 and 3 with respect to Aβ-induced neuronal toxicity has been well documented (Farrer et al., 1997; Sen et al., 2012). For example, APOE3 but not APOE4 protects against synaptic loss through increased expression of synaptogenic PKC (Sen et al. 2012). Almost all GWAS have found a significant link between APOE variants and AD. A very recent GWAS on age at onset of LOAD found a much stronger association between LOAD and APOE variants than with other gene loci, including CR1, BIN1, and PICALM (Naj et al., 2014). The TOMM40 gene with a variable-length poly-T sequence polymorphism (rs10524523) in combination with APOE alleles (E2, E3, E4) significantly influences LOAD risk (Roses et al., 2010).

Worldwide attributable risk due to APOE4 in the LOAD population varies by ethnicity and region (Crean et al., 2011). By meta-analytic estimation, the proportion of APOE4 carriers in AD patients increased in the following order Asia < Southern Europe < Central Europe < North America < Northern Europe (Mayeux 2003). Among European ethnic groups, APOE4 has a significant association with LOAD; however, these associations vary in strength based on ethnicity (Evans et al., 2003; Mayeux 2003; Kalaria et al., 2008). APOE4 does not play as strong a role in AD risk in African Americans (Reitz et al., 2013). Because the degree of association of APOE4 in risk of AD varies based on not only ethnicity (Evans et al. 2003; Mayeux 2003; Kalaria et al., 2008; Reitz et al. 2013), but also age (Farrer et al., 1997) and sex (Farrer et al., 1997), the APOE4 allele is not a suitable AD biomarker, but may be considered as a disease risk factor in certain populations. Several correlation studies have presented significant associations between the APOE4 allele and smaller hippocampal volume measured by neuroimaging AD biomarkers (Soininen et al., 1995; Lehtovirta et al., 1995, 1996; Plassman et al., 1997; den Heijer et al., 2002; Morra et al., 2009; Schuff et al., 2009; Shen et al., 2010; Lu et al., 2011; O'Dwyer et al., 2012; Liu et al., 2015); however, same association was not observed in other studies (Reiman et al., 1998; Ferencz et al., 2013; Khan et al., 2014). A multicentered [Alzheimer's Disease Neuroimaging Initiative (ADNI)] GWAS found that APOE4 was associated with multiple structural magnetic resonance imaging (sMRI) biomarkers, more specifically, it was associated with hippocampal volume loss ($p = 0.9 \times 10^{-14}$), amygdala volume loss ($p = 3.6 \times 10^{-11}$), loss of entorhinal cortex thickness ($p = 8.7 \times 10^{-7}$), loss of parahippocampal gyrus cortex thickness ($p = 3.3 \times 10^{-4}$), and loss of temporal pole cortex thickness ($p = 0.002$) (Biffi et al., 2010).

Other important genes associated with AD that were discovered using the GWAS approach are summarized in Table 4.5.

4.4.1.2 *BRIDGING INTEGRATOR 1 (BIN1) (SNP: RS6733839; CHROMOSOMAL LOCATION: 2Q14.3)*

Several GWAS have independently reported association of BIN1 with AD (Lambert et al., 2013; Chapuis et al., 2013; Tan et al., 2013b; Naj et al., 2014) (Table 4.5). BIN1 contains up to 20 exons and spans 59.3 kb (Nicot et al., 2007). The BIN1 protein (70 kDa) plays important roles in AD-related pathology, including production and clearance of Aβ and cellular signaling. The BIN1 gene variant increases the LOAD risk ~1.4-fold by modulating tau pathology (Tan et al., 2013a). GWAS have identified the BIN1 gene as the most important genetic susceptibility locus in LOAD after APOE4 for individuals of European ancestry (Chapuis et al., 2013), as well as for Han Chinese individuals (Tan et al., 2014a). BIN1 is expressed in many tissues including blood, and is overexpressed in AD brains. In addition, it

Table 4.5 Other Important Genes in LOAD

Gene Identified	Methodology	Comments	References
CLU (also known as APOJ) ($p = 7.5 \times 10^{-9}$) CR1 ($p = 3.7 \times 10^{-9}$)	GWAS. Patient population: France: AD ($n = 2032$), control ($n = 5328$). Belgium, Finland, Italy, and Spain: AD ($n = 3978$), control ($n = 3297$)	Both CLU and CR1 are associated with Aβ clearance	Lambert et al. (2009), Harold et al. (2009)
Already known: ABCA7, BIN1, CD33, CLU, CR1, CD2AP, EPHA1, MS4A6A-MS4A4E, and PICALM Other new genes: CASS4, CELF1, FERMT2, INPP5D, MEF2C, NME8, PTK2B, SLC24A4, SORL1, ZCWPW1	GWAS with two-stage meta-analysis. Stage I: AD ($n = 17,008$), control ($n = 37,154$); 7,055,881 SNPs. $P < 5 \times 10^{-8}$. Stage II: AD ($n = 8\,572$), control ($n = 11,312$); 211,632 SNPs. $P < 1 \times 10^{-5}$	Newly discovered genes are associated with immunocompetence and histocompatibility	Lambert et al. (2013)
CLU, PICALM, CR1	GWAS International plus 3-stage sequential analysis, and replicated in a separate cohort.	Largest sample size in any genetic analysis of AD	Seshadri et al. (2010)
SORL1	Genotyping using GenomeLab SNP stream system. Genotyping samples: LOAD ($n = 12$), EOAD ($N = 12$), control ($n = 2$); RT-PCR: LOAD ($n = 17$), control ($n = 16$). Confirmed by GWAS in different ethnicity with larger population.	Confirmed by various studies	Rogaeva et al. (2007)
TOMM40	Long-range PCR and DNA sequencing. Two independent cohorts. AD ($n = 191$), control ($n = 131$), $P = 0.02$	Increases precision in estimating LOAD risk. Contradictory results in later studies	Roses et al. (2010)

Table 4.5	Other Important Genes in LOAD (*cont.*)		
Gene Identified	**Methodology**	**Comments**	**References**
ABCA7, MS4A6A/MS4A4E/MS4A4A;EPHA1, CD33, CD2AP	Combined analysis of four GWAS. AD ($n = 25{,}900$), control ($n = 41{,}584$) $P = 5.0 \times 10^{-21}$ (for ABC7); AD ($n = 28{,}183$), control ($n = 46{,}911$) $P = 1.2 \times 10^{-16}$ (for MS4A6A); AD ($n = 28{,}484$), control ($n = 46{,}850$) $P = 1.1 \times 10^{-16}$ (for MS4A4E) AD ($n = 6{,}992$), control ($n = 13{,}472$) $P = 3.4 \times 10^{-4}$ (for EHPA1); $P = 2.2 \times 10^{-4}$ (for CD33); $P = 8.0 \times 10^{-4}$ (for CD2AP)	The study found common variants at ABC7, MS4A6A/MS4A4E, EPHA1, CD33 and CD2AP in AD	Hollingworth et al. (2011)
APOE, CR1, BIN1, PICALM	The Alzheimer Disease Genetics Consortium: GWAS of 14 case-control, prospective, and family-based populations. $n = 9162$ participants APOE: $P = 3.3 \times 10^{-96}$; CR1: $P = 7.2 \times 10^{-4}$; BIN1: $P = 4.8 \times 10^{-4}$; PICALM: $P = 2.2 \times 10^{-3}$.	Associations of CR1, BIN1, and PICALM with age at onset of AD. APOE locus contributes to risk of LOAD.	Naj et al. (2014).

ABCA7, ATP-binding cassette subfamily A member 7; AD, Alzheimer's disease; APOJ, apolipoprotein J; BIN1, bridging integrator 1; CASS4, Cas scaffolding protein family member 4; CD33, CD33 antigen; CD2AP, CD2-associated protein; CELF1, CUGBP Elav-like family member1; CLU, clusterin; CR1, complement cell-surface receptor; EPHA1, ephrin type-A receptor 1; FERMT2, fermitin family member 2; GWAS, genome-wide association study; INPP5D, inositol polyphosphate-5-phosphatase; MEF2C, myocyte enhancer factor 2; MS4A, membrane-spanning 4-domains subfamily A gene cluster; MS4A6A, membrane-spanning 4-domains subfamily A gene cluster 6A; MS4A4A, membrane-spanning 4-domains subfamily A gene cluster 4A; MS4A4E, membrane-spanning 4-domains subfamily A gene cluster 4E; NME8, NME/ NM23 family member8; PICALM, phosphatidylinositol binding clathrin assembly protein; PTK2B, encoding protein tyrosine kinase 2β; RTPCR, real time polymerase chain reaction; SLC24A4, solute carrier family 24 (sodium/potassium/calcium exchanger; member 4); SNP, single nucleotide polymorphism; SORL1, sortilin-related receptor; Tomm40, translocase of outer mitochondrial membrane 40 homolog; ZCWPW1, zinc finger CW type with PWWP domain1.

has been reported that BIN1 expression increases with age at onset and shorter disease duration in AD patients (Karch et al., 2012). Very recently, another GWAS also reported that BIN1 is associated with age LOAD onset (Naj et al., 2014); however, the effect was less significant than that of APOE4. BIN1 was also associated with smaller hippocampal volume as measured by structural magnetic resonance imaging (sMRI) in a study of crosscorrelation of results from GWAS (Chauhan et al., 2015). A multicentered (ADNI) GWAS found that BIN1 was associated with multiple sMRI biomarkers, more specifically, BIN1 was associated with loss of entorhinal cortex thickness ($p = 0.004$) and loss of temporal pole cortex thickness ($p = 0.02$) (Biffi et al., 2010).

4.4.1.3 CLUSTERIN (CLU) (SNP: RS11136000; CHROMOSOMAL LOCATION: 8P21-12)

Several GWAS have independently reported association of CLU with AD (Lambert et al., 2009, 2013; Harold et al., 2009; Seshadri et al., 2010) (Table 4.5). The CLU gene contains nine exons and spans 16 kb of DNA (Wong et al., 1994). The CLU protein (75 kDa), also named apolipoprotein J (APOJ), is a lipoprotein that is highly expressed in the mammalian brain as well as in peripheral tissues of the body. CLU is involved in a variety of neuronal signaling processes, such as synaptic turnover, apoptosis, lipid transport, membrane protection, endocrine secretion, promotion of cell interactions, and chaperone activities. The mRNA levels of CLU are higher in the brain areas affected by AD, including the hippocampus and entorhinal cortex, and it is found in amyloid plaques and cerebrospinal fluid (CSF) (McGeer et al., 1992; Nuutinen et al., 2009). CLU expression is inversely related to APOE4 levels. Similar to APOE, CLU is involved in several AD-related pathways, including Aβ clearance, the immune response, and lipid metabolism. It acts as a chaperone protein. Involvement of CLU in transport of Aβ from plasma to brain and Aβ fibrillation have been reported (DeMattos et al., 2004; Nuutinen et al., 2009).

4.4.1.4 ATP-BINDING CASSETTE SUBFAMILY A MEMBER 7 (ABCA7) (SNP: RS 3764650; CHROMOSOMAL LOCATION: 19P13.3)

Several GWAS have independently reported association of ABCA7 with AD (Hollingworth et al., 2011; Lambert et al., 2013) (Table 4.5). ABCA7 is highly expressed in hippocampal CA1 neurons and microglia. Gene expression levels of ABCA7 increase with AD pathology. The ABCA7 gene is involved in cholesterol and lipid regulation and the transport of various molecules across extra- and intracellular membranes. With respect to AD-related processes, ABCA7 regulates APP processing and inhibits Aβ secretion in APP-overexpressed cells (Jehle et al., 2006). It is also involved in Aβ clearance, immune response, and lipid metabolism. ABCA7 is a definitive genetic risk factor for LOAD in African Americans and it has an effect on disease risk comparable to that of APOE4 (Reitz et al. 2013).

4.4.1.5 COMPLEMENT CELL-SURFACE RECEPTOR (CR1) (SNP: RS3818361; CHROMOSOMAL LOCATION: 1Q32)

Several GWAS have independently reported association of CR1 with AD (Lambert et al., 2009, 2013; Harold et al., 2009; Naj et al., 2014) (Table 4.5). CR1 has been reported to be associated with amyloid plaques. Consistent with increased complement cascade activity in AD pathology, CR1 gene expression level was found to be high in AD brains. The AD-related pathways in which CR1 is involved are Aβ clearance and the immune response. Erythrocytes express CR1, and this cell type plays an important role in peripheral clearance of $Aβ_{1-42}$ via CR1. Very recently, CR1 was found to be associated with age at onset in LOAD (Naj et al., 2014); however, the effect was less significant than that of APOE4. Multicentered GWAS that correlated genetic and neuroimaging biomarkers found that CR1 was associated with loss of entorhinal cortex thickness ($p = 0.03$) (Biffi et al., 2010).

4.4.1.6 *PHOSPHATIDYLINOSITOL BINDING CLATHRIN ASSEMBLY PROTEIN (PICALM) (SNP: RS561655; CHROMOSOMAL LOCATION: 11Q14.2)*

Several GWAS have independently confirmed association of PICALM with AD (Seshadri et al., 2010; Lambert et al., 2013; Naj et al., 2014) (Table 4.5). PICALM plays an important role in clathrin-mediated endocytosis. The role of PICALM may be related to AD via three different mechanisms that affect endocytosis and neurotransmission. Endocytosis reduces APP internalization and Aβ production. PICALM is expressed prominently in neurons and is nonselectively distributed in pre- and postsynaptic terminals, where it plays an important role in fusion of synaptic vesicles to the presynaptic membrane in neurotransmitter release. Loss of synapses in the early stages of AD might be associated with dysfunctional PICALM. In addition, PICALM plays a critical role in iron homeostasis, offering a new perspective on iron-related AD pathogenesis. PICALM is involved in AD pathways related to APP trafficking, Aβ clearance, synaptic function, and endocytosis. Very recently, PICALM was found to be associated with age at onset in LOAD (Naj et al., 2014); again, the effect was less significant than that of APOE4. Multicentered GWAS correlating genetic and neuroimaging biomarkers found that PICALM was associated multiple sMRI biomarkers, such as loss of hippocampal volume ($p = 0.04$) and loss of entorhinal cortex thickness ($p = 0.01$) (Biffi et al., 2010).

4.4.1.7 *CD33 ANTIGEN (CD33) (SNP: RS3865444; CHROMOSOMAL LOCATION: 19Q13.3)*

Several GWAS have independently reported the CD33 locus and its association with AD (Hollingworth et al., 2011; Lambert et al., 2013) (Table 4.5). In AD, increased cell surface expression of CD33 in monocytes decreases internalization of Aβ peptide (Bradshaw et al., 2013). Overexpression of CD33 leads to the impairment of microglia-mediated clearance of Aβ, which results in the formation of amyloid plaques in the brain (Jiang et al., 2014). CD33 is involved in Aβ clearance and immune response. CD33 is the only gene other than APOE that has been associated with AD in both case–control and family-based GWAS. CD33 was associated with smaller intracranial volume measured by sMRI in a crosscorrelation of GWAS results (Chauhan et al., 2015).

4.4.1.8 *MEMBRANE-SPANNING 4-DOMAINS SUBFAMILY A (MS4A) GENE CLUSTER (CHROMOSOMAL LOCATION: 11Q12.2)*

Several GWAS analyses have independently reported an association between the MS4A gene cluster and AD (Antúnez et al., 2011; Hollingworth et al., 2011; Lambert et al., 2013) (Table 4.5). There are three members of the MS4A family: MS4A4A (rs4938933), MS4A4E (rs670139), and MS4A6A (rs983392) that have been identified in GWAS analyses. These genes encode transmembrane proteins that are involved in signal transduction and have immunological functions (Liang et al., 2001; Ma et al., 2015). The MS4A family contains 16 genes within a 600 kb cluster that encode proteins with four or more transmembrane domains (Ishibashi et al., 2001; Liang et al., 2001). The MS4A6A gene contains four exons and spans 5 kb, and MS4A4 contains seven exons and spans 23 kb (Liang et al., 2001).

4.4.1.9 EPHRIN TYPE-A RECEPTOR 1 (EPHA1) (RS11767557; CHROMOSOMAL LOCATION: 7Q34)

Several GWAS have independently reported an association of EPH1 with AD (Hollingworth et al., 2011; Lambert et al., 2013) (Table 4.5). It contains 18 exons and spans ~18 kb (Owshalimpur and Kelley, 1999). EPHA1 is a membrane receptor of the ephrin subfamily and is expressed mainly in epithelial tissues and the nervous system. EPH1 (~108 kDa) belongs to family of receptor tyrosine kinases (Zhou, 1998; Coulthard et al., 2001) and is involved in variety of cellular signaling pathways regulating neural development (Martínez and Soriano, 2005; Lai and Ip, 2009), cell proliferation, angiogenesis, innate immunity, and apoptosis (Kullander and Klein, 2002).

4.4.1.10 CD2-ASSOCIATED PROTEIN (CD2AP) (SNP: RS10948363; CHROMOSOMAL LOCATION: 6P12.3)

Several GWAS have independently reported an association between CD2AP and AD (Hollingworth et al., 2011; Lambert et al., 2013) (Table 4.5). The CD2AP protein has 639 amino acid residues with ~70 kDa molecular mass (Kirsch et al., 1999). As a scaffolding protein, CD2AP regulates actin cytoskeleton dynamics in neurons, including pathways related to synaptic function and endocytosis. CD2AP acts as a link between epidermal growth factor receptor endocytosis and the actin cytoskeleton (Lynch et al., 2003) and membrane trafficking. CD2AP is also associated with end-stage renal disease in patients with type 1 diabetes (Hyvönen et al., 2013).

4.4.1.11 SORTILIN-RELATED RECEPTOR (SORL1) (SNP: RS11218343; CHROMOSOMAL LOCATION: 11Q23–24)

Association of the SORL1 gene (a neuronal sortilin-related receptor) with AD was reported by an international study (Rogaeva et al., 2007) and was subsequently supported by an independent GWAS (Seshadri et al., 2010) (Table 4.5). SORL1 is also known as LR11. It is highly expressed in brain and directs trafficking of APP into recycling pathways. Reduced levels of SORL1 in AD patient brains or preclinical AD cases might lead to increased Aβ levels. Underexpression of SORL1 protein would sort APP into Aβ-generating compartments. SORL1 binds to APOE as well as APP. The association of SORL1 with AD has been assessed in several studies with conflicting results; however, a meta-analysis conducted by the Genetic and Environmental Risk in Alzheimer Disease 1 Consortium detected a significant association of polymorphisms in SORL1 and AD in both Caucasians and Asians (Reitz et al., 2011). Association of SORL1 with LOAD has also been confirmed in patients of various ethnicities (Japanese, Koreans, and Caucasians) (Miyashita et al., 2013).

4.4.1.12 *TRANSLOCASE OF OUTER MITOCHONDRIAL MEMBRANE 40 HOMOLOG (TOMM40) (SNP: RS10524523; CHROMOSOMAL LOCATION: 19Q13)*

There is a possibility that mitochondrial dysfunction might be the earliest pathophysiological event in AD. A variable-length polymorphism in TOMM40 has been shown to predict the age of LOAD onset (Roses et al., 2010). The TOMM40 polymorphism is in linkage disequilibrium with APOE and greatly increases precision of the estimated risk of LOAD onset in APOE3/4 patients (Roses et al., 2010). However, some reports have not been able to reproduce this association (Jun et al. 2012; Cruchaga et al., 2011). Roses et al. challenged these negative findings and claimed that the experiments were not replicated properly, leading to different results (Roses et al., 2013). An ADNI study found that TOMM40 was associated with lower MRI volumetry in AD cases (Potkin et al., 2009; Shen et al., 2010).

4.5 GENETIC VARIABILITY OF LOAD RISK BASED ON ETHNICITY

Genetic susceptibility of LOAD may vary among different ethnic populations. Cumulative risk assessment for LOAD examined in the Risk Evaluation and Education for Alzheimer Disease study suggested that the extent of risk of LOAD depends on ethnicity, gender, and APOE genotype (Christensen et al., 2008). Recent GWAS have also revealed variability of LOAD risk with ethnicity. For example, AKAP9 variants are associated with AD in African Americans (Logue et al., 2014), but not in white Europeans. Variants in ABCA7 are also more strongly associated with AD in African Americans than in white Europeans (Reitz et al., 2013). This GWAS examined more than 2000 LOAD patients and 4000 control cases in more than 20 different locations.

4.6 GENETICS BACKGROUND AND PRECLINICAL ALZHEIMER'S DISEASE

The ability to diagnose AD early—ideally in its preclinical stage, before widespread loss of dendritic spines and synapses has occurred—is essential for the development of AD therapeutics that can slow down or even prevent the disease. There are well-established preclinical genetic variants in patients with EOAD, but none for LOAD. While research has identified a genetic component (heritability) of sporadic AD, the genes underlying the development of LOAD have yet to be deciphered. To date, only APOE4 has been consistently associated with an increased risk of LOAD, with APOE4-containing MCI cases having a greater likelihood of converting to AD compared to MCI cases without the gene variant. For these reasons, research has turned to other molecular markers that may interact with or modify genetic factors to increase the risk of developing LOAD.

4.7 TRANSCRIPTOME BIOMARKERS IN ALZHEIMER'S DISEASE

The transcriptome is the set of messenger RNAs (mRNAs) produced by specific cells or tissues. Several attempts have been made to explore the utility of transcriptomics to distinguish AD patients from non-AD cases (Table 4.6) (Booij et al., 2011; Rye et al., 2011; Fehlbaum-Beurdeley et al., 2012; Lunnon et al., 2013). A simple blood-based transcriptome profiling test was developed by AclarusDx in a cohort of 177 individuals (90 AD patients and 87 age-matched controls) using the human Genome-Wide Splice Array (Fehlbaum-Beurdeley et al., 2012). In a validation study in a cohort of 209 individuals (111 AD patients and 98 age-matched controls) showed high sensitivity (81.3%) but low specificity (67.1%) (Fehlbaum-Beurdeley et al., 2012). Transcriptomic profiles as markers of AD have yet to be validated in very large-scale cohorts. Blood-based transcriptomic AD biomarkers achieved lower accuracies than more expensive MRI markers for detecting AD (Lunnon et al., 2013).

4.8 MICRO RNA (MIRNA) IN ALZHEIMER'S DISEASE

The lack of specific genetic links in AD in the context of clear evidence of heritability of the disease led researchers to suggest the involvement of epigenetic factors and/or miRNA-controlled gene expression influenced by environmental factors and age. Small noncoding regulatory RNAs (micro RNAs or miRNAs, 18–25 nucleotides) bind the 3'UTR of target genes and have been identified in wide range of human tissues, including the central nervous system (CNS) (Eacker et al., 2009). miRNA are the products of a series of mRNA cleavages. They generally form RNA silencing complexes by binding complementary mRNA target sequences. These target sequences may only involve 2–7 complementary nucleotides, called a seed sequence. miRNAs regulate ~60% of all known genes through posttranscriptional gene silencing.

The roles of miRNA in the pathogenesis of AD have been studied since 2008. Spatial and stage-specific abnormalities in brain miRNA expression have been revealed in AD-related pathogenic processes, such as Aβ processing, innate immunity, neurodegeneration, and insulin resistance (Cogswell et al., 2008; Hébert et al., 2008; Patel et al., 2008; Wang et al., 2008; Geekiyanage et al., 2012). Alterations in protein production are key events in neurodegeneration, including in AD. For example, miRNA miR-29a or miR-29b interacts with BACE1 and suppresses accumulation of soluble toxic Aβ oligomers (Hébert et al., 2008). miR-107 has been reported to be reduced in the AD brain neocortex (Nelson and Wang, 2010) and miR-137/181c has been shown to regulate serine palmitoyltransferase and, in turn, Aβ, in LOAD (Geekiyanage and Chan, 2011).

Circulating miRNAs are potential AD peripheral diagnostic markers that could be detected in blood. miRNA is easily isolated from blood samples and circulating miRNA profiling has shown promising results in discriminating between different tumors. Studies of blood-based miRNA biomarker signatures of AD are summarized in Table 4.7. One of the most promising studies

Table 4.6	Transcriptomic Biomarkers in Alzheimer's Disease (AD)		
Method	**Study Design**	**Biomarker Performance**	**References**
Micro-array gene expression AB1700 Whole Genome Survey Microarrays	*Experiment set:* AD (*n*= 94), age-matched control (*n*= 73), young control (*n*= 21)	*Experiment set:* Accuracy = 87%, Sensitivity = 85%, Specificity = 88%	Booij et al. (2011)
	Validation: AD (*n*= 31), age-matched control (*n*= 25), young control (*n*= 7), PD (*n* = 27), MCI (*n* = 10)	*Validation:* Accuracy = 87%, Sensitivity = 84%, Specificity = 91% (95% confidence interval)	
Real Time PCR 96-gene array set for AD, ADtect® (DiaGenic ASA, Oslo, Norway)	*Experiment set:* AD (*n*= 103), age-matched control (*n*= 105)	*Experiment set:* Accuracy = 71.6 ± 6.1%, Sensitivity = 71.8 ± 8.7%, Specificity = 71.4 ± 8.6%	Rye et al. (2011)
	Validation 1: AD (*n*= 32), age-matched control (*n*= 42)	*Validation 1:* Accuracy = 71.6 ± 10.3%, Sensitivity = 71.9 ± 15.6%, Specificity = 71.4 ± 13.7%	
	Validation 2: AD (*n*= 68), age-matched control (*n*= 62)	*Validation 2:* Accuracy = 71.5 ± 7.8%, Sensitivity = 71.4 ± 6.2%, Specificity = 71.8 ± 6.1%. (95% confidence interval)	
AclarusDx Transcriptomic signature using human Genome-Wide Splice Array	*Test study:* AD (*n* = 90), age-matched control (*n* = 87)		Fehlbaum-Beurdeley et al. (2012)
	Validation study: AD (*n* = 111), age-matched control (*n* = 98)	*Validation study:* Sensitivity = 81.3%, Specificity = 67.1% (95% confidence interval)	
HT-12v3 BeadChips	AddNeuroMed cohort *Test study:* AD (*n* = 78), age-matched control (*n* = 78)		Lunnon et al. (2013)
	Validation study: AD (*n* = 26), age-matched control (*n* = 26), MCI (converter) (*n* = 41), MCI (converter) (*n*= 77); total MCI (*n* = 118)	Validation: Accuracy = 75% (AD vs control)	

AD, Alzheimer's disease; MCI, mild cognitive impairment; PCR, polymerase chain reaction; PD, Parkinson's disease.

Table 4.7	Micro-RNA (miRNA) Based Alzheimer's Disease Biomarkers in Peripheral Blood		
mi-RNA Identified	**Methodology**	**Comments**	**References**
↑ brain-miR-112, brain-miR-161, hsa-let-7d-3p, hsa-miR-5010-3p, hsa-miR-26a-5p, hsamiR-1285-5p, and hsa-miR-151a-3p ↓ hsamiR-103a-3p, hsa-miR-107, hsa-miR-532-5p, hsa-miR-26b-5p, hsa-let-7f-5p	Sample: whole blood next-generation sequencing technology and RT-qPCR. Patient population: discovery stage: AD ($n = 48$), control ($n = 22$). Accuracy = 93%, Specificity =95%, Sensitivity= 92%. Validation stage: AD ($n = 58$), MCI ($n = 18$), multiple sclerosis ($n = 16$), Parkinson's disease ($n = 9$), major depression ($n = 15$), bipolar disease ($n = 15$), and schizophrenia ($n = 14$). Accuracy = 74–78%; specificity and sensitivity were not reported	Patients were treated with other drugs. Chronic antidepressant drug treatment has effects on blood miRNA profile Lower accuracy at validation	Leidinger et al. (2013)
↑hsa-let-7d-5p, hsa-let-7g-5p, hsa-miR-15b-5p, hsa-miR-142-3p, hsa-miR-191-5p, hsa-miR-301a-3p, hsa-miR-545-3p)	Sample: blood plasma Patient population: discovery (cohort 1): AD ($n = 11$), control ($n = 9$), MCI ($n = 20$). Validation (cohort 2): AD ($n = 20$), control ($n = 17$). Accuracy =95%	The study included no non-AD dementia patients. No comparison with crucial non-AD dementia cases	Kumar et al. (2013)
↓miR-125b, miR-23a, miR-26b	Sample: blood serum Discovery stage: micro-array. Validation stage RT-q-PCR. Patient population: AD ($n = 22$), control ($n = 24$), frontotemporal dementia ($n = 10$). Accuracy =82%	Results are correlated with those identified using CSF. No report of specificity and sensitivity	Galimberti et al. (2014)
↑miR-9 ↓miR-125b, miR-181c	Sample: blood serum Specificity= 68.3%, Sensitivity of 80.8%	Low specificity miR-125b was correlated with the MMSE	Tan et al. (2014b)

↑, up regulation; ↓, down regulation; RT-qPCR, real-time quantitative polymerase chain reaction; MCI, mild cognitive impairment; CSF, cerebrospinal fluid; MMSE, mini-mental state examination.

found a 12-biomarker signature associated with AD that had high diagnostic sensitivity, specificity, and accuracy (accuracy = 93%, specificity = 95%, and sensitivity = 92%) (Leidinger et al., 2013). A validation study including non-AD dementia patients found this miRNA marker panel to be considerably less accurate (74–78%), but did not report specificity and sensitivity (Table 4.7). This latter study was also limited by the fact that the some of the patients in the study population were treated with antidepression drugs and it is known that chronic antidepressant drug treatment affects the blood miRNA profile (Bocchio-Chiavetto et al., 2013). Prediction of functional miRNA target sites is also challenging. Although there are a substantial number of miRNA studies in AD-related pathologies, the findings must be validated in studies with larger sample sizes.

4.9 CROSSCORRELATION OF ALZHEIMER'S DISEASE GENETIC BIOMARKERS WITH OTHER BIOMARKERS

The relationship between all types of AD biomarkers is an important topic, as diagnosis of LOAD will likely require multiple biomarkers that reflect the complex pathophysiology of the disease. To avoid repetition, a discussion of the crosscorrelation of genetic AD biomarkers with other AD biomarker types is included in later chapters. Crosscorrelation between AD genetic biomarkers and AD CSF biomarkers is presented in Chapter 5 (see Section 5.13); crosscorrelation with AD blood plasma biomarkers is presented in Chapter 6 (see Section 6.7); and crosscorrelation with AD neuroimaging biomarkers is presented in Chapter 3 (see Section 3.11).

4.10 COMMERCIAL PATH FOR GENETIC BIOMARKER TESTS FOR ALZHEIMER'S DISEASE

The AD gene signature test that is closest to the market is AclarusDx (a microarray-transcriptomic based test) developed by Exonhit Therapeutics (France). The test is based on a gene expression signature of inflammatory and immune markers. The diagnostic sensitivity is moderate (81%), and specificity is quite low (67%) based on an evaluation study in 167 cases by the company (Exonhit. Approval of US pilot clinical trial to evaluate AclarusDx, the Exonhit investigational test for Alzheimer's disease diagnosis. Press release October 26, 2011). The study was conducted in French specialized memory clinics and published in 2012 (Fehlbaum-Beurdeley et al., 2012). DiaGenic ASA (Oslo, Norway) has been engaged to develop a novel whole blood RNA-based gene expression test for AD (ADtect). The company assayed sets of genes and transformed the resulting gene expression profile into a simple and convenient diagnostic test for AD. The agreement with clinical diagnosis of AD and the ADtect test was 72% ($n = 412$), whereas agreement with CSF biomarkers (low CSF Aβ_{1-42}, high tau and phospho-tau-181) was much higher (85%, $n = 30$) (Rye et al., 2011).

4.11   CONCLUSIONS

While EOAD has a well-established genetic profile (APP, PSEN1, and PSEN2), there is no clear complete genetic picture of the most common form of AD, LOAD, with the exception of the APOE4 gene variant. No other gene locus has been found to have a consistently strong association with LOAD. GWAS has not revealed any high-frequency SNPs that associate with LOAD at a strength equal to that of APOE. However, association of APOE4 with LOAD is also dependent on ethnicity, which limits its use as a clinical biomarker of LOAD.

SNP analysis has found several gene variants associated with LOAD, but most studies reported a diagnostic accuracy that was too low to be useful clinically. As these gene variants affect the fundamental biological mechanisms underlying AD, it is clear that we must look at other mechanisms that affect gene activity, including posttranslational modification and miRNA, as potential biomarkers of AD. A major shift in the approach to AD diagnosis and therapeutic intervention is required because of inability of presently available drugs to alter the course of AD. One future strategy could be personal genome sequence analysis to clarify AD risk or identify protective genes. Another approach that is already underway is to compare genetic and other information from families with multiple LOAD cases with information from families without LOAD. There is no evidence of association between genes identified from GWAS and plasma Aβ AD biomarkers (Chouraki et al., 2014). Pharmacogenomics profiling in association with next-generation GWAS sequencing might be a future step toward discovery of AD genetic biomarkers. A set of intricate networks of genes rather than individual genes is responsible for complex and heterogeneous diseases such as LOAD. Modeling of next-generation sequence-wide association (SWA) data to discover gene–gene interactions and/or gene–environment interactions will be necessary for developing diagnostic genetic tests for multifactorial diseases like LOAD.

Bibliography

Ando, K., Brion, J.P., Stygelbout, V., Suain, V., Authelet, M., Dedecker, R., Chanut, A., Lacor, P., Lavaur, J., Sazdovitch, V., Rogaeva, E., Potier, M.C., Duyckaerts, C., 2013. Clathrin adaptor CALM/PICALM is associated with neurofibrillary tangles and is cleaved in Alzheimer's brains. Acta Neuropathol. 125, 861–878.

Antúnez, C., Boada, M., González-Pérez, A., Gayán, J., Ramírez-Lorca, R., Marín, J., Hernández, I., Moreno-Rey, C., Morón, F.J., López-Arrieta, J., Mauleón, A., Rosende-Roca, M., Noguera-Perea, F., Legaz-García, A., Vivancos-Moreau, L., Velasco, J., Carrasco, J.M., Alegret, M., Antequera-Torres, M., Manzanares, S., Romo, A., Blanca, I., Ruiz, S., Espinosa, A., Castaño, S., García, B., Martínez-Herrada, B., Vinyes, G., Lafuente, A., Becker, J.T., Galán, J.J., Serrano-Ríos, M., Vázquez, E., Tárraga, L., Sáez, M.E., López, O.L., Real, L.M., Ruiz, A., Alzheimer's Disease Neuroimaging Initiative, 2011. The membrane-spanning 4-domains, subfamily A (MS4A) gene cluster contains a common variant associated with Alzheimer's disease. Genome Med. 3, 33.

Bergem, A.L., Engedal, K., Kringlen, E., 1997. The role of heredity in late-onset Alzheimer disease and vascular dementia. A twin study. Arch. Gen. Psychiatry 54, 264–270.

Bertram, L., Tanzi, R.E., 2009. Genome-wide association studies in Alzheimer's disease. Hum. Mol. Genet. 18 (R2), R137–R145.

Bertram, L., Lange, C., Mullin, K., Parkinson, M., Hsiao, M., Hogan, M.F., Schjeide, B.M., Hooli, B., Divito, J., Ionita, I., Jiang, H., Laird, N., Moscarillo, T., Ohlsen, K.L., Elliott, K., Wang, X., Hu-Lince, D., Ryder, M., Murphy, A., Wagner, S.L., Blacker, D., Becker, K.D., Tanzi, R.E., 2008. Genome-wide association analysis reveals putative Alzheimer's disease susceptibility loci in addition to APOE. Am. J. Hum. Genet. 83, 623–632.

Biffi, A., Anderson, C.D., Desikan, R.S., Sabuncu, M., Cortellini, L., Schmansky, N., Salat, D., Rosand, J., Alzheimer's Disease Neuroimaging Initiative (ADNI), 2010. Genetic variation and neuroimaging measures in Alzheimer disease. Arch. Neurol. 67, 677–685.

Bocchio-Chiavetto, L., Maffioletti, E., Bettinsoli, P., Giovannini, C., Bignotti, S., Tardito, D., Corrada, D., Milanesi, L., Gennarelli, M., 2013. Blood microRNA changes in depressed patients during antidepressant treatment. Eur. Neuropsychopharmacol. 23, 602–611.

Booij, B.B., Lindahl, T., Wetterberg, P., Skaane, N.V., Sæbø, S., Feten, G., Rye, P.D., Kristiansen, L.I., Hagen, N., Jensen, M., Bårdsen, K., Winblad, B., Sharma, P., Lönneborg, A., 2011. A gene expression pattern in blood for the early detection of Alzheimer's disease. J. Alzheimers Dis. 23, 109–119.

Bradshaw, E.M., Chibnik, L.B., Keenan, B.T., Ottoboni, L., Raj, T., Tang, A., Rosenkrantz, L.L., Imboywa, S., Lee, M., Von Korff, A., Morris, M.C., Evans, D.A., Johnson, K., Sperling, R.A., Schneider, J.A., Bennett, D.A., De Jager, P.L., Alzheimer Disease Neuroimaging Initiative, 2013. CD33 Alzheimer's disease locus: altered monocyte function and amyloid biology. Nat. Neurosci. 16, 848–850.

Cacabelos, R., 2007. Pharmacogenetic basis for therapeutic optimization in Alzheimer's disease. Mol. Diagn. Ther. 11, 385–405.

Cacabelos, R., 2009. Pharmacogenomics and therapeutic strategies for dementia. Expert Rev. Mol. Diagn. 9, 567–611.

Cacabelos, R., Llovo, R., Fraile, C., Fernández-Novoa, L., 2007. Pharmacogenetic aspects of therapy with cholinesterase inhibitors: the role of CYP2D6 in Alzheimer's disease pharmacogenetics. Curr. Alzheimer Res. 4, 479–500.

Cacabelos, R., Martínez, R., Fernández-Novoa, L., Carril, J.C., Lombardi, V., Carrera, I., Corzo, L., Tellado, I., Leszek, J., McKay, A., Takeda, M., 2012. Genomics of dementia: APOE- and CYP2D6-related pharmacogenetics. Int. J. Alzheimers Dis. 2012, 518901.

Calero, M., Rostagno, A., Matsubara, E., Zlokovic, B., Frangione, B., Ghiso, J., 2000. Apolipoprotein J (clusterin) and Alzheimer's disease. Microsc. Res. Tech. 50, 305–315.

Castellano, J.M., Kim, J., Stewart, F.R., Jiang, H., DeMattos, R.B., Patterson, B.W., Fagan, A.M., Morris, J.C., Mawuenyega, K.G., Cruchaga, C., Goate, A.M., Bales, K.R., Paul, S.M., Bateman, R.J., Holtzman, D.M., 2011. Human apoE isoforms differentially regulate brain amyloid-β peptide clearance. Sci. Transl. Med. 3 (89ra57), 1–11.

Chan, S.L., Kim, W.S., Kwok, J.B., Hill, A.F., Cappai, R., Rye, K.A., Garner, B., 2008. ATP-binding cassette transporter A7 regulates processing of amyloid precursor protein in vitro. J. Neurochem. 106, 793–804.

Chapuis, J., Hansmannel, F., Gistelinck, M., Mounier, A., Van Cauwenberghe, C., Kolen, K.V., Geller, F., Sottejeau, Y., Harold, D., Dourlen, P., Grenier-Boley, B., Kamatani, Y., Delepine, B., Demiautte, F., Zelenika, D., Zommer, N., Hamdane, M., Bellenguez, C., Dartigues, J.F., Hauw, J.J., Letronne, F., Ayral, A.M., Sleegers, K., Schellens, A., Broeck, L.V., Engelborghs, S., De Deyn, P.P., Vandenberghe, R., O'Donovan, M., Owen, M., Epelbaum, J., Mercken, M., Karran, E., Bantscheff, M., Drewes, G., Joberty, G., Campion, D., Octave, J.N., Berr, C., Lathrop, M., Callaerts, P., Mann, D., Williams, J., Buée, L., Dewachter, I., Van Broeckhoven, C., Amouyel, P., Moechars, D., Dermaut, B., Lambert, J.C., GERAD consortium, 2013. Increased expression of BIN1 mediates Alzheimer genetic risk by modulating tau pathology. Mol. Psychiatry 18, 1225–1234.

Chauhan, G., Adams, H.H., Bis, J.C., Weinstein, G., Yu, L., Töglhofer, A.M., Smith, A.V., van der Lee, S.J., Gottesman, R.F., Thomson, R., Wang, J., Yang, Q., Niessen, W.J., Lopez, O.L., Becker, J.T., Phan, T.G., Beare, R.J., Arfanakis, K., Fleischman, D., Vernooij, M.W., Mazoyer, B., Schmidt,

H., Srikanth, V., Knopman, D.S., Jack, Jr., C.R., Amouyel, P., Hofman, A., DeCarli, C., Tzourio, C., van Duijn, C.M., Bennett, D.A., Schmidt, R., Longstreth, Jr., W.T., Mosley, T.H., Fornage, M., Launer, L.J., Seshadri, S., Ikram, M.A., Debette, S., 2015. Association of Alzheimer's disease GWAS loci with MRI markers of brain aging. Neurobiol. Aging 36, 1765.e7–1765.e16.

Chouraki, V., De Bruijn, R.F., Chapuis, J., Bis, J.C., Reitz, C., Schraen, S., Ibrahim-Verbaas, C.A., Grenier-Boley, B., Delay, C., Rogers, R., Demiautte, F., Mounier, A., Fitzpatrick, A.L., Berr, C., Dartigues, J.F., Uitterlinden, A.G., Hofman, A., Breteler, M., Becker, J.T., Lathrop, M., Schupf, N., Alpérovitch, A., Mayeux, R., van Duijn, C.M., Buée, L., Amouyel, P., Lopez, O.L., Ikram, M.A., Tzourio, C., Lambert, J.C., Alzheimer's Disease Neuroimaging Initiative, 2014. A genome-wide association meta-analysis of plasma Aβ peptides concentrations in the elderly. Mol. Psychiatry 19, 1326–1335.

Christensen, K.D., Roberts, J.S., Royal, C.D., Fasaye, G.A., Obisesan, T., Cupples, L.A., Whitehouse, P.J., Butson, M.B., Linnenbringer, E., Relkin, N.R., Farrer, L., Cook-Deegan, R., Green, R.C., 2008. Incorporating ethnicity into genetic risk assessment for Alzheimer disease: the REVEAL study experience. Genet. Med. 10, 207–214.

Chung, S.J., Lee, J.H., Kim, S.Y., You, S., Kim, M.J., Lee, J.Y., Koh, J., 2013. Association of GWAS top hits with late-onset Alzheimer disease in Korean population. Alzheimer Dis. Assoc. Disord. 27, 250–257.

Cogswell, J.P., Ward, J., Taylor, I.A., Waters, M., Shi, Y., Cannon, B., Kelnar, K., Kemppainen, J., Brown, D., Chen, C., Prinjha, R.K., Richardson, J.C., Saunders, A.M., Roses, A.D., Richards, C.A., 2008. Identification of miRNA changes in Alzheimer's disease brain and CSF yields putative biomarkers and insights into disease pathways. J. Alzheimers Dis. 14, 27–41.

Coulthard, M.G., Lickliter, J.D., Subanesan, N., Chen, K., Webb, G.C., Lowry, A.J., Koblar, S., Bottema, C.D., Boyd, A.W., 2001. Characterization of the Epha1 receptor tyrosine kinase: expression in epithelial tissues. Growth Factors 18, 303–317.

Crean, S., Ward, A., Mercaldi, C.J., Collins, J.M., Cook, M.N., Baker, N.L., Arrighi, H.M., 2011. Apolipoprotein E ε4 prevalence in Alzheimer's disease patients varies across global populations: a systematic literature review and meta-analysis. Dement. Geriatr. Cogn. Disord. 31, 20–30.

Crehan, H., Holton, P., Wray, S., Pocock, J., Guerreiro, R., Hardy, J., 2012. Complement receptor 1 (CR1) and Alzheimer's disease. Immunobiol 217, 244–250.

Cruchaga, C., Nowotny, P., Kauwe, J.S., Ridge, P.G., Mayo, K., Bertelsen, S., Hinrichs, A., Fagan, A.M., Holtzman, D.M., Morris, J.C., Goate, A.M., Alzheimer's Disease Neuroimaging Initiative, 2011. Association and expression analyses with single-nucleotide polymorphisms in TOMM40 in Alzheimer disease. Arch. Neurol. 68, 1013–1019.

Cruts, M., Backhovens, H., Wang, S.Y., Van Gassen, G., Theuns, J., De Jonghe, C.D., Wehnert, A., De Voecht, J., De Winter, G., Cras, P., et al., 1995. Molecular genetic analysis of familial early-onset Alzheimer's disease linked to chromosome 14q24. 3. Hum. Mol. Genet. 4, 2363–2371.

Daborg, J., Andreasson, U., Pekna, M., Lautner, R., Hanse, E., Minthon, L., Blennow, K., Hansson, O., Zetterberg, H., 2012. J. Neural Transm. 119, 789–797.

Deane, R., Sagare, A., Hamm, K., Parisi, M., Lane, S., Finn, M.B., Holtzman, D.M., Zlokovic, B.V., 2008. apoE isoform-specific disruption of amyloid beta peptide clearance from mouse brain. J. Clin. Invest. 118, 4002–4013.

DeMattos, R.B., Cirrito, J.R., Parsadanian, M., May, P.C., O'Dell, M.A., Taylor, J.W., Harmony, J.A., Aronow, B.J., Bales, K.R., Paul, S.M., Holtzman, D.M., 2004. ApoE and clusterin cooperatively suppress Ab levels and deposition: evidence that ApoE regulates extracellular Ab metabolism in vivo. Neuron 41, 193–202.

den Heijer, T., Oudkerk, M., Launer, L.J., van Duijn, C.M., Hofman, A., Breteler, M.M., 2002. Hippocampal, amygdalar, and global brain atrophy in different apolipoprotein E genotypes. Neurology 59, 746–748.

Deng, Y.L., Liu, L.H., Wang, Y., Tang, H.D., Ren, R.J., Xu, W., Ma, J.F., Wang, L.L., Zhuang, J.P., Wang, G., Chen, S.D., 2012. The prevalence of CD33 and MS4A6A variant in Chinese Han population with Alzheimer's disease. Hum. Genet. 131, 1245–1249.

Eacker, S.M., Dawson, T.M., Dawson, V.L., 2009. Understanding microRNAs in neurodegeneration. Nat. Rev. Neurosci. 10, 837–841.

Elbein, S.C., 1997. The genetics of human noninsulin-dependent (type 2) diabetes mellitus. J. Nutr. 127, 1891S–1896S.

Evans, D.A., Bennett, D.A., Wilson, R.S., Bienias, J.L., Morris, M.C., Scherr, P.A., Hebert, L.E., Aggarwal, N., Beckett, L.A., Joglekar, R., Berry-Kravis, E., Schneider, J., 2003. Incidence of Alzheimer disease in a biracial urban community: relation to apolipoprotein E allele status. Arch. Neurol. 60, 185–189.

Farrer, L.A., Cupples, L.A., Haines, J.L., Hyman, B., Kukull, W.A., Mayeux, R., Myers, R.H., Pericak-Vance, M.A., Risch, N., van Duijn, C.M., 1997. Effects of age, sex, and ethnicity on the association between apolipoprotein E genotype and Alzheimer disease. A meta-analysis. APOE and Alzheimer Disease Meta Analysis Consortium. JAMA 278, 1349–1356.

Fehlbaum-Beurdeley, P., Sol, O., Désiré, L., Touchon, J., Dantoine, T., Vercelletto, M., Gabelle, A., Jarrige, A.C., Haddad, R., Lemarié, J.C., Zhou, W., Hampel, H., Einstein, R., Vellas, B., EHTAD/002 study group, 2012. Validation of AclarusDx™, a blood-based transcriptomic signature for the diagnosis of Alzheimer's disease. J. Alzheimers Dis. 32, 169–181.

Ferencz, B., Laukka, E.J., Lovden, M., Kalpouzos, G., Keller, L., Graff, C., Wahlund, L.O., Fratiglioni, L., Backman, L., 2013. The influence of APOE and TOMM40 polymorphisms on hippocampal volume and episodic memory in old age. Front. Hum. Neurosci. 7, 198.

Galimberti, D., Villa, C., Fenoglio, C., Serpente, M., Ghezzi, L., Cioffi, S.M., Arighi, A., Fumagalli, G., Scarpini, E., 2014. Circulating miRNAs as potential biomarkers in Alzheimer's disease. J. Alzheimers Dis. 42, 1261–1267.

Gatz, M., Pedersen, N.L., Berg, S., Johansson, B., Johansson, K., Mortimer, J.A., Posner, S.F., Viitanen, M., Winblad, B., Ahlbom, A., 1997. Heritability for Alzheimer's disease: the study of dementia in Swedish twins. J. Gerontol. A Biol. Sci. Med. Sci. 52, M117–M125.

Gatz, M., Reynolds, C.A., Fratiglioni, L., Johansson, B., Mortimer, J.A., Berg, S., Fiske, A., Pedersen, N.L., 2006. Role of genes and environments for explaining Alzheimer disease. Arch. Gen. Psychiatry 63, 168–174.

Geekiyanage, H., Chan, C., 2011. MicroRNA-137/181c regulates serine palmitoyltransferase and in turn amyloid β, novel targets in sporadic Alzheimer's disease. J. Neurosci. 31, 14820–14830.

Geekiyanage, H., Jicha, G.A., Nelson, P.T., Chan, C., 2012. Blood serum miRNA: non-invasive biomarkers for Alzheimer's disease. Exp. Neurol. 235, 491–496.

Glenner, G.G., Wong, C.W., 1984. Alzheimer's disease: initial report of the purification and characterization of a novel cerebrovascular amyloid protein. Biochem. Biophys. Res. Commun. 120, 885–890.

Goate, A., 2006. Segregation of a missense mutation in the amyloid beta-protein precursor gene with familial Alzheimer's disease. J. Alzheimers Dis. 9 (3 Suppl), 341–347.

Goate, A., Chartier-Harlin, M.C., Mullan, M., Brown, J., Crawford, F., Fidani, L., Giuffra, L., Haynes, A., Irving, N., James, L., et al., 1991. Segregation of a missense mutation in the amyloid precursor protein gene with familial Alzheimer's disease. Nature 349, 704–706.

Goldgaber, D., Lerman, M.I., McBride, O.W., Saffiotti, U., Gajdusek, D.C., 1987. Characterization and chromosomal localization of a cDNA encoding brain amyloid of Alzheimer's disease. Science 235, 877–880.

Griciuc, A., Serrano-Pozo, A., Parrado, A.R., Lesinski, A.N., Asselin, C.N., Mullin, K., Hooli, B., Choi, S.H., Hyman, B.T., Tanzi, R.E., 2013. Alzheimer's disease risk gene CD33 inhibits microglial uptake of amyloid beta. Neuron 78, 631–643.

Grupe, A., Abraham, R., Li, Y., Rowland, C., Hollingworth, P., Morgan, A., Jehu, L., Segurado, R., Stone, D., Schadt, E., Karnoub, M., Nowotny, P., Tacey, K., Catanese, J., Sninsky, J., Brayne, C., Rubinsztein, D., Gill, M., Lawlor, B., Lovestone, S., Holmans, P., O'Donovan, M., Morris, J.C., Thal, L., Goate, A., Owen, M.J., Williams, J., 2007. Evidence for novel susceptibility genes for late-onset Alzheimer's disease from a genome-wide association study of putative functional variants. Hum. Mol. Genet. 16, 865–873.

Gusella, J.F., Wexler, N.S., Conneally, P.M., Naylor, S.L., Anderson, M.A., Tanzi, R.E., Watkins, P.C., Ottina, K., Wallace, M.R., Sakaguchi, A.Y., et al., 1983. A polymorphic DNA marker genetically linked to Huntington's disease. Nature 306, 234–238.

Hardy, J., Selkoe, D.J., 2002. The amyloid hypothesis of Alzheimer's disease: progress and problems on the road to therapeutics. Science 297, 353–356.

Harold, D., Abraham, R., Hollingworth, P., Sims, R., Gerrish, A., Hamshere, M.L., Pahwa, J.S., Moskvina, V., Dowzell, K., Williams, A., et al., 2009. Genome-wide association study identifies variants at CLU and PICALM associated with Alzheimer's disease. Nat. Genet. 41, 1088–1093.

Hébert, S.S., Horré, K., Nicolaï, L., Papadopoulou, A.S., Mandemakers, W., Silahtaroglu, A.N., Kauppinen, S., Delacourte, A., De Strooper, B., 2008. Loss of microRNA cluster miR-29a/b-1 in sporadic Alzheimer's disease correlates with increased BACE1/beta-secretase expression. Proc. Natl. Acad. Sci. USA 105, 6415–6420.

Hollingworth, P., Harold, D., Sims, R., Gerrish, A., Lambert, J.C., Carrasquillo, M.M., Abraham, R., et al., 2011. Common variants at ABCA7, MS4A6A/MS4A4E, EPHA1, CD33 and CD2AP are associated with Alzheimer's disease. Nat. Genet. 43, 429–435.

Hyvönen, M.E., Ihalmo, P., Sandholm, N., Stavarachi, M., Forsblom, C., McKnight, A.J., Lajer, M., Maestroni, A., Lewis, G., Tarnow, L., Maestroni, S., Zerbini, G., Parving, H.H., Maxwell, A.P., Groop, P.H., Lehtonen, S., 2013. CD2AP is associated with end-stage renal disease in patients with type 1 diabetes. Acta Diabetol. 50, 887–897.

Ishibashi, K., Suzuki, M., Sasaki, S., Imai, M., 2001. Identification of a new multigene four-transmembrane family (MS4A) related to CD20, HTm4 and beta subunit of the high-affinity IgE receptor. Gene 264, 87–93.

Jehle, A.W., Gardai, S.J., Li, S., Linsel-Nitschke, P., Morimoto, K., Janssen, W.J., Vandivier, R.W., Wang, N., Greenberg, S., Dale, B.M., Qin, C., Henson, P.M., Tall, A.R., 2006. ATP-binding cassette transporter A7 enhances phagocytosis of apoptotic cells and associated ERK signaling in macrophages. J. Cell Biol. 174, 547–556.

Jiang, T., Yu, J.T., Hu, N., Tan, M.S., Zhu, X.C., Tan, L., 2014. CD33 in Alzheimer's disease. Mol. Neurobiol. 49, 529–535.

Jun, G., Vardarajan, B.N., Buros, J., Yu, C.E., Hawk, M.V., Dombroski, B.A., Crane, P.K., Larson, E.B., Mayeux, R., Haines, J.L., Lunetta, K.L., Pericak-Vance, M.A., Schellenberg, G.D., Farrer, L.A., Alzheimer's Disease Genetics Consortium, 2012. Comprehensive search for Alzheimer disease susceptibility loci in the APOE region. Arch. Neurol. 69, 1270–1279.

Kalaria, R.N., Maestre, G.E., Arizaga, R., Friedland, R.P., Galasko, D., Hall, K., Luchsinger, J.A., Ogunniyi, A., Perry, E.K., Potocnik, F., Prince, M., Stewart, R., Wimo, A., Zhang, Z.X., Antuono, P., World Federation of Neurology Dementia Research Group, 2008. Alzheimer's disease and vascular dementia in developing countries: prevalence, management, and risk factors. Lancet Neurol. 7, 812–826.

Kamino, K., Orr, H.T., Payami, H., Wijsman, E.M., Alonso, M.E., Pulst, S.M., Anderson, L., O'dahl, S., Nemens, E., White, J.A., et al., 1992. Linkage and mutational analysis of familial Alzheimer disease kindreds for the APP gene region. Am. J. Hum. Genet. 51, 998–1014.

Kang, J., Lemaire, H.G., Unterbeck, A., Salbaum, J.M., Masters, C.L., Grzeschik, K.H., Multhaup, G., Beyreuther, K., Muller-Hill, B., 1987. The precursor of Alzheimer's disease amyloid A4 protein resembles a cell-surface receptor. Nature 325, 733–736.

Karch, C.M., Jeng, A.T., Nowotny, P., Cady, J., Cruchaga, C., Goate, A.M., 2012. Expression of novel Alzheimer's disease risk genes in control and Alzheimer's disease brains. PLoS One 7, e50976.

Kehoe, P.G., Russ, C., McIlory, S., Williams, H., Holmans, P., Holmes, C., Liolitsa, D., Vahidassr, D., Powell, J., McGleenon, B., Liddell, M., Plomin, R., Dynan, K., Williams, N., Neal, J., Cairns, N.J., Wilcock, G., Passmore, P., Lovestone, S., Williams, J., Owen, M.J., 1999. Variation in DCP1, encoding ACE, is associated with susceptibility to Alzheimer disease. Nat. Genet. 21, 71–72.

Khan, W., Giampietro, V., Ginestet, C., Dell'Acqua, F., Bouls, D., Newhouse, S., Dobson, R., Banaschewski, T., Barker, G.J., Bokde, A.L., Büchel, C., Conrod, P., Flor, H., Frouin, V., Garavan, H., Gowland, P., Heinz, A., Ittermann, B., Lemaître, H., Nees, F., Paus, T., Pausova, Z., Rietschel, M., Smolka, M.N., Ströhle, A., Gallinat, J., Westman, E., Schumann, G., Lovestone, S., Simmons, A., IMAGEN consortium, 2014. No differences in hippocampal volume between carriers and non-carriers of the ApoE ε4 and ε2 alleles in young healthy adolescents. J. Alzheimers Dis. 40, 37–43.

Khan, T.K., Sen, A., Hongpaisan, J., Lim, C.S., Nelson, T.J., Alkon, D.L., 2015. PKCε deficits in Alzheimer's disease brains and skin fibroblasts. J. Alzheimers Dis. 43, 491–509.

Kim, J., Basak, J.M., Holtzman, D.M., 2009a. The role of apolipoprotein E in Alzheimer's disease. Neuron 63, 287–303.

Kim, J., Castellano, J.M., Jiang, H., Basak, J.M., Parsadanian, M., Pham, V., Mason, S.M., Paul, S.M., Holtzman, D.M., 2009b. Overexpression of low-density lipoprotein receptor in the brain markedly inhibits amyloid deposition and increases extracellular A beta clearance. Neuron 64, 632–644.

Kirsch, K.H., Georgescu, M.M., Ishimaru, S., Hanafusa, H., 1999. CMS: an adapter molecule involved in cytoskeletal rearrangements. Proc. Natl. Acad. Sci. USA 96, 6211–6216.

Klimkowicz-Mrowiec, A., Sado, M., Dziubek, A., Dziedzic, T., Pera, J., Szczudlik, A., Slowik, A., 2013. Lack of association of CR1, PICALM and CLU gene polymorphisms with Alzheimer disease in a Polish population. Neurol. Neurochir. Pol. 47, 157–160.

Kullander, K., Klein, R., 2002. Mechanisms and functions of Eph and ephrin signalling. Nat. Rev. Mol. Cell. Biol. 3, 475–486.

Kumar, P., Dezso, Z., MacKenzie, C., Oestreicher, J., Agoulnik, S., Byrne, M., Bernier, F., Yanagimachi, M., Aoshima, K., Oda, Y., 2013. Circulating miRNA biomarkers for Alzheimer's disease. PLoS One 8, e69807.

Lai, K.O., Ip, N.Y., 2009. Synapse development and plasticity: roles of ephrin/Eph receptor signaling. Curr. Opin. Neurobiol. 19, 275–283.

Lambert, J.C., Amouyel, P., 2011. Genetics of Alzheimer's disease: new evidences for an old hypothesis? Curr. Opin. Genet. Dev. 3, 295–301.

Lambert, J.-C., Heath, S., Even, G., Campion, D., Sleegers, K., Hiltunen, M., Combarros, O., Zelenika, D., Bullido, M.J., Tavernier, B., European Alzheimer's Disease Initiative Investigators, et al., 2009. Genome-wide association study identifies variants at CLU and CR1 associated with Alzheimer's disease. Nat. Genet. 41, 1094–1099.

Lambert, J.C., Ibrahim-Verbaas, C.A., Harold, D., Naj, A.C., Sims, R., Bellenguez, C., DeStafano, A.L., Bis, J.C., Beecham, G.W., Grenier-Boley, B., Russo, G., Thorton-Wells, T.A., Jones, N., Smith, A.V., Chouraki, V., Thomas, C., Ikram, M.A., Zelenika, D., Vardarajan, B.N., Kamatani, Y., Lin, C.F., Gerrish, A., Schmidt, H., Kunkle, B., Dunstan, M.L., Ruiz, A., Bihoreau, M.T., Choi, S.H., Reitz, C., Pasquier, F., Cruchaga, C., Craig, D., Amin, N., Berr, C., Lopez, O.L., De Jager, P.L., Deramecourt, V., Johnston, J.A., Evans, D., Lovestone, S., Letenneur, L., Morón, F.J., Rubinsztein, D.C., Eiriksdottir, G., Sleegers, K., Goate, A.M., Fiévet, N., Huentelman, M.W., Gill, M., Brown, K., Kamboh, M.I., Keller, L., Barberger-Gateau, P., McGuiness, B., Larson, E.B., Green, R., Myers, A.J., Dufouil, C., Todd, S., Wallon, D., Love, S., Rogaeva, E., Gallacher, J., St George-Hyslop, P., Clarimon, J., Lleo, A., Bayer, A., Tsuang, D.W., Yu, L., Tsolaki, M., Bossù, P., Spalletta, G., Proitsi, P., Collinge, J., Sorbi, S., Sanchez-Garcia, F., Fox, N.C., Hardy, J., Deniz Naranjo, M.C., Bosco, P., Clarke, R., Brayne, C., Galimberti, D., Mancuso, M., Matthews, F., Moebus, S., Mecocci, P., Del Zompo, M., Maier, W., Hampel, H., Pilotto, A., Bullido, M., Panza, F., Caffarra, P., Nacmias, B., Gilbert, J.R., Mayhaus, M., Lannefelt, L., Hakonarson, H., Pichler, S., Carrasquillo, M.M., Ingelsson, M., Beekly, D., Alvarez, V., Zou, F., Valladares, O., Younkin, S.G., Coto, E., Hamilton-Nelson,

K.L., Gu, W., Razquin, C., Pastor, P., Mateo, I., Owen, M.J., Faber, K.M., Jonsson, P.V., Combarros, O., O'Donovan, M.C., Cantwell, L.B., Soininen, H., Blacker, D., Mead, S., Mosley, Jr., T.H., Bennett, D.A., Harris, T.B., Fratiglioni, L., Holmes, C., de Bruijn, R.F., Passmore, P., Montine, T.J., Bettens, K., Rotter, J.I., Brice, A., Morgan, K., Foroud, T.M., Kukull, W.A., Hannequin, D., Powell, J.F., Nalls, M.A., Ritchie, K., Lunetta, K.L., Kauwe, J.S., Boerwinkle, E., Riemenschneider, M., Boada, M., Hiltuenen, M., Martin, E.R., Schmidt, R., Rujescu, D., Wang, L.S., Dartigues, J.F., Mayeux, R., Tzourio, C., Hofman, A., Nöthen, M.M., Graff, C., Psaty, B.M., Jones, L., Haines, J.L., Holmans, P.A., Lathrop, M., Pericak-Vance, M.A., Launer, L.J., Farrer, L.A., van Duijn, C.M., Van Broeckhoven, C., Moskvina, V., Seshadri, S., Williams, J., Schellenberg, G.D., Amouyel, P., European Alzheimer's Disease Initiative (EADI); Genetic and Environmental Risk in Alzheimer's Disease; Alzheimer's Disease Genetic Consortium; Cohorts for Heart and Aging Research in Genomic Epidemiology, 2013. Meta-analysis of 74,046 individuals identifies 11 new susceptibility loci for Alzheimer's disease. Nat. Genet. 45, 1452–1458.

Lehtovirta, M., Laakso, M.P., Soininen, H., Helisalmi, S., Mannermaa, A., Helkala, E.L., Partanen, K., Ryynanen, M., Vainio, P., Hartikainen, P., Riekkinen, P.J., 1995. Volumes of hippocampus, amygdala and frontal lobe in Alzheimer patients with different apolipoprotein E genotypes. Neuroscience 67, 65–72.

Lehtovirta, M., Soininen, H., Laakso, M.P., Partanen, K., Helisalmi, S., Mannermaa, A., Ryynanen, M., Kuikka, J., Hartikainen, P., Riekkinen@@Sr., P.J., 1996. SPECT and MRI analysis in Alzheimer's disease: relation to apolipoprotein E epsilon 4 allele. J. Neurol. Neurosurg. Psychiatry 60, 644–649.

Leidinger, P., Backes, C., Deutscher, S., Schmitt, K., Mueller, S.C., Frese, K., Haas, J., Ruprecht, K., Paul, F., Stähler, C., Lang, C.J., Meder, B., Bartfai, T., Meese, E., Keller, A., 2013. A blood based 12-miRNA signature of Alzheimer disease patients. Genome Biol. 14, R78.

Levy-Lahad, E., Wasco, W., Poorkaj, P., Romano, D.M., Oshima, J., Pettingell, W.H., Yu, C.E., Jondro, P.D., Schmidt, S.D., Wang, K., et al., 1995. Candidate gene for the chromosome 1 familial Alzheimer's disease locus. Science 269, 973–977.

Liang, Y., Buckley, T.R., Tu, L., Langdon, S.D., Tedder, T.F., 2001. Structural organization of the human MS4A gene cluster on Chromosome 11q12. Immunogenetics 53, 357–368.

Liu, Y., Yu, J.T., Wang, H.F., Han, P.R., Tan, C.C., Wang, C., Meng, X.F., Risacher, S.L., Saykin, A.J., Tan, L., 2015. APOE genotype and neuroimaging markers of Alzheimer's disease: systematic review and meta-analysis. J. Neurol. Neurosurg. Psychiatry 86, 127–134.

Logue, M.W., Schu, M., Vardarajan, B.N., Farrell, J., Bennett, D.A., Buxbaum, J.D., Byrd, G.S., Ertekin-Taner, N., Evans, D., Foroud, T., Goate, A., Graff-Radford, N.R., Kamboh, M.I., Kukull, W.A., Manly, J.J., Alzheimer Disease Genetics Consortium; Alzheimer Disease Genetics Consortium, 2014. Two rare AKAP9 variants are associated with Alzheimer's disease in African Americans. Alzheimers Dement. 10, 609–618.

Lu, P.H., Thompson, P.M., Leow, A., Lee, G.J., Lee, A., Yanovsky, I., Parikshak, N., Khoo, T., Wu, S., Geschwind, D., Bartzokis, G., 2011. Apolipoprotein E genotype is associated with temporal and hippocampal atrophy rates in healthy elderly adults: a tensor-based morphometry study. J. Alzheimers Dis. 23, 433–442.

Lunnon, K., Sattlecker, M., Furney, S.J., Coppola, G., Simmons, A., Proitsi, P., Lupton, M.K., Lourdusamy, A., Johnston, C., Soininen, H., Kłoszewska, I., Mecocci, P., Tsolaki, M., Vellas, B., Geschwind, D., Lovestone, S., Dobson, R., Hodges, A., dNeuroMed Consortium, 2013. A blood gene expression marker of early Alzheimer's disease. J. Alzheimers Dis. 33, 737–753.

Lynch, D.K., Winata, S.C., Lyons, R.J., Hughes, W.E., Lehrbach, G.M., Wasinger, V., Corthals, G., Cordwell, S., Daly, R.J., 2003. A Cortactin-CD2-associated protein (CD2AP) complex provides a novel link between epidermal growth factor receptor endocytosis and the actin cytoskeleton. J. Biol. Chem. 278, 21805–21813.

Ma, J., Yu, J.T., Tan, L., 2015. MS4A cluster in Alzheimer's disease. Mol. Neurobiol. 51, 1240–1248.

Martinelli-Boneschi, F., Giacalone, G., Magnani, G., Biella, G., Coppi, E., Santangelo, R., Brambilla, P., Esposito, F., Lupoli, S., Clerici, F., Benussi, L., Ghidoni, R., Galimberti, D., Squitti, R., Confaloni, A., Bruno, G., Pichler, S., Mayhaus, M., Riemenschneider, M., Mariani, C., Comi, G., Scarpini, E., Binetti, G., Forloni, G., Franceschi, M., Albani, D., 2013. Pharmacogenomics in Alzheimer's disease: a genome-wide association study of response to cholinesterase inhibitors. Neurobiol. Aging 34, 1711.e7–1711.e13.

Martínez, A., Soriano, E., 2005. Functions of ephrin/Eph interactions in the development of the nervous system: emphasis on the hippocampal system. Brain Res. Brain Res. Rev. 49, 211–226.

Mayeux, R., 2003. Apolipoprotein E, Alzheimer disease, and African Americans. Arch. Neurol. 60, 161–163.

McGeer, P.L., Kawamata, T., Walker, D.G., 1992. Distribution of clusterin in Alzheimer brain tissue. Brain Res. 579, 337–341.

Mines, M., Ding, Y., Fan, G., 2007. The many roles of chemokine receptors in neurodegenerative disorders: emerging new therapeutical strategies. Curr. Med. Chem. 14, 2456–2470.

Miyashita, A., Koike, A., Jun, G., Wang, L.S., Takahashi, S., Matsubara, E., Kawarabayashi, T., Shoji, M., et al., 2013. SORL1 is genetically associated with late-onset Alzheimer's disease in Japanese, Koreans and Caucasians. PLoS One 8 (4), e58618.

Morra, J.H., Tu, Z., Apostolova, L.G., Green, A.E., Avedissian, C., Madsen, S.K., Parikshak, N., Toga, A.W., Jack, Jr., C.R., Schuff, N., Weiner, M.W., Thompson, P.M., Alzheimer's Disease Neuroimaging Initiative, 2009. Automated mapping of hippocampal atrophy in 1-year repeat MRI data from 490 subjects with Alzheimer's disease, mild cognitive impairment, and elderly controls. Neuroimage 45 (1 Suppl), S3–15.

Naj, A.C., Jun, G., Reitz, C., Kunkle, B.W., Perry, W., Park, Y.S., Beecham, G.W., Rajbhandary, R.A., Hamilton-Nelson, K.L., Wang, L.S., Kauwe, J.S., Huentelman, M.J., Myers, A.J., Bird, T.D., Boeve, B.F., Baldwin, C.T., Jarvik, G.P., Crane, P.K., Rogaeva, E., Barmada, M.M., Demirci, F.Y., Cruchaga, C., Kramer, P.L., Ertekin-Taner, N., Hardy, J., Graff-Radford, N.R., Green, R.C., Larson, E.B., St George-Hyslop, P.H., Buxbaum, J.D., Evans, D.A., Schneider, J.A., Lunetta, K.L., Kamboh, M.I., Saykin, A.J., Reiman, E.M., De Jager, P.L., Bennett, D.A., Morris, J.C., Montine, T.J., Goate, A.M., Blacker, D., Tsuang, D.W., Hakonarson, H., Kukull, W.A., Foroud, T.M., Martin, E.R., Haines, J.L., Mayeux, R.P., Farrer, L.A., Schellenberg, G.D., Pericak-Vance, M.A., Albert, M.S., Albin, R.L., Apostolova, L.G., Arnold, S.E., Barber, R., Barnes, L.L., Beach, T.G., Becker, J.T., Beekly, D., Bigio, E.H., Bowen, J.D., Boxer, A., Burke, J.R., Cairns, N.J., Cantwell, L.B., Cao, C., Carlson, C.S., Carney, R.M., Carrasquillo, M.M., Carroll, S.L., Chui, H.C., Clark, D.G., Corneveaux, J., Cribbs, D.H., Crocco, E.A., DeCarli, C., DeKosky, S.T., Dick, M., Dickson, D.W., Duara, R., Faber, K.M., Fallon, K.B., Farlow, M.R., Ferris, S., Frosch, M.P., Galasko, D.R., Ganguli, M., Gearing, M., Geschwind, D.H., Ghetti, B., Gilbert, J.R., Glass, J.D., Growdon, J.H., Hamilton, R.L., Harrell, L.E., Head, E., Honig, L.S., Hulette, C.M., Hyman, B.T., Jicha, G.A., Jin, L.W., Karydas, A., Kaye, J.A., Kim, R., Koo, E.H., Kowall, N.W., Kramer, J.H., LaFerla, F.M., Lah, J.J., Leverenz, J.B., Levey, A.I., Li, G., Lieberman, A.P., Lin, C.F., Lopez, O.L., Lyketsos, C.G., Mack, W.J., Martiniuk, F., Mash, D.C., Masliah, E., McCormick, W.C., McCurry, S.M., McDavid, A.N., McKee, A.C., Mesulam, M., Miller, B.L., Miller, C.A., Miller, J.W., Murrell, J.R., Olichney, J.M., Pankratz, V.S., Parisi, J.E., Paulson, H.L., Peskind, E., Petersen, R.C., Pierce, A., Poon, W.W., Potter, H., Quinn, J.F., Raj, A., Raskind, M., Reisberg, B., Ringman, J.M., Roberson, E.D., Rosen, H.J., Rosenberg, R.N., Sano, M., Schneider, L.S., Seeley, W.W., Smith, A.G., Sonnen, J.A., Spina, S., Stern, R.A., Tanzi, R.E., Thornton-Wells, T.A., Trojanowski, J.Q., Troncoso, J.C., Valladares, O., Van Deerlin, V.M., Van Eldik, L.J., Vardarajan, B.N., Vinters, H.V., Vonsattel, J.P., Weintraub, S., Welsh-Bohmer, K.A., Williamson, J., Wishnek, S., Woltjer, R.L., Wright, C.B., Younkin, S.G., Yu, C.E., Yu, L., Alzheimer Disease Genetics Consortium, 2014. Effects of multiple genetic loci on age at onset in late-onset Alzheimer disease: a genome-wide association study. JAMA Neurol. 71, 1394–1404.

Nelson, P.T., Wang, W.X., 2010. MiR-107 is reduced in Alzheimer's disease brain neocortex: validation study. J. Alzheimers Dis. 21, 75–79.

Nicot, A.S., Toussaint, A., Tosch, V., Kretz, C., Wallgren-Pettersson, C., Iwarsson, E., Kingston, H., Garnier, J.M., Biancalana, V., Oldfors, A., Mandel, J.L., Laporte, J., 2007. Mutations in amphiphysin 2 (BIN1) disrupt interaction with dynamin 2 and cause autosomal recessive centronuclear myopathy. Nat. Genet. 39, 1134–1139.

Nuutinen, T., Suuronen, T., Kauppinen, A., Salminen, A., 2009. Clusterin: a forgotten player in Alzheimer's disease. Brain Res. Rev. 61, 89–104.

O'Dwyer, L., Lamberton, F., Matura, S., Tanner, C., Scheibe, M., Miller, J., Rujescu, D., Prvulovic, D., Hampel, H., 2012. Reduced hippocampal volume in healthy young ApoE4 carriers: an MRI study. PLoS One 7, e48895.

Owshalimpur, D., Kelley, M.J., 1999. Genomic structure of the EPHA1 receptor tyrosine kinase gene. Mol. Cell Probes 13, 169–173.

Patel, N., Hoang, D., Miller, N., Ansaloni, S., Huang, Q., Rogers, J.T., Lee, J.C., Saunders, A.J., 2008. MicroRNAs can regulate human APP levels. Mol. Neurodegener. 3, 10.

Pericak-Vance, M.A., Bebout, J.L., Gaskell, Jr., P.C., Yamaoka, L.H., Hung, W.Y., Alberts, M.J., Walker, A.P., Bartlett, R.J., Haynes, C.A., Welsh, K.A., et al., 1991. Linkage studies in familial Alzheimer disease: evidence for chromosome 19 linkage. Am. J. Hum. Genet. 48, 1034–1050.

Plassman, B.L., Welsh-Bohmer, K.A., Bigler, E.D., Johnson, S.C., Anderson, C.V., Helms, M.J., Saunders, A.M., Breitner, J.C., 1997. Apolipoprotein E epsilon 4 allele and hippocampal volume in twins with normal cognition. Neurology 48, 985–989.

Potkin, S.G., Guffanti, G., Lakatos, A., Turner, J.A., Kruggel, F., Fallon, J.H., Saykin, A.J., Orro, A., Lupoli, S., Salvi, E., Weiner, M., Macciardi, F., Alzheimer's Disease Neuroimaging Initiative, 2009. Hippocampal atrophy as a quantitative trait in a genome-wide association study identifying novel susceptibility genes for Alzheimer's disease. PLoS One 4, e6501.

Rademakers, R., Cruts, M., Sleegers, K., Dermaut, B., Theuns, J., Aulchenko, Y., Weckx, S., De Pooter, T., Van den Broeck, M., Corsmit, E., De Rijk, P., Del-Favero, J., van Swieten, J., van Duijn, C.M., van Broeckhoven, C., 2005. Linkage and association studies identify a novel locus for Alzheimer disease at 7q36 in a Dutch population-based sample. Am. J. Hum. Genet. 77, 643–652.

Rademakers, R., Dermaut, B., Peeters, K., Cruts, M., Heutink, P., Goate, A., Van Broeckhoven, C., 2003. Tau (MAPT) mutation Arg406Trp presenting clinically with Alzheimer disease does not share a common founder in Western Europe. Hum. Mutat. 22, 409–411.

Reiman, E.M., Chen, K., Alexander, G.E., Caselli, R.J., Bandy, D., Osborne, D., Saunders, A.M., Hardy, J., 2005. Correlations between apolipoprotein E epsilon4 gene dose and brain-imaging measurements of regional hypometabolism. Proc. Natl. Acad. Sci. USA 102, 8299–8302.

Reiman, E.M., Uecker, A., Caselli, R.J., Lewis, S., Bandy, D., de Leon, M.J., De Santi, S., Convit, A., Osborne, D., Weaver, A., Thibodeau, S.N., 1998. Hippocampal volumes in cognitively normal persons at genetic risk for Alzheimer's disease. Ann. Neurol. 44, 288–291.

Reiman, E.M., Webster, J.A., Myers, A.J., Hardy, J., Dunckley, T., Zismann, V.L., Joshipura, K.D., Pearson, J.V., Hu-Lince, D., Huentelman, M.J., Craig, D.W., Coon, K.D., Liang, W.S., Herbert, R.H., Beach, T., Rohrer, K.C., Zhao, A.S., Leung, D., Bryden, L., Marlowe, L., Kaleem, M., Mastroeni, D., Grover, A., Heward, C.B., Ravid, R., Rogers, J., Hutton, M.L., Melquist, S., Petersen, R.C., Alexander, G.E., Caselli, R.J., Kukull, W., Papassotiropoulos, A., Stephan, D.A., 2007. GAB2 alleles modify Alzheimer's risk in APOE 14 carriers. Neuron 54, 713–720.

Reitz, C., Cheng, R., Rogaeva, E., Lee, J.H., Tokuhiro, S., Zou, F., Bettens, K., Sleegers, K., Tan, E.K., Kimura, R., Shibata, N., Arai, H., Kamboh, M.I., Prince, J.A., Maier, W., Riemenschneider, M., Owen, M., Harold, D., Hollingworth, P., Cellini, E., Sorbi, S., Nacmias, B., Takeda, M., Pericak-Vance, M.A., Haines, J.L., Younkin, S., Williams, J., van Broeckhoven, C., Farrer, L.A., St George-Hyslop, P.H., Mayeux, R., Genetic and Environmental Risk in Alzheimer Disease 1 Consortium, 2011. Meta-analysis of the association between variants in SORL1 and Alzheimer disease. Arch. Neurol. 68, 99–106.

Reitz, C., Jun, G., Naj, A., Rajbhandary, R., Vardarajan, B.N., Wang, L.S., Valladares, O., Lin, C.F., Larson, E.B., Alzheimer Disease Genetics Consortium, et al., 2013. Variants in the ATP-binding cassette transporter (ABCA7), apolipoprotein E ε4, and the risk of late-onset Alzheimer disease in African Americans. JAMA 309, 1483–1492.

Robakis, N.K., Ramakrishna, N., Wolfe, G., Wisniewski, H.M., 1987. Molecular cloning and characterization of a cDNA encoding the cerebrovascular and the neuritic plaque amyloid peptides. Proc. Natl. Acad. Sci. USA 84, 4190–4194.

Rogaev, E.I., Sherrington, R., Rogaeva, E.A., Levesque, G., Ikeda, M., Liang, Y., Chi, H., Lin, C., Holman, K., Tsuda, T., et al., 1995. Familial Alzheimer's disease in kindreds with missense mutations in a gene on chromosome 1 related to the Alzheimer's disease type 3 gene. Nature 376, 775–778.

Rogaeva, E., Meng, Y., Lee, J.H., Gu, Y., Kawarai, T., Zou, F., Katayama, T., Baldwin, C.T., Cheng, R., Hasegawa, H., Chen, F., Shibata, N., Lunetta, K.L., Pardossi-Piquard, R., Bohm, C., Wakutani, Y., Cupples, L.A., Cuenco, K.T., Green, R.C., Pinessi, L., Rainero, I., Sorbi, S., Bruni, A., Duara, R., Friedland, R.P., Inzelberg, R., Hampe, W., Bujo, H., Song, Y.Q., Andersen, O.M., Willnow, T.E., Graff-Radford, N., Petersen, R.C., Dickson, D., Der, S.D., Fraser, P.E., Schmitt-Ulms, G., Younkin, S., Mayeux, R., Farrer, L.A., St George-Hyslop, P., 2007. The neuronal sortilin-related receptor SORL1 is genetically associated with Alzheimer disease. Nat. Genet. 39, 168–177.

Roses, A.D., 1996. Apolipoprotein E and Alzheimer's disease. A rapidly expanding field with medical and epidemiological consequences. Ann. NY Acad. Sci. 802, 50–57.

Roses, A.D., Lutz, M.W., Amrine-Madsen, H., Saunders, A.M., Crenshaw, D.G., Sundseth, S.S., Huentelman, M.J., Welsh-Bohmer, K.A., Reiman, E.M., 2010. A TOMM40 variable-length polymorphism predicts the age of late-onset Alzheimer's disease. Pharmacogenomics J. 10, 375–384.

Roses, A.D., Lutz, M.W., Crenshaw, D.G., Grossman, I., Saunders, A.M., Gottschalk, W.K., 2013. TOMM40 and APOE: Requirements for replication studies of association with age of disease onset and enrichment of a clinical trial. Alzheimers Dement. 9, 132–136.

Royer-Pokora, B., Kunkel, L.M., Monaco, A.P., Goff, S.C., Newburger, P.E., Baehner, R.L., Cole, F.S., Curnutte, J.T., Orkin, S.H., 1986. Cloning the gene for an inherited human disorder-chronic granulomatous disease-on the basis of its chromosomal location. Nature 322, 32–38.

Rye, P.D., Booij, B.B., Grave, G., Lindahl, T., Kristiansen, L., Andersen, H.M., Horndalsveen, P.O., Nygaard, H.A., Naik, M., Hoprekstad, D., Wetterberg, P., Nilsson, C., Aarsland, D., Sharma, P., Lönneborg, A., 2011. A novel blood test for the early detection of Alzheimer's disease. J. Alzheimers Dis. 23, 121–129.

Saunders, A.M., Strittmatter, W.J., Schmechel, D., George-Hyslop, P.H., Pericak-Vance, M.A., Joo, S.H., Rosi, B.L., Gusella, J.F., Crapper-MacLachlan, D.R., Alberts, M.J., et al., 1993. Association of apolipoprotein E allele epsilon4 with late-onset familial and sporadic Alzheimer's disease. Neurology 43, 1467–1472.

Schuff, N., Woerner, N., Boreta, L., Kornfield, T., Shaw, L.M., Trojanowski, J.Q., Thompson, P.M., Jack, Jr., C.R., Weiner, M.W., 2009. MRI of hippocampal volume loss in early Alzheimer's disease in relation to ApoE genotype and biomarkers. Brain 132, 1067–1077.

Sen, A., Alkon, D.L., Nelson, T.J., 2012. Apolipoprotein E3 (ApoE3) but not ApoE4 protects against synaptic loss through increased expression of protein kinase C epsilon. J. Biol. Chem. 287, 15947–15958.

Seshadri, S., Fitzpatrick, A.L., Ikram, M.A., DeStefano, A.L., Gudnason, V., Boada, M., Bis, J.C., Smith, A.V., Consortium; EADI1 Consortium, et al., 2010. Genome-wide analysis of genetic loci associated with Alzheimer disease. JAMA 303, 1832–1840.

Shen, L., Kim, S., Risacher, S.L., Nho, K., Swaminathan, S., West, J.D., Foroud, T., Pankratz, N., Moore, J.H., Sloan, C.D., Huentelman, M.J., Craig, D.W., Dechairo, B.M., Potkin, S.G., Jack, Jr., C.R., Weiner, M.W., Saykin, A.J., Alzheimer's Disease Neuroimaging Initiative, 2010. Whole genome association study of brain-wide imaging phenotypes for identifying quantitative trait loci in MCI and AD: a study of the ADNI cohort. Neuroimage 53, 1051–1063.

Sherrington, R., Rogaev, E.I., Liang, Y., Rogaeva, E.A., Levesque, G., Ikeda, M., Chi, H., Lin, C., Li, G., et al., 1995. Cloning of a gene bearing missense mutations in early-onset familial Alzheimer's disease. Nature 375, 754–760.

Shi, H., Belbin, O., Medway, C., Brown, K., Kalsheker, N., Carrasquillo, M., Proitsi, P., Powell, J., Lovestone, S., Goate, A., Younkin, S., Passmore, P., 2012. Genetic variants influencing human aging from late-onset Alzheimer's disease (LOAD) genome-wide association studies (GWAS). Neurobiol. Aging Aug. 33 (1849), e5–18.

Soininen, H., Partanen, K., Pitkanen, A., Hallikainen, M., Hanninen, T., Helisalmi, S., Mannermaa, A., Ryynanen, M., Koivisto, K., Riekkinen@@Sr., P., 1995. Decreased hippocampal volume asymmetry on MRIs in nondemented elderly subjects carrying the apolipoprotein E epsilon 4 allele. Neurology 45, 391–392.

Strittmatter, W.J., Saunders, A.M., Schmechel, D., Pericak-Vance, M., Enghild, J., Salvesen, G.S., Roses, A.D., 1993. Apolipoprotein E: high-avidity binding to beta-amyloid and increased frequency of type 4 allele in late-onset familial Alzheimer disease. Proc. Natl. Acad. Sci. USA 90, 1977–1981.

Sullivan, P.F., Kendler, K.S., Neale, M.C., 2003. Schizophrenia as a complex trait: evidence from a meta-analysis of twin studies. Arch. Gen. Psychiatry 60, 1187–1192.

Tan, M.S., Yu, J.T., Jiang, T., Zhu, X.C., Guan, H.S., Tan, L., 2014a. Genetic variation in BIN1 gene and Alzheimer's disease risk in Han Chinese individuals. Neurobiol. Aging 35 (1781), e1–8.

Tan, L., Yu, J.T., Liu, Q.Y., Tan, M.S., Zhang, W., Hu, N., Wang, Y.L., Sun, L., Jiang, T., Tan, L., 2014b. Circulating miR-125b as a biomarker of Alzheimer's disease. J. Neurol. Sci. 336, 52–56.

Tan, M.S., Yu, J.T., Tan, L., 2013a. Bridging integrator 1 (BIN1): form, function, and Alzheimer's disease. Trends Mol. Med. 19, 594–603.

Tan, L., Yu, J.T., Zhang, W., Wu, Z.C., Zhang, Q., Liu, Q.Y., Wang, W., Wang, H.F., Ma, X.Y., Cui, W.Z., 2013b. Association of GWAS-linked loci with late-onset Alzheimer's disease in a northern Han Chinese population. Alzheimers Dement. 9, 546–553.

Tanzi, R.E., Gusella, J.F., Watkins, P.C., Bruns, G.A., St George-Hyslop, P., Van Keuren, M.L., Patterson, D., Pagan, S., Kurnit, D.M., Neve, R.L., 1987. Amyloid beta protein gene: cDNA, mRNA distribution, and genetic linkage near the Alzheimer locus. Science 235, 880–884.

Tanzi, R.E., Kovacs, D.M., Kim, T.W., Moir, R.D., Guenette, S.Y., Wasco, W., 1996. The gene defects responsible for familial Alzheimer's disease. Neurobiol. Dis. 3, 159–168.

Tanzi, R.E., Vaula, G., Romano, D.M., Mortilla, M., Huang, T.L., Tupler, R.G., Wasco, W., Hyman, B.T., Haines, J.L., Jenkins, B.J., et al., 1992. Assessment of amyloid beta-protein precursor gene mutations in a large set of familial and sporadic Alzheimer disease cases. Am. J. Hum. Genet. 51, 273–282.

Thompson, P.M., Stein, J.L., Medland, S.E., Hibar, D.P., Vasquez, A.A., Renteria, M.E., Toro, R., Jahanshad, N., Alzheimer's Disease Neuroimaging Initiative, EPIGEN Consortium, IMAGEN Consortium, Saguenay Youth Study (SYS) Group, et al., 2014. The ENIGMA Consortium: large-scale collaborative analyses of neuroimaging and genetic data. Brain Imaging Behav. 8, 153–182.

Wang, W.X., Rajeev, B.W., Stromberg, A.J., Ren, N., Tang, G., Huang, Q., Rigoutsos, I., Nelson, P.T., 2008. The expression of microRNA miR-107 decreases early in Alzheimer's disease and may accelerate disease progression through regulation of beta-site amyloid precursor protein-cleaving enzyme 1. J. Neurosci. 28, 1213–1223.

Warner, T.T., Schapira, A.H., 2003. Genetic and environmental factors in the cause of Parkinson's disease. Ann. Neurol. 53, S16–S23.

Wilson, R.S., Barral, S., Lee, J.H., Leurgans, S.E., Foroud, T.M., Sweet, R.A., Graff-Radford, N., Bird, T.D., Mayeux, R., Bennett, D.A., 2011. Heritability of different forms of memory in the Late Onset Alzheimer's Disease Family Study. J. Alzheimers Dis. 23, 249–255.

Wong, P., Taillefer, D., Lakins, J., Pineault, J., Chader, G., Tenniswood, M., 1994. Molecular characterization of human TRPM-2/clusterin, a gene associated with sperm maturation, apoptosis and neurodegeneration. Eur. J. Biochem. 221, 917–925.

Xu, W., Xu, J., Wang, Y., Tang, H., Deng, Y., Ren, R., Wang, G., Niu, W., Ma, J., Wu, Y., Zheng, J., Chen, S., Ding, J., 2013. The genetic variation of SORCS1 is associated with late-onset Alzheimer's disease in Chinese Han population. PLoS One 8, e63621.

Yu, J.T., Tan, L., 2012. The role of clusterin in Alzheimer's disease: pathways, pathogenesis, and therapy. Mol. Neurobiol. 45, 314–326.

Zhou, R., 1998. The Eph family receptors and ligands. Pharmacol. Ther. 77, 151–181.

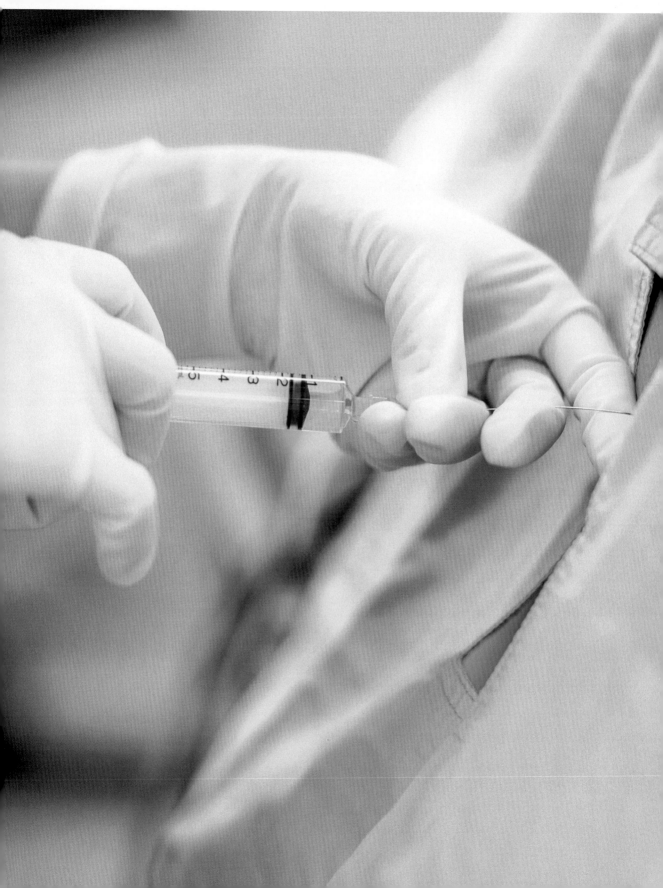

CHAPTER 5

Alzheimer's Disease Cerebrospinal Fluid (CSF) Biomarkers

Chapter Outline

Biomarkers in Alzheimer's Disease. http://dx.doi.org/10.1016/B978-0-12-804832-0.00005-5
Copyright © 2016 Elsevier Inc. All rights reserved.

5.1 PATHOPHYSIOLOGY OF Aβ PRODUCTION IN THE ALZHEIMER'S DISEASE BRAIN AND DETECTION IN THE CSF

The major hypothesis regarding the pathogenesis of AD involves an increase in Aβ production (by an amyloidogenic pathway) and accumulation (due to inefficient degradation and clearance of toxic oligomeric Aβ from the brain), leading to the deposition of amyloid plaques and neurodegeneration that ultimately disrupts cognitive function. Aβ plaques are aggregates of Aβ peptides (4 kDa, mostly $A\beta_{1-40}$ and $A\beta_{1-42}$) that are formed upon enzymatic cleavage of Aβ by amyloid precursor protein (APP; Fig. 5.1). APP is subjected to posttranslational processing by three major enzymes (α-, β-, and γ-secretase). Normally, APP is cleaved by α-secretase (a member of the ADAM family of proteases), which releases nontoxic, neuro-protective, soluble (s)APPα into the extracellular fluid and cerebrospinal fluid (CSF) (Lammich et al., 1999; Fig. 5.1A). In the abnormal amyloidogenic pathway, APP is cleaved by β-secretase (β-site APP-cleaving enzyme 1, BACE-1; Vassar et al., 1999), which releases sAPPβ into the extracellular fluid and eventually into the CSF (Blennow et al., 2010; Portelius et al., 2011; Beher et al., 2002; Fig. 5.1B). The γ-secretase complex, consisting of four components: presenilin, presenilin enhancer 2 (PEN2), nicastrin (Nct), and anterior pharynx-defective (Aph; Selkoe and Wolfe, 2007), acts on the remaining extracellular carboxy-terminated fragment (CTFβ) in the plasma membrane and generates $A\beta_{1-42}$ and several carboxy-terminal truncated Aβ-peptides ($A\beta_{1-40,}$ $A\beta_{1-17,}$ and others; Blennow et al., 2010; Fig. 5.1C). In an alternative pathway, cleavage by β-secretase is followed by α-secretase to produce several short forms of Aβ-peptides ($A\beta_{1-16}$– $A\beta_{1-13}$; Blennow et al., 2010). Taken together, activation of α-secretase or inhibition of β- and/or γ-secretase decreases Aβ production in vitro and in vivo (Fig. 5.1B). For this reason, targeting the secretase enzymes is a primary area of research for AD therapeutic intervention.

Pathologically elevated oligomeric Aβ has been found to be neurotoxic and correlates with cognitive dysfunction (Hongpaisan et al., 2011), eliciting abnormal patterns of activity in neuronal network circuits in transgenic animal models of AD (Mucke and Selkoe, 2012). Individuals with early-onset, or familial, AD have an overproduction of Aβ, whereas those with late-onset, or sporadic, AD show a dysregulation of Aβ clearance from AD brain (Selkoe, 2001; Hardy and Selkoe, 2002). Among Aβ peptides, $A\beta_{1-42}$ is the most hydrophobic in nature, aggregating into toxic Aβ oligomers and fibrils that accumulate as amyloid plaques in the brain. As $A\beta_{1-42}$ aggregates into amyloid plaques, the level of circulating $A\beta_{1-42}$ decreases, resulting in lower levels of $A\beta_{1-42}$ detected in the CSF of AD patients. Several studies have shown that accumulation of Aβ occurs early in AD progression, and reduced levels of CSF Aβ have been measured before the onset of AD symptoms (Oddo et al., 2003; Braak et al., 2013). The differences in CSF $A\beta_{1-42}$ levels between AD and healthy control cases are within 50%, making CSF $A\beta_{1-42}$ alone one of the less accurate AD biomarkers. While most studies have found lower levels

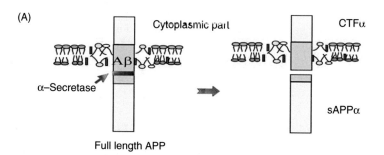

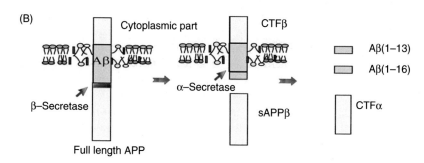

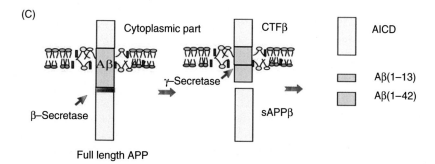

FIGURE 5.1

Generation of Aβ peptides via amyloid precursor protein *(APP)* metabolism. (A) In the normal nonamyloidogenic pathway, APP is cleaved in the middle of the Aβ sequence by α-secretase, resulting nontoxic soluble sAPPα in the extracellular fluid and cerebrospinal fluid (CSF), and a carboxy terminal fragment *(CTFα)* in the cytoplasm. (B) In an alternate pathway, cleavage of full-length APP by β-secretase (BACE1) followed by α-secretase activity results in several short Aβ peptides. (C) In the amyloidogenic pathway, APP is first cleaved by β-secretase or BACE1 producing sAPPβ that is released into extracellular fluid and CSF. In the second step, the γ-secretase complex (presenilin, nicastrin, PEN2, and APH1) acts on the remaining fragment in the plasma membrane to generate $A\beta_{1-40}$, $A\beta_{1-42}$, and several other Aβ peptides.

of CSF $A\beta_{1-42}$ in AD cases compared to healthy control cases (Sunderland et al., 2003), two studies reported different results (Fukuyama et al., 2000; Csernansky et al., 2002). The diagnostic sensitivity (80%) and specificity (82%) of CSF $A\beta_{1-42}$ for AD are lower than those of CSF total tau (t-tau), as discussed later (Bloudek et al., 2011).

5.2 PATHOPHYSIOLOGY OF ELEVATED TAU AND PHOSPHORYLATED-TAU IN THE ALZHEIMER'S DISEASE BRAIN AND CSF

tau is an axonal protein that binds microtubules to promote assembly and stability. As part of its normal function in microtubule organization, the tau protein becomes phosphorylated at multiple serine and threonine binding sites. Neurofibrillary tangles are formed as a result of abnormal hyperphosphorylation of tau protein, which disrupts microtubule organization. Compared to $A\beta$ pathophysiology, tau-related pathology occurs later in AD pathoprogression (Oddo et al., 2003), with elevated t-tau and p-tau detected at the onset and during progression of AD (Braak et al., 2013). Iqbal and Grundke-Iqbal (2008), the discoverers of hyperphosphorylated tau in AD brain, proposed that the possible cause of tau hyperphophoryaltion is the imbalance of regulation of kinases and phosphatases in AD brain.

tau is released upon neuronal death and as such is thought to be general biomarker of neurodegeneration. Elevated levels of tau have been reported in the brains of individuals with various types of neurodegenerative disease, with the following order of increasing tau: healthy control < Lewy body disease (LBD), frontotemporal dementia (FTD) and vascular dementia (VaD) < taupathy and AD < Creutzfeldt–Jakob disease (CJD). A meta-analysis by Sunderland et al. (2003) showed significantly higher CSF tau levels in AD cases compared to healthy control cases, and the difference in CSF tau levels between AD and healthy control cases is much higher compared to CSF $A\beta_{1-42}$ levels. The diagnostic sensitivity (82%) and specificity (90%) of CSF t-tau for AD are reasonably good (Bloudek et al., 2011).

Consistent with tau pathophysiology, higher levels of phosphorylated tau (p-tau) in the CSF have been associated with formation of neurofibrillary tangles in AD brain. There are approximately 80 phosphorylation sites on the tau protein (Mitchell, 2009; Blennow and Humpel, 2003), but the most commonly used CSF p-tau biomarker assay detects phosphorylation at the threonine 181 (p-tau-181) residue. Phosphorylation at the threonine 231 residue (p-tau-231) has also been investigated extensively as a potential AD biomarker. Assays for both residues show similar levels of sensitivies and specificities for detecting AD. Diagnostic sensitivity (78%) and specificity (83%) of CSF p-tau (p-tau-181) for AD and healthy control are lower compared to those of CSF t-tau (Bloudek et al., 2011).

5.3 RATIONALE FOR USING CSF BIOMARKERS OF NEURODEGENERATIVE DISEASES

During the last two decades, several groups have reported that CSF from patients with AD has lower $A\beta_{1-42}$ (~50% reduction) and higher t-tau and p-tau-181 (~300% elevation) compared with age-matched control cases (Motter et al., 1995; Zetterberg et al., 2003; Wahlund and Blennow, 2003; Sunderland et al., 2003; Blennow and Humpel, 2003; Georganopoulou et al., 2005; Blennow et al., 2010; Table 5.1). CSF is a sensitive, but nonspecific indicator of central nervous system (CNS) pathology. The scientific rationale for using CSF biomarkers to diagnose AD is based on (1) the direct physical contact between the CSF and interstitial brain fluid and (2) changes in CSF composition that reflect biochemical and pathological changes in the brain associated with AD. Blood-based assays of $A\beta$ also have greater variability compared to CSF-based $A\beta$ assays (Rissman et al., 2012), in part because the source of $A\beta_{1-42}$, t-tau, and p-tau in CSF is different than in blood. $A\beta_{1-42}$ detected in the CSF comes from abnormal amyloidogenic cleavage of APP and inefficient clearance of toxic $A\beta$ peptides in the CNS. In contrast, $A\beta_{1-42}$ detected in blood comes from a wide array of sources including platelets, erythrocytes, vascular wall, skin cells, liver, kidney, skeletal muscles, intestine, and several glands. CSF thus represents a single, uniform tissue source for assaying CNS biomarkers. Moreover, blood plasma contains a greater number of $A\beta$-interacting factors that can affect quantification, such as serum amyloid P component, complement factors, immunoglobulins, α-2 macroglobulin, apolipoprotein A-I, A-IV, E and J, transthyretin, and apoferritin.

Furthermore, changes in CSF biomarkers are consistent with what is known about AD pathophysiology and CSF biomarkers have been shown to predict the conversion of MCI to AD. While synaptic loss may have started prior to deposition of plaques and tangles in MCI and early-stage AD (Scheff et al., 2006), older individuals may have evidence of molecular pathophysiology of AD prior

Table 5.1	CSF Core Biomarkers in Alzheimer's Disease	
Method	**CNS Biomarkers**	**Criteria for AD**
CSF collection by lumbar puncture and analyzed by ELISA, multianalyte Luminex xMAP assay, electrochemoluminescence	$A\beta_{1-42}$[a] Total tau[a] p-tau-181[a]	Low CSF $A\beta_{1-42}$ High CSF total tau High CSF p-tau-181

$A\beta$, beta-amyloid; AD, Alzheimer's disease; CNS, central nervous system; CSF, cerebrospinal fluid; ELISA, enzyme-linked immunosorbent assay; p-tau-181, phosphorylated tau at threonine 181.

[a]These CSF biomarkers are included in National Institute on Aging-Alzheimer's Association (NIA-AA) 2011 criteria for Alzheimer's disease diagnosis for research purposes. The International Working Group 2 (IGW-2) in 2014 recognized the use of these three CSF biomarkers as an integral part of Alzheimer's disease diagnosis.

to experiencing cognitive impairment. In one study, older non-AD (cognitively normal) cases had lower $A\beta_{1-42}$ and higher p-tau-181 levels compared with younger non-AD patients (Bouwman et al., 2009). This supports the utility of CSF biomarkers in early diagnosis of AD or assessment of AD risk, before clinical symptoms manifest. It is important to note that several studies have reported inconsistent differences in CSF biomarkers in patients with familial AD and sporadic AD (Green et al., 1999; Brunnström et al., 2010; Mitchell et al., 2010; Williams et al., 2011).

5.4 CSF BIOMARKER MEASUREMENT TECHNOLOGIES AND STANDARDIZATION

Poorly developed and validated analytical methods lead to reduced sensitivity and specificity of the biomarker and increase the number of false-positive and false-negative results. In recent years, there has been an explosive increase in the discovery, validation, and application of CSF AD biomarkers for disease diagnosis, prognosis, and assessment of therapeutic efficacy in clinical trials (Humpel 2011; Humpel et al., 2012; Rosén et al., 2013; Kang et al., 2013; Blennow et al., 2014; Ferreira et al., 2014; Lleó et al., 2015; Khan and Alkon, 2015b). Once a CSF AD biomarker is identified that reflects AD pathology in preclinical research, development and validation of more advanced analytical methods are needed to ensure the high sensitivity, specificity, and accuracy of the biomarker in the clinical setting in a high-throughput manner. New ELISA-based technologies have been introduced to reduce intra- and interassay variability (Fujirebio Europe, Belgium; previously Innogenetics, Ghent, Belgium). Two new ELISA platforms, INNO-BIA, and INNOTEST, have been used in AD autopsy cohorts (Table 5.2). Luminex xMAP (Innogenetics) is a semiautomated assay platform with reduced intrasample variance that has been used as an alternative to ELISA-based assays (Wang et al., 2012). xMAP technology is more appropriate in CLIA-approved clinical laboratories. In the US, Meso Scale Discovery (Rockville, MD) introduced an electrochemiluminescence-based single-analyte and multiplex assay platform to detect $A\beta_{1-42}$ and tau. The technology is based on antibody binding reactions that can be detected by electrodes. Due to differences between laboratories and a wide variety of protocols and technologies available to detect CSF biomarkers, identification of a universal cut-off value for each of the three core CSF biomarkers for the diagnosis of AD or to distinguish between AD and non-AD dementias has been difficult (Mattsson et al., 2013). Cut-off values for CSF AD biomarkers identified by several important studies are presented in Table 5.2. According to the Alzheimer's Disease Neuroimaging Initiative (ADNI), the cut-off values of CSF biomarkers for a diagnosis of AD are: $A\beta_{1-42}$ <192 pg/mL; total tau >93 pg/mL; and p-tau-181 >23 pg/mL; with threshold values of tau/$A\beta_{1-42}$ = 0.39 and p-tau-181/$A\beta_{1-42}$ = 0.1 (Shaw et al., 2009). Manufacturer-defined cut-off values for CSF AD biomarkers for the INNOTEST ELISA kits (Innogenetics) are higher than those for the Luminex xMAP assay (INNO-BIA AlzBio3, Innogenetics; Struyfs et al., 2014). Cut-off values for CSF AD biomarkers reported in the Oxford Project to Investigate Memory and Ageing (OPTIMA)

Table 5.2 Cut-Off Values of Core CSF Biomarkers to Distinguish Alzheimer's Disease and Other Non-Alzheimer's Disease Cases

Study Design and Method	Cut-Off Values (pg/mL)			References
	$A\beta_{1-42}$	t-tau	p-tau-181	
Validation study: ELISA-kits (INNOTEST, Innogenetics) Sample: AD (n = 51), autopsy confirmed; age-matched control (n = 95)[a]; Biobank of the Institute Born-Bunge (Antwerp, Belgium)	539.44	422.21	50.50	Struyfs et al. (2014)
Validation study: Multianalyte Luminex xMAP assay (INNO-BIA AlzBio3, Innogenetics) Sample: AD (n = 51), autopsy confirmed; age-matched control (n = 95)[a]; Biobank of the Institute Born-Bunge (Antwerp, Belgium)	159.05	83.60	46.833	Struyfs et al. (2014)
ELISA-kits (INNOTEST, Innogenetics) Sample: AD (n = 51), autopsy confirmed; non-AD dementia (n = 15), autopsy confirmed; age-matched control (n = 95)[a]; Biobank of the Institute Born-Bunge (Antwerp, Belgium)	539	422	50	Le Bastard et al. (2013)
Multianalyte Luminex xMAP assay (INNO-BIA AlzBio3, Innogenetics) Sample: AD (n = 51), autopsy confirmed; non-AD dementia (n = 15), autopsy confirmed; age-matched control (n = 95)[a]; Biobank of the Institute Born-Bunge (Antwerp, Belgium)	159	84	47	Le Bastard et al. (2013)[a]
Multicenter ADNI Multiplex xMAP Luminex platform (Luminex Corp, Austin, TX) Sample: AD (n = 56), autopsy confirmed; age-matched control (n = 52)[a]	192	93	23	Shaw et al. (2009)
OPTIMA Cohort INNOTEST Sample: AD (n = 91), autopsy confirmed; age-matched control (n = 74)[a]	503	330	65	Seeburger et al. (2015)

AD, Alzheimer's disease; ADNI, Alzheimer's Disease Neuroimaging Initiative; OPTIMA cohort, Oxford Project to Investigate Memory and Ageing cohort; p-tau-181, tau phosphorylated on threonine 181; t-tau, total tau.
[a]Age-matched control cases were clinically confirmed (not autopsy validated).

cohort were slightly lower than those for the INNOTEST ELISA kits (Innogenetics; Seeburger et al., 2015). To reduce interlaboratory variability, three options have been considered: (a) using individual laboratory standard reference values (Ferreira et al., 2014), (b) introduction of a numerical normalization to account for variability (Hansson et al., 2006), and (c) presenting data in terms of normalized indexing values (AD-CSF index; Molinuevo et al., 2013).

5.5 DIAGNOSTIC ACCURACY OF CSF BIOMARKERS IN AUTOPSY- VERSUS CLINICALLY VALIDATED ALZHEIMER'S DISEASE COHORTS

The diagnostic performance of core CSF biomarkers is satisfactory for discriminating between autopsy-confirmed AD cases and healthy controls (Table 5.3). Using state-of-the-art technologies for detecting CSF biomarkers, similar moderate-level sensitivities, specificities, and accuracy were obtained for CSF biomarkers across several autopsy registry studies that included AD and nondemented control cases (Le Bastard et al., 2013; Struyfs et al., 2014; note that non-AD dementia patients were not included in most of these studies; Table 5.3). Some studies suggest that a combination of all three CSF AD biomarkers may be able to distinguish between AD and MCI. A combination of CSF biomarkers and multimodal neuroimaging techniques may achieve higher sensitivity and specificity (Lista et al., 2014); however, such combination testing approaches are more expensive, time consuming, and may not be suitable in the clinical settings. It should be noted, however, that most CSF biomarker studies included cohorts validated using clinical confirmation of an AD diagnosis. Clinical diagnoses show high accuracy for diagnosing autopsy-confirmed AD in patients more than 4 years of the onset of dementia symptoms, but are not as accurate within the first few years of the onset of symptoms of dementia (Khan and Alkon, 2010; Hogervorst et al., 2003). The suboptimal accuracy of CSF markers in these studies may be due to the use of clinically validated rather than autopsy-validated AD cohorts, as 25–30% of cognitively normal older individuals are likely to have some degree of AD pathology despite no clinical signs of the disease.

A recent review of peripheral biomarker studies showed similar levels of accuracy with respect to autopsy confirmation (Khan and Alkon, 2015a). However, another study found that CSF levels of $A\beta_{1-42}$, t-tau, and p-tau-181 were not associated with APOE4 (widely regarded as one of the main risk factors of sporadic AD), tangle, or plaque burden in 50 autopsy-confirmed AD patients (Le Bastard et al., 2010).

5.6 DISCRIMINATION BETWEEN ALZHEIMER'S DISEASE AND NON-ALZHEIMER'S DISEASE DEMENTIAS BY CSF BIOMARKERS

Though the diagnostic performance of CSF biomarkers is generally satisfactory for discriminating between AD patients versus healthy controls (Table 5.3), and the combination of the CSF biomarkers, particularly with neuroimaging

Table 5.3 Sensitivity, Specificity, and Accuracy of Alzheimer's Disease Core CSF Biomarkers to Detect Autopsy-Validated Alzheimer's Disease

Study Design and Method	Results			References
	$A\beta_{1-42}$ (%)	t-tau (%)	p-tau-181 (%)	
Validation study: ELISA-kits (INNOTEST, Innogenetics) AD (n = 51), autopsy confirmed Sample: age-matched control (n = 95)[b]; Biobank of the Institute Born-Bunge (Antwerp, Belgium)	SN = 94 SP = 88 ACU = 90	SN = 69 SP = 94 ACU = 90	SN = 75 SP = 79 ACU = 77	Struyfs et al. (2014)
Validation study: Multianalyte Luminex xMAP assay (INNO-BIA AlzBio3, Innogenetics) Sample: AD (n = 51), autopsy confirmed; age-matched control (n = 95)[b]; Biobank of the Institute Born-Bunge (Antwerp, Belgium)	SN = 88 SP = 92 ACU = 90	SN = 82 SP = 87 ACU = 86	SN = 69 SP = 91 ACU = 83	Struyfs et al. (2014)
ELISA-kits (INNOTEST, Innogenetics) Sample: AD (n = 51), autopsy confirmed; non-AD dementia (n = 15), autopsy confirmed; age-matched control (n = 95)[b]; Biobank of the Institute Born-Bunge (Antwerp, Belgium)	SN = 94 SP = 88 ACU = 90	SN = 69 SP = 94 ACU = 85	SN = 77 SP = 78 ACU = 77	Le Bastard et al. (2013)[a]
Multianalyte Luminex xMAP assay (INNO-BIA AlzBio3, Innogenetics) Sample: AD (n = 51), autopsy confirmed; non-AD dementia (n = 15), autopsy confirmed; age-matched control (n = 95)[b]; Biobank of the Institute Born-Bunge (Antwerp, Belgium)	SN = 88 SP = 92 ACU = 90	SN = 82 SP = 87 ACU = 86	SN = 69 SP = 91 ACU = 83	Le Bastard et al. (2013)[a]

(Continued)

Table 5.3 Sensitivity, Specificity, and Accuracy of Alzheimer's Disease Core CSF Biomarkers to Detect Autopsy-Validated Alzheimer's Disease (cont.)

Study Design and Method	Results			References
	$A\beta_{1-42}$ (%)	t-tau (%)	p-tau-181 (%)	
Multicenter ADNI Multiplex xMAP Luminex platform (Luminex Corp, Austin, TX) Sample: AD (n = 56), autopsy confirmed; age-matched control (n = 52)[b]	SN = 96.4, SP = 76.9, ACU = 87	SN = 69.6, SP = 92.3, ACU = 80.6	SN = 67.9, SP = 73.1, ACU = 70.3	Shaw et al. (2009)
OPTIMA cohort INNOTEST Sample: AD (n = 91), autopsy confirmed; age-matched control (n = 74)[b]	SN = 807.6, SP = 88.1, ACU = 94.1	SN = 86.1, SP = 81.6, ACU = 92	SN = 65.7, SP = 93.2, ACU = 86.9	Seeburger et al. (2015)

AD, Alzheimer's disease; ACU, accuracy; ADNI, Alzheimer's Disease Neuroimaging Initiative; OPTIMA cohort, Oxford Project to Investigate Memory and Ageing cohort; p-tau-181, tau phosphorylated on threonine 181; SN, sensitivity; SP, specificity; t-tau, total tau.
[a]Non-AD dementia patients were included in these studies.
[b]Age-matched control cases were clinically confirmed (not autopsy validated).

biomarkers, may be able to predict MCI progression to AD, CSF biomarkers alone or in combination generally fail to distinguish AD from other non-AD dementias with high sensitivity and specificity (Table 5.4). Most published studies have reported that CSF concentrations of $A\beta_{1-42}$ are not significantly different in AD and non-AD dementias (VaD, FTP, and LBD). Moreover, no significant differences in CSF $A\beta_{1-42}$ levels were found in CJD versus AD (Otto et al., 2000; Shoji et al., 2002) or in amyotrophic lateral sclerosis (ALS) versus AD (Shoji et al., 2002; Sjögrena et al., 2002; Table 5.4). However, progressive supra nuclear palsy patients have elevated $A\beta_{1-42}$ compared to age-matched nondemented controls and Parkinson's disease (PD) cases (Süssmuth et al., 2010). It is well known that $A\beta_{1-42}$ is lowered in AD patients' CSF compared to age-matched nondemented controls. Therefore, progressive supranuclear palsy patients can easily be distinguished from AD cases by examining CSF $A\beta_{1-42}$ levels. Tau levels in LBD, FTD, and VaD are increased compared to healthy control cases, but lower than those seen in AD. Tau levels in CJD are significantly higher than in AD. There were no significant differences in CSF levels of $A\beta_{1-42}$ and t-tau concentrations in patients with AD and ALS or amyloid angiopathy (AA; Sjögrena et al., 2002; Shoji et al., 2002), and most studies have found that CSF concentrations of t-tau are not significantly different in AD and non-AD dementia cases, and reported low sensitivity (Table 5.4). A systematic meta-analysis of CSF $A\beta_{1-42}$ alone showed a sensitivity of 73% and a specificity of 67% for detection of AD versus non-AD dementias, and CSF tau alone showed a sensitivity of 78% and a specificity of 75% (Bloudek et al., 2011). Among the three core CSF biomarkers, p-tau appears to be the best able to discriminate between AD and non-AD dementias, including LBD, FTD, and VaD, but not CJD (Le Bastard et al., 2010, 2013). A large cohort study by the French Alzheimer's Plan showed that p-tau outperformed other CSF biomarkers in distinguishing between AD and non-AD dementia patients (Gabelle et al., 2013). That study included 272 patients with AD and 370 patients with a wide variety of non-AD dementia disorders ($n = 9$ AA, $n = 60$ FTD, $n = 26$ LBD, $n = 22$ with psychiatric disorders, $n = 181$ with other nondegenerative diseases, and $n = 72$ with other degenerative diseases). The discrepancy in accuracy between t-tau and p-tau could be due to the fact that p-tau is closely related to formation of neurofibrillary tangles whereas tau is related to neuronal damage in general. Though measurement of p-tau phosphorylated at threonine 181 (p-tau-181) may be useful for the differential diagnosis of AD and non-AD dementia, the existing data are inconsistent (Humpel et al., 2004; Le Bastard et al., 2007, 2010; Table 5.4). For example, Bloudek et al. (2011) found slightly higher sensitivity (79%) and specificity (80%) for p-tau-181 for distinguishing AD from non-AD dementia cases, whereas Mitchell (2009) found lower sensitivity (72%) and specificity (78%). A combination of the three CSF biomarkers improved sensitivity (86%) but specificity remained quite low (67%; Bloudek et al., 2011).

5.6.1 AD Versus VaD

Several studies have shown that CSF tau concentrations in VaD patients are at intermediate levels, between those of control groups and AD patient groups (Sjögren

Table 5.4 CSF Biomarkers for Differential Diagnosis of Alzheimer's Disease and Non- Alzheimer's Disease Dementia

Study Design AD Versus Non-AD Dementia	CSF Biomarker Measured Values (Average ± Standard Deviation)	Biomarker Performance (As Reported)	Reference
AD versus VaD AD (n = 64); VaD (n = 21) 76 of 85 autopsy confirmed Sample: Biobank of the Institute Born-Bunge (Antwerp, Belgium)	$A\beta_{1-42}$ (pg/mL) 318 ± 135 (AD) 492 ± 273 (VaD) p-tau-181 (pg/mL) 103 ± 94 (AD) 91 ± 139 (VaD)	No statistical difference between AD and non-AD dementia SN: 97–50% SP: 38–81% SN and SP varied with cut-off value	Le Bastard et al. (2007)
AD versus VaD Probable AD (n = 105) Possible AD (n = 58) VaD (n = 23) Sample: Community population-based study of the Piteå River Valley Hospital, Piteå, Sweden	$A\beta_{1-42}$ (pg/mL) 523 ± 180 (probable AD) 572 ± 225 (possible AD) 704 ± 321 (VaD) p-tau-181 (pg/mL) 759 ± 417 (probable AD) 699 ± 275 (possible AD) 461 ± 280 (VaD)	SN: Not determined SP: 48% P = 0.247	Andreasen et al. (2001)
AD versus VaD AD (n = 47) VaD (n = 44) Sample: Neurochemistry Laboratory at the Department of Neurology, University Medical School, Göttingen	$A\beta_{1-42}$ (pg/mL) 580 ± 211 (AD) 701 ± 341 (VaD) t-tau (pg/mL) 391 ± 232 (AD) 302 ± 252 (VaD)	No statistically significant change P = 0.579; no statistically significant change	Kaerst et al. (2013)

Sample	Biomarker	Significance	Reference
AD versus FTD AD (n = 72) VaD (n = 42) Clinical confirmation Sample: prospective study (start at 2006- end at 2009) by Department of Neurology, IRCCS Multimedica, Milan, Italy	$A\beta_{1-42}$ (pg/mL) 372.9 ± 20.8[#] (AD) 482.4 ± 42.9[#] (FTD)	$P = 0.03$; statistically significant change	de Rino et al. (2012)
	t-tau (pg/mL) 1175.8 ± 609.4[#] (AD) 452.9 ± 83.9[#] (FTD)	$P = 0.36$; no statistically significant change	
	p-tau-181 (pg/mL) 97.7 ± 5.5[#] (AD) 67.1 ± 7.2[#] (FTD)	$P = 0.011$; statistically significant change	
AD versus FTD AD (n = 35) FTD (n = 37) Clinical confirmation Sample: longitudinal study database of the University of Eastern Finland, Finland	$A\beta_{1-42}$ (pg/mL) 580 ± 250 (AD) 632 ± 205 (FTD)	$P = 0.448$; no statistically significant change	Muñoz-Ruiz et al. (2013)
	t-tau (pg/mL) 465 ± 241(AD) 350 ± 250 (FTD)	$P = 0.27$; no statistically significant change	
	p-tau-181 (pg/mL) 69 ± 23 (AD) 51 ± 30 (FTD)	$P = 0.032$; statistically significant change	
AD versus FTD AD (n = 272) FTD (n = 60) Clinical confirmation Sample: Montpellier neurological and CMRR (Centres Mémoire de Res-sources et de Recherche), France	$A\beta_{1-42}$ (pg/mL) 428 (334–499)[a] (AD) 665(472–800)[a] (FTD)	$P < 0.001$; statistically significant change	Gabelle et al. (2013)
	t-tau (pg/mL) 546 (398–850)[a] (AD) 301(191–412)[a] (FTD)	$P < 0.001$; statistically significant change	
	p-tau-181 (pg/mL) 77 (63–103)[a] (AD) 42(26–51)[a] (FTD)	$P < 0.001$; statistically significant change	

(Continued)

Table 5.4 CSF Biomarkers for Differential Diagnosis of Alzheimer's Disease and Non-Alzheimer's Disease Dementia (cont.)

Study Design AD Versus Non-AD Dementia	CSF Biomarker Measured Values (Average ± Standard Deviation)	Biomarker Performance (As Reported)	Reference
AD versus LBD Probable AD (n = 105) Possible AD (n = 58) LBD (n = 9). Sample: Community population-based study of the Piteå River Valley Hospital, Piteå, Sweden	$A\beta_{1-42}$ (pg/mL) 523 ± 180 (probable AD) 572 ± 225 (possible AD) 568 ± 183 (LBD)	No significant change SN: Not determined SP: 67%	Andreasen et al. (2001)
AD versus LBD AD (n = 272) LBD (n = 26) Clinical confirmation Sample: Montpellier neurological and CMRR (Centres Mémoire de Ressources et de Recherche), France	$A\beta_{1-42}$ (pg/mL) 428 (334–499)[a] (AD) 432(358–659)[a] (LBD) t-tau (pg/mL) 546 (398–850)[a] (AD) 269 (200–401)[a] (LBD) p-tau-181 (pg/mL) 77 (63–103)[a] (AD) 44(35–56)[a] (LBD)	$P < 0.05$; statistically significant change $P < 0.01$; statistically significant change $P < 0.001$; statistically significant change	Gabelle et al. (2013)
AD versus non-AD dementia Severe AD (n = 123) Moderate AD (n = 145) Early-stage AD (n = 98) Non-AD dementia (n = 33)	t-tau (pg/mL) 460 ± 263 (severe AD) 508 ± 268 (moderate AD) 463 ± 273 (early-stage AD) 271 ± 203 (non-AD dementia) 213 ± 172 (VaD)	Early-stage AD versus non-AD dementia: SN = 59.1 SP = 80.4	Shoji et al. (2002)

Sample / Population	Biomarker values	Statistical significance	Reference
AD versus CJD, AA, ALS, FTD, LBD Severe AD (n = 123) Moderate AD (n = 145) Early-stage AD (n = 98) CJD (n = 6) AA (n = 2) ALS (n = 8) FTD (n = 14) LBD (n = 14)	t-tau (pg/mL) 460 ± 263 (severe AD) 508 ± 268 (moderate AD) 463 ± 273 (early-stage AD) 410 ± 400 (CJD) 493 ± 441 (AA) 410 ± 147 (ALS) 331 ± 124 (FTD) 330 ± 204 (LBD)	SN and SP were not determined	Shoji et al. (2002)
AD versus non-AD dementia (autopsy confirmed)	$A\beta_{1-42}$ (pg/mL) 304 (137–557)[a] (AD) 519 (327–581)[a] (non-ADD)	$P = 0.409$; no statistically significant change	Le Bastard et al. (2010)
AD (n = 16) Non-AD dementia (n = 6)	t-tau (pg/mL) 532 (219–1094)[a] (AD) 489 (198–1071)[a] (non-ADD)	$P = 0.94$; no statistically significant change	
Sample: Biobank facilities of the Institute Born-Bunge (Antwerp, Belgium)	p-tau-181 (pg/mL) 66.2(40.7–102.5)[a] (AD) 36.9 (25.4–49.2)[a] (non-ADD)	$P = 0.029$[b]	
AD versus non-AD dementia (autopsy confirmed)	$A\beta_{1-42}$ (pg/mL) 355 (68–774)[a] (AD) 610 (327–581)[a] (non-AD dementia)	$P < 0.05$; statistically significant change	Le Bastard et al. (2013)
AD (n = 51) Non-AD dementia (n = 15)By	t-tau (pg/mL) 564 (92–1194)[a] (AD) 351 (194–1028)[a] (non-AD dementia)	$P < 0.05$; statistically significant change	
INNOTEST® platform Sample: Biobank of the Institute Born-Bunge (Antwerp, Belgium)	p-tau-181 (pg/mL) 66 (21–121)[a] (AD) 41 (27–66))[a] (non-AD dementia)	$P < 0.05$; statistically significant change	

(Continued)

Table 5.4   CSF Biomarkers for Differential Diagnosis of Alzheimer's Disease and Non- Alzheimer's Disease Dementia (*cont.*)

Study Design AD Versus Non-AD Dementia	CSF Biomarker Measured Values (Average ± Standard Deviation)	Biomarker Performance (As Reported)	Reference
AD versus non-AD dementia (2008–2012 French Alzheimer's Plan)			Gabelle et al. (2013)
AD (n = 272)	$A\beta_{1-42}$ (pg/mL)	AD versus AA; NS	
AA (n = 9)	428 (334–449)[a] (AD)	AD versus FTD; $P < 0.001$	
FTD (n = 60)	349 (263–491)[a] (AA)	AD versus LBD; $P < 0.05$	
LBD (n = 26)	665 (472–800)[a] (FTD)	AD versus ODD; $P < 0.001$	
ODD (n = 72)	432 (358–659)[a] (LBD)	AD versus OND; $P < 0.001$	
OND (n = 181)	639 (437–836)[a] (ODD)	AD versus PSY; $P < 0.001$	
PSY (n = 22)	741 (543–867)[a] (OND)		
	654 (557–859)[a] (PSY)		
By INNOTEST® platform	t-tau (pg/mL)	AD versus AA; NS	
	546 (398–850)[a] (AD)	AD versus FTD; $P < 0.001$	
	348 (229–613)[a] (AA)	AD versus LBD; $P < 0.01$	
	301 (191–412)[a] (FTD)	AD versus ODD; $P < 0.001$	
	269 (200–401)[a] (LBD)	AD versus OND; $P < 0.001$	
	264 (187–405)[a] (ODD)	AD versus PSY; $P < 0.001$	
	212 (161–284)[a] (OND)		
	153 (104–198)[a] (PSY)		

Sample: Montpellier neurological and CMRR (Centres Mémoire de Ressources et de Recherche), France	p-tau (pg/mL)	
	77 (63–103)[a] (AD)	AD versus AA; $P < 0.05$
	54 (41–64)[a] (AA)	AD versus FTD; $P < 0.001$
	42 (26–51)[a] (FTD)	AD versus LBD; $P < 0.01$
	44 (35–56)[a] (LBD)	AD versus ODD; $P < 0.001$
	43 (31–57)[a] (ODD)	AD versus OND; $P < 0.001$
	36 (26–46)[a] (OND)	AD versus PSY; $P < 0.001$
	35 (25–46)[a] (PSY)	All statistically significant change

AA, amyloid angiopathy; ALS, amyotrophic lateral sclerosis; AD, Alzheimer's disease; CJD, Creutzfeldt–Jackob disease; CSF, cerebrospinal fluid; FTD, frontotemporal dementia; LBD, Lewy body disease; p-tau-181, ODD, other degenerative diseases; OND, other nondegenerative diseases; PSY, psychiatric disorders; SN, sensitivity; SP, specificity; tau phosphorylated on threonine 181; t-tau, total tau; VaD, vascular dementia.

[a] Data range (standard deviation was not reported; " Standard error of mean).

[b] Significant (although the number of non-AD dementia patients was small).

et al., 2001; Shoji et al., 2002; Schönknecht et al., 2003; de Jong et al., 2006; Spies et al., 2009). In these studies, sensitivity and specificity are low, with poor accuracy when trying to distinguish between AD and VaD. A meta-analysis of published articles that included VaD, AD, and control groups found statistically significant differences in CSF tau between VaD and control groups ($P < 0.01$; van Harten et al., 2011). The same study also estimated sensitivity to be 70% (60–86%) and specificity 86% (80–94%) for detecting VaD versus AD. In general, p-tau values in VaD are lower than in AD, resulting in high sensitivity (88%), but moderate specificity (78%) for distinguishing AD versus VaD (van Harten et al., 2011). Yet several other studies have shown that there is no correlation between CSF p-tau-181 with the extent of Braak neurofibrillary tangles and neuritic plaques, the gold standard for autopsy diagnosis of AD (Engelborghs et al., 2007).

CSF biomarkers are more efficient for the differential diagnosis of AD versus subcortical VaD (SVaD). SVaD is a very common type of VaD that shares many of the same neuropsychological deficits as AD, making the two disorders difficult to diagnose with neuropsychological tests alone. The primary risk factors for SVaD are age and infarcts caused by strokes. Reduced blood flow through small vessels in the brain causes nerve fiber damage and disruption of neuronal signaling. Unlike AD, the pathophysiology of SVaD is not related to elevated tau and hyperphosphorylated tau, which allows SVaD to be distinguished from AD based on t-tau and p-tau biomarker levels. Furthermore, in SVaD, there is no evidence of increased CSF tau compared to controls; therefore, MCI–SVaD patients can also be differentiated from patients with MCI–AD based on CSF tau levels (Bjerke et al., 2009). A study using multivariate analysis and a combination of CSF t-tau, p-tau, $A\beta_{1-42}$, matrix metalloproteinases, and tissue inhibitors of metalloproteinases was able to distinguish between AD and SVaD with high sensitivity, specificity, and accuracy (Bjerke et al., 2011).

5.6.2 AD Versus LBD

Lower tau levels are expected in LBD patients compared with AD patients; however, studies have found that mean values of CSF tau were higher for LBD patients than control cases. A meta-analysis of 19 studies (208 LBD and 473 AD patients) did show lower tau values in LBD compared with AD and that tau could differentiate between the two groups with low sensitivity (73%) but high specificity (90%; van Harten et al., 2011). The same study found that CSF p-tau could differentiate between LBD and AD with the same sensitivity (74%) but lower specificity, likely because the difference in p-tau levels in AD patients versus LBD patients is smaller than the difference in p-tau between AD and other non-AD dementias such as VaD and FTD.

5.6.3 AD Versus FTD

In general, FTD occurs before the age of 65 years. Therefore, a comparison of CSF markers in FTD with those in early-onset AD is reasonable. The sensitivity (66–82%) and specificity (66–81%) of CSF t-tau to distinguish between AD and FTD are relatively low (van Harten et al., 2011). However, the sensitivity (79%) and specificity (83%) are higher for CSF p-tau to distinguish between AD and FTD (van Harten et al., 2011). In general CSF tau and p-tau are lowered in FTD cases compared to AD patients (Table 5.4). Several studies found statistically significant difference of CSF $A\beta_{1-42}$ and total tau biomarkers between AD patients groups and FTD cases (Gabelle et al., 2013; de Rino et al., 2012) but other studies are unable to find such differences in CSF tau levels (Table 5.4; Muñoz-Ruiz et al., 2013). However, p-tau was significantly lowered in FTD cases by all studies (Table 5.4).

5.6.4 AD Versus CJD

Lower $A\beta_{1-42}$ and elevated tau have been reported in the CSF of patients with CJD (Otto et al., 1997, 2000; Shoji et al., 2002). CJD has extremely high CSF tau values compared to AD, resulting in high sensitivity (91%) and specificity (98%) to differentially diagnose the two disorders (van Harten et al., 2011). Unlike other non-AD dementias, p-tau concentrations are less able to distinguish between AD and CJD.

In summary, CSF tau levels in FTD, VaD, and LBD are at intermediate levels between age-matched controls and AD patients, with low diagnostic sensitivity and specificity. After reviewing most of the articles related to CSF biomarkers in non-AD dementia, we can conclude that, compared to age-matched controls, tau concentrations are moderately elevated in LBD, FTD, and VaD, while p-tau-181 concentrations are slightly elevated in LBD but not in FTD and VaD (Rosso et al., 2003; Humpel et al., 2004; Le Bastard et al., 2007, 2010). CSF tau levels are also elevated after traumatic brain injury, stroke, and in taupathies. No significant differences have been reported in the core CSF biomarkers in patients with stroke versus patients with AD (Kaerst et al., 2013). Some cohort studies have shown a low sensitivity and specificity of CSF biomarkers to distinguish AD from non-AD dementias. Suboptimal differential diagnostic accuracy of CSF biomarkers may be related to the fact that a substantial number of patients with AD present the comorbid age-related dementias.

5.7 LONGITUDINAL ASSESSMENT OF ALZHEIMER'S DISEASE USING CSF BIOMARKERS

AD involves more than decade of progression through prodromal stages, and understanding the molecular pathologic changes that occur throughout the AD continuum is critical in order to diagnose the disease in its early stages and initiate therapeutic intervention. CSF biomarkers have been examined in longitudinal cohorts to track disease progression. Based on what is known about AD pathophysiology, a decrease in CSF $A\beta_{1-42}$ and increase in total tau and p-tau as a function of time would be expected as the disease progresses. Indeed, CSF $A\beta_{1-42}$ levels start to decrease even before clinical symptoms appear, while the values of total tau and p-tau increase over the course of AD progression (Seppälä et al., 2011; Toledo et al., 2013; Fagan et al., 2014). Longitudinal studies have shown that a combination of all three core CSF biomarkers can predict conversion from MCI to AD with a high accuracy (Table 5.5). A combination of

Table 5.5 Longitudinal Studies of CSF Biomarkers in Alzheimer's Disease

Study Design	Changes of Biomarkers	Comments	Reference
Follow-up: 18 months MCI ($n = 28$) Outcome: progress to AD ($n = 10$), progress to FTD ($n = 2$), progressive MCI ($n = 6$), MCI (stable; $n = 10$)	CSF $A\beta_{1-42}$ ↓, tau ↑, p-tau ↑ for MCI cases who converted to AD	CSF biomarkers profile can distinguish AD with MCI cases	Riemenschneider et al. (2002)
Follow-up MCI: 4–6 years, control: 3 years MCI (converter) to AD ($n = 57$), MCI (converter) to non-AD dementia ($n = 21$), MCI (stable; $n = 56$), control ($n = 39$)	Detection of progression of MCI (converter) to AD was highly accurate (specificity = 83%, sensitivity = 95%)	Total tau, p-tau, and $A\beta_{1-42}$ in CSF are strongly associated with the conversion of MCI (converter) to development of AD	Hansson et al. (2006)
Follow-up: 3 years DESCRIPA (development of screening guidelines and criteria for predementia Alzheimer's disease) study SCI ($n = 60$), aMCI ($n = 71$), non-aMCI ($n = 37$), control ($n = 89$)	Baseline CSF $A\beta_{1-42}$ ↓, tau ↑, p-tau ↑ for all kind of MCI and SCI cases	CSF biomarkers profile was associated with cognitive decline	Visser et al. (2009)

Study	Findings	Conclusion	Reference
Multicenter study Follow-up: 2 years MCI ($n = 271$)	CSF $A\beta_{1-42}$ ↓, tau ↑, p-tau ↑ for MCI cases who converted to AD	CSF biomarkers profile was associated with the conversion of MCI to AD	Mattsson et al. (2009)
Follow-up: two lumbar punctures with a median interval of 3 years AD ($n = 56$), MCI ($n = 57$), non-AD dementia ($n = 10$)	Baseline CSF $A\beta_{1-42}$ ↓, tau ↑, p-tau ↑	Well correlated MMSE with decline rate of baseline lowest $A\beta(42)$, highest tau, and highest p-tau-181 in CSF	Seppälä et al. (2011)
ADNI study Follow-up: 4 years AD ($n = 18$), MCI ($n = 74$), control ($n = 50$)	Baseline CSF $A\beta_{1-42}$ ↓, low baseline $A\beta_{1-42}$ values were associated with longitudinal increases in p-tau	CSF $A\beta_{1-42}$ changes before p-tau	Toledo et al. (2013)
DIAN study (longitudinal DIAN subcohort) MCI (stable; $n = 17$), MCI (converter; $n = 9$), control ($n = 11$)	Baseline CSF $A\beta_{1-42}$ ↓, tau ↑, p-tau ↑ for all MCI cases compared to control	Study was conducted on autosomal dominant AD cases	Fagan et al. (2014)

AD, Alzheimer's disease; ADNI, Alzheimer's Disease Neuroimaging Initiative; DIAN, dominantly inherited Alzheimer network; FTD, frontotemporal dementia; MCI, mild cognitive impairment; aMCI, amnestic MCI; non-aMCI, nonamnestic MCI; MMSE, mini-mental score examination; MRI, magnetic resonance imaging; SCI, subjective cognitive impairment.

CSF core biomarkers can also differentiate between MCI patients who progress to AD (MCI-C) and MCI patients who remain stable over time (MCI-S; Diniz et al., 2008; Mitchell, 2009; Schmand et al., 2010; Ferreira et al., 2014). Patients with MCI-C who show a decrease in $A\beta_{1-42}$ are more likely to develop AD later, suggesting that CSF $A\beta_{1-42}$ may be predictive of early AD. CSF biomarkers may also be superior to neuroimaging biomarkers for tracking disease progression (Humpel et al., 2008; Anoop et al., 2010).

5.8   LIMITATIONS OF LUMBAR PUNCTURE TO THE WIDESPREAD APPLICATION OF CSF-BASED DIAGNOSTIC ASSAYS IN ALZHEIMER'S DISEASE

Collection of CSF samples can only be done by lumbar puncture (LP), which is a much more invasive procedure than a blood draw or even punch biopsy for skin sample collection. While the LP procedure is fairly routine and consistent across centers, there are risks and minor side effects that may limit its use in diagnostic AD testing, particularly for elderly patients. Post-LP headache (PLPH) is one of the main side effects of the procedure, caused by inadvertent rupture of blood vessels (Hart et al., 1988; Strupp and Katsarava, 2009; de Almeida et al., 2011). PLPH typically begins within a few hours after the procedure and can persist for few days, though in some cases, symptoms can last several weeks or months. The amount of CSF removed at a single LP does not appear to influence the frequency of headache (Strupp and Katsarava, 2009). However, one study that included patients from a wide range of age groups found that only 2.6% of patients ($n = 1089$; aged 23–89 years) reported PLPH without other local or general complications (Kuntz et al., 1992). A multicenter, 13-week study of CSF cholinesterase activity in AD patients reported a favorable safety profile of LP procedures and that fewer than 2% of patients experienced a PLPH (Zetterberg et al., 2010). Although LP is a relatively safe procedure, several other minor and major complications can occur, even when standard infection control measures and good techniques are used. In addition, low CSF pressure or volume in elderly patients may increase the possibility of an unsuccessful LP (Peskind et al., 2009). LP for sample collection for the diagnosis of AD is not yet approved in US, though it is routinely performed in European countries. Due to pain and post-LP effects, negative perceptions of LP pose a challenge to its use in routine AD diagnostic testing. Furthermore, the multiple LP procedures that must be performed in drug efficacy clinical trials or to track disease progression present considerable logistical challenges.

5.9   INSTABILITY OF BASELINE Aβ IN CSF

Fluctuations in Aβ levels are a major concern that may limit the use of CSF Aβ as a diagnostic biomarker for AD in clinical pathology laboratories. Aβ levels vary due to circadian fluctuations and patient activity levels, including sleep patterns. One seminal CSF Aβ baseline study reported that CSF Aβ levels fluctuate 1.5- to 4-fold over a period 12–36 h, and appear to be dependent on the time of day or activity level (Bateman et al., 2007). The sleep cycles for AD patients are not regular. Another in vivo microdialysis study in mice described that Aβ levels in brain interstitial fluid (ISF) correlated with wakefulness, and that Aβ levels significantly increased during acute sleep deprivation (Kang et al., 2009). The study showed clear evidence of diurnal fluctuations in Aβ levels in the CSF of young healthy male volunteers over a 33-h period ($N = 10$). The study also found that Aβ levels increased throughout the day and peaked in the evening, then decreased at overnight, and increased throughout the next day. On the other hand, two studies found no obvious diurnal variability in Aβ levels (Bjerke et al., 2010; Slats, 2010); particularly when only 6 mL of CSF sample was collected every hour for 36 h (Slats, 2010). The Alzheimer's Biomarkers Standardization Initiative (SBSI) concluded that there is no clear evidence to support a diurnal variation in CSF biomarkers, though SBSI members are conducting an ongoing study of baseline diurnal variation (Vanderstichele et al., 2012).

Blood–brain barrier (BBB) dysfunction in AD may cause instability in the levels of CSF biomarkers. Approximately 500 mL of CSF is exchanged daily into the circulation in humans. Proteins in blood plasma such as albumin, α2-macroglobulin, and low-density lipoprotein receptor-related proteins can also bind to Aβ, which may lead to an underestimation of CSF Aβ levels. Because the blood–brain barrier becomes dysfunctional in AD, there is also a greater likelihood of blood contamination of CSF samples during LP (Petzold et al., 2006).

5.10 INTERLABORATORY VARIATION IN CSF BIOMARKER ASSAYS

Another factor limiting the application of AD CSF biomarker assays to clinical care is the high interlaboratory variation. The specificity and sensitivity of CSF biomarkers are reasonably good in single-site cohort studies. By contrast, in a multicenter study of the diagnostic accuracy of the three core CSF biomarkers, conducted at 12 sites in Europe and the US in 750 individuals with MCI, 529 with AD, and 304 control cases, diagnostic sensitivity was 83%, specificity was 72%, positive predictive value was 62%, and negative predictive value was 88% (Mattsson et al., 2009). Variations in assay results across the 12 sites may have contributed to the low diagnostic accuracy in the study. The source of variations can be divided into two categories: methodological factors and biological factors. Methodological factors refer to variability in assay materials and techniques, including collection tube materials, sample handling and storage, dilution and buffer composition, heat treatment, plasma contamination, and immunoassay procedures. Biological factors include variations in patient selection, inclusion of patients with comorbid AD and non-AD dementias, and assessment of AD CSF biomarker performance based on clinically confirmed AD instead of autopsy-confirmed AD. Inclusion of non-AD cases as AD and inclusion of preclinical AD cases as controls (eg, due to a long prodromal stage) would contribute to a lower diagnostic accuracy in any individual study. Inclusion of comorbid cases may shift cut-off values and underestimate sensitivity and specificity. Our own autopsy-validated AD cohort study (Khan and Alkon, 2010) found that approximately 33% of patients with AD also had comorbid non-AD dementias, such as PD, FTD, VaD, LBD, and taupathy.

Substantial interlaboratory variations of CSF biomarker levels also make comparisons of data from different laboratories problematic, and this issue must be resolved before these assays can be routinely used for AD diagnosis in the clinic setting. The accuracy of CSF Aβ measurements can be confounded by interlaboratory variations in the immunoassay materials and methods, including the type of sample and assay tubes used, the number of freeze/thaw cycles, storage and incubation temperatures, sample preparation protocols, and antibody selection. Between studies, there is considerable variation in the reported levels of CSF $A\beta_{1-42}$, total tau, and p-tau-181. For CSF $A\beta_{1-42}$, the variation among laboratories ranges from 13% to 36% (CV = 26.5%; Paris: Standardization a Hurdle for Spinal Fluid, Imaging Markers; http://www.alzforum.org/news/conference-coverage/paris-standardization-hurdle-spinal-fluid-imaging-markers). The Alzheimer's Association Quality Control Program Work Group found a 20–30% CV between laboratories within a consortium of total 84 laboratories (Mattsson et al., 2013). An international quality control survey of 14 laboratories in Germany, Austria, and Switzerland found ever higher CVs for each CSF biomarker (CSF $A\beta_{1-42}$ = 29%, t-tau = 26%, and p-tau-181 = 27%; Lewczuk et al., 2006). Reduction in variation between studies and laboratories, through standardization of analytical protocols, preanalytical sample collection, and sample processing procedures, is essential. To address this issue, international scientists working on CSF biomarkers have established a working group called the Alzheimer's Biomarkers Standardization Initiative (ABSI; Vanderstichele et al., 2012). To reduce interlaboratory variability, the ABSI has reached a consensus on various preanalytical issues such as CSF collection and storage tubes, the effect of fasting, storage temperature, length of storage time, centrifugation speed, and storage concentrations of CSF $A\beta_{1-42}$, total tau, and p-tau-181. A standard protocol for CSF preparation and immunoassay, universal cut-off values, internationally recognized reference standards, and a mechanism to evaluate assay performance are still needed. Ongoing standardization efforts have been introduced to harmonize good laboratory practice (GLP), standard operating procedures (SOPs), defined procedures on CSF collection and handling, and assay calibration for different technology platforms, with the ultimate goal of reducing interlaboratory variability in CSF biomarker assays (Mattsson and Zetterberg, 2012; Andreasson et al., 2012; Mattsson et al., 2012; Teunissen et al., 2010).

5.11 PERFORMANCE OF CSF BIOMARKERS IN ASSESSING DRUG EFFICACY IN ALZHEIMER'S DISEASE CLINICAL TRIALS

Biomarkers are incorporated into AD clinical trial designs to establish the homogeneity of the recruited patient group, assess drug response, provide surrogate endpoints for drug efficacy, and give insights into the mechanisms of drug action. The European Medicines Agency has endorsed the use of CSF $A\beta_{1-42}$ and tau in clinical trials in patients with AD. Along with neuropsychological tests, CSF biomarkers and MRI volumetric measurements are used for patient selection in most AD clinical trials. CSF biomarkers have been incorporated into the majority of AD clinical trials to demonstrate target engagement, to enrich and stratify treatment groups, and to find evidence of disease modification by treatment. In general, CSF biomarker endpoints in AD clinical trials include changes in CSF t-tau and p-tau-181 (due to neuronal injuries) and $A\beta_{1-42}$ (due to amyloid plaque deposition in brain), with drug treatment efficacy detected as decreased tau and increased $A\beta_{1-42}$. Animal model studies showed γ-secretase inhibitor treatment resulted reduction of $A\beta$ in cortical and CSF samples (Lanz et al., 2004; Anderson et al., 2005), but despite promising preclinical results with anti-$A\beta$ immunotherapies, as well as β- and γ-secretase inhibitors, all of these approaches have failed in recent AD clinical trials (Fleisher et al., 2008; Blennow et al., 2012; Salloway et al., 2014; Doody et al., 2014; Table 5.6).

The performance of CSF biomarkers in longitudinal studies to track AD progression has encouraged their use in selecting patients for inclusion in clinical trials, and changes in biomarker data with respect to trial dose/time may ultimately lead to correlation of biomarkers to clinical benefits such as reduced neurodegeneration (Holtzman, 2011). Most of the $A\beta$ immunization clinical trials resulted in clearance of plaques in AD; however, no improvement in neurodegeneration was seen (Table 5.6). Furthermore, changes in CSF biomarkers did not correlate with cognitive test results (Table 5.6). An ideal biomarker would predict clinical trial benefits and act as surrogate endpoint marker of neurodegeneration. Yet, there is no evidence to support the use of CSF biomarkers as surrogate markers for AD treatment efficacy in clinical trials (Coley et al., 2009).

5.12   CROSSCORRELATION OF CSF BIOMARKERS WITH OTHER ALZHEIMER'S DISEASE BIOMARKERS MODALITIES

CSF core biomarkers can track pathophysiological changes in the brain in AD that have been correlated with other biomarkers such as neuroimaging, amyloid plaques and fibrillary tangles at brain autopsy (the gold standard of AD diagnosis), and neuro-psychometric tests. Low levels of CSF $A\beta_{1-42}$ were well correlated with amyloid plaques in the AD brain autopsy (Strozyk et al., 2003). Both the neuroimaging biomarkers ^{11}C-labeled Pittsburgh compound B positron emission tomography (^{11}C-PiB PET) and[18]F-2-fluoro-2-deoxy-D-glucose (FDG) PET showed significant correlations with CSF $A\beta_{1-42}$ and tau (Tolboom et al., 2009; Table 5.7). $A\beta$ plaques and fibrillary tau aggregates in the cortical brain biopsies of AD patients were correlated with CSF $A\beta_{1-42}$, t-tau, and p-tau levels measured by ELISA (Seppälä et al., 2012). Another recent study found that CSF $A\beta_{1-42}$ correlated significantly with amyloid plaques in AD brain measured by ^{18}F-flutemetamol PET (Palmqvist et al., 2014). On the other hand, several studies found that changes in CSF biomarker levels were not related to changes in either the mini-mental state examination (MMSE) or atrophy rate (Sluimer et al., 2010;

Table 5.6	Performance of Core CSF Biomarkers to Measure Treatment Efficacy in Alzheimer's Disease Clinical Trials			
Study Design	**Cognitive Measure**	**Biomarker Results**	**Conclusion**	**References**
Aβ aggregation				
Phase II trial Tramiposate	No significant improvement in cognitive function	▪ CSF $A\beta_{1-42}$ was significantly lower after treatment ▪ No change in CSF $A\beta_{1-40}$	▪ Side effects were nausea, vomiting, and diarrhea	Aisen et al. (2006)
Phase II trial PBT2: A compound affects Zn2+ and Cu2+ mediated Aβ aggregation	Improvement in cognitive tests scores	▪ Lower CSF Aβ after treatment	▪ No serious adverse effect	Lannfelt et al. (2008)
Phase II trial ELND005: scyllo-inositol	No statistically significant improvement in cognitive test results and functional ability	▪ Lower CSF Aβ after treatment	▪ Insufficient evidence to support or refute a benefit	Salloway et al. (2011)
Aβ-immunization				
Phase III trial: Aβ-immunotherapy with bapineuzumab	No statistically significant improvement in cognitive test results	▪ CSF p-tau low in APOE ε4 group only ▪ Slight decrease in amount of Aβ plaques measured by ^{11}C-PIB-PET imaging in APOE ε4 group	▪ No significant clinical improvement in this trial ▪ There were differences in CSF and neuroimaging biomarkers in APOE ε4 carriers	Salloway et al. (2014)

Phase III trial: Aβ-immunotherapy with solanezumab (a humanized monoclonal antibody that binds Aβ)	No statistically significant improvement in cognitive test results or functional ability	■ Total CSF $A\beta_{1-42}$ was significantly higher after treatment	■ No significant clinical improvement with treatment ■ No correlation of CSF biomarker results and the clinical outcome of the trial	Doody et al. (2014)
Phase II trial of bapineuzumab	Not reported	■ No significant difference in CSF $A\beta_{1-42}$ between placebo and treatment group	■ Significant decrease of CSF tau ($P = 0.03$) ■ Significant decrease of CSF p-tau ($P = 0.001$)	Blennow et al. (2012)
APP processing				
Phase I trial Posiphen	Tolerability test in MCI patients	■ Decreased CSF total tau ■ Decreased CSF p-tau ■ No significant change in CSF $A\beta_{1-42}$	■ Target engagement, and pharmacokinetics of Posiphen for MCI and AD was positive	Maccecchini et al. (2012)
β-Secretase inhibitor				
Phase III trial MK-8931	Not reported	■ Significant reduction in CSF $A\beta_{1-42}$ ■ Significant reduction in CSF $A\beta_{1-40}$	■ No side effect reported	NCT 01496170 NCT 01739348 NCT 01953601
Phase I trial E2609	Not reported	■ Significant reduction in CSF $A\beta_{1-42}$	■ Headache and dizziness	NCT 01511783 NCT 01600859

(Continued)

Table 5.6	Performance of Core CSF Biomarkers to Measure Treatment Efficacy in Alzheimer's Disease Clinical Trials *(cont.)*			
Study Design	**Cognitive Measure**	**Biomarker Results**	**Conclusion**	**References**
γ-Secretase inhibitor				
Phase II trial LY450139 (Semagacestat)	No statistically significant changes in cognitive test results or measures of functional ability	■ No significant reduction of CSF Aβ	■ Lowered plasma Aβ consistent with the action of γ-secretase activity	Fleisher et al. (2008)
Phase III trial LY450139 (Semagacestat)	No improvement in cognitive status	■ Significant reduction in CSF Aβ$_{1-42}$ ■ Significant reduction in CSF Aβ$_{1-40}$	■ Study discontinued. Patients lost weight and had skin cancers and infections during treatment	Doody et al. (2013)
Phase II trial BMS-708163 (Avagacestat)	Trends for worsening cognition	■ Significant reduction in CSF Aβ$_{1-42}$ ■ Significant reduction in CSF Aβ$_{1-40}$	■ Study discontinued for adverse events, gastrointestinal and dermatologic	Coric et al. (2012)
Nonamyloid based therapeutic methods				
Cholesterol lowering statins trial	No report of cognitive state examination	■ No significant reduction of CSF Aβ ■ No change in CSF tau	■ Simvastatin, but not pravastatin, reduced CSF levels of p-tau	Riekse et al. (2006)
Long-term treatment with inhibitors of AChE	Temporary cognitive improvement	■ No significant reduction of CSF biomarkers	■ AChE inhibitors induced different effects on CSF AChE activity	Parnetti et al. (2002)
Phase II trial Intranasal insulin	Improved delayed memory	■ No significant reduction of CSF biomarkers	■ No treatment-related severe adverse effects occurred	Craft et al. (2012)

Aβ, amyloid beta; AChE, acetylcholinesterase; AD, Alzheimer's disease; APP, amyloid precursor protein; CSF, cerebrospinal fluid; MCI, mild cognitive impairment; ^{11}C-PIB, [11C]-Pittsburgh compound B; p-tau-181, tau phosphorylated on threonine 181.

Table 5.7	**Crosscorrelation of CSF and Neuroimaging Biomarkers in Alzheimer's Disease**		
Study Design	**Biomarkers**	**Comments**	**References**
ADNI AD (n = 10), control (n = 11), and MCI (n= 34)	FDG-PET versus CSF biomarkers	FDG-PET was modestly related to CSF biomarkers	Jagust et al. (2009)
AD (n = 88), FTD (n = 97) 32 of 185 has autopsy confirmed	MRI versus CSF biomarkers	75% overlap between MRI and CSF biomarkers	McMillan et al. (2013)
AD (n = 47), MCI (n = 29), control (n = 23)	MRI versus CSF biomarkers	No association of CSF $A\beta_{1-42}$ and tau biomarkers with whole brain atrophy. CSF levels of p-tau-181 showed a mild association with whole-brain atrophy rate in AD but not in controls or MCI cases	Sluimer et al. (2010)
AD (n = 15), control (n = 10), and MCI (n= 12)	FDG-PET and [11C] PIB PET, CSF biomarkers	Positive association between FDG-PET and CSF biomarkers. Inverse association between [11C] PIB PET and CSF biomarkers	Tolboom et al. (2009)
MCI (n = 33), AD (n = 35; clinically confirmed)	CSF biomarkers versus [11C] PIB PET	No correlation	Leuzy et al. (2015)
Control (n = 103), EMCI (n = 187), LMCI (n = 62), AD (n = 22)	CSF biomarkers versus Florbetapir PET	CSF biomarkers andorbetapir PET were consistent	Landau et al. (2013)
ADNI study Healthy control (n = 97) MCI (n = 226) AD (n = 21; clinically confirmed)	CSF biomarkers versus Florbetapir PET	Good correlation between CSF biomarkers with Florbetapir PET	Hake et al. (2015)

AD, Alzheimer's disease; ADNI, Alzheimer's Disease Neuroimaging Initiative; CSF, cerebrospinal fluid; MCI, mild cognitive impairment; MMSE, mini-mental score examination MRI, magnetic resonance imaging; PET, position emission tomography; [11C]PIB, [11C] Pittsburgh compound B PET; FDG-PET, [18F] fluorodeoxyglucose-PET.

van Der Flier and Scheltens, 2009). There are several conflicting reports regarding crosscorrelation of various AD biomarkers. For example, a patient with clinically and CSF-positive AD was negative for plaque burden by ^{11}C-PIB PET (Cairns et al., 2009), whereas in another study, normal individuals with cortical amyloid deposition had higher CSF levels of tau and p-tau (Fagan et al., 2009). Very recently, an AD autopsy report found only neurofibrillary tangles, but no amyloid plaques (Crary et al., 2014).

Some earlier investigations of cross correlation of CSF and APOE4 found substantial association (Galasko et al., 1998; Sunderland et al., 2004; Prince et al., 2004), while another report using an ApoE genotyping method found no association with CSF biomarkers $A\beta_{1-42}$ and p-tau-181 (Engelborghs et al., 2007). All 50 AD patients in the study had autopsy-confirmed disease. A genome-wide association study (GWAS) of CSF biomarkers ($A\beta_{1-42}$, t-tau, p-tau-181, p-tau-181/ $A\beta_{1-42}$, and t-tau/$A\beta_{1-42}$) found sound correlation with sporadic AD risk genes like APOE4 and TOMM40 (Kim et al., 2011). However, the most commonly identified variants in BIN1, CLU, CR1, and PICALM that show replicable association with risk for disease by GWAS showed no correlation with CSF biomarkers (Kauwe et al., 2011; Table 5.8). Crosscorrelation of CSF AD biomarkers with blood plasma AD biomarkers will be presented in Chapter 6.

5.13 EARLY DIAGNOSIS OF ALZHEIMER'S DISEASE USING CSF BIOMARKERS

Currently, the search for preclinical AD biomarkers is dominating the research effort. Therapeutic interventions for AD are likely to have the greatest effect if initiated in the early, preclinical stages of the disease, before synaptic loss and neuronal death occur. There are several lines of evidence that CSF biomarkers can be used as part of a preclinical signature that can predict the likelihood of developing AD. Several studies have described changes in CSF biomarkers in patients with dominantly inherited AD mutations but who are cognitively normal—these patients would be considered at-risk individuals (Rosén et al., 2013).

The NIA-AA working group defined preclinical AD as a prodromal phase consisting of three stages (Fig. 5.2). In the first stage, a patient is positive for amyloid plaques by PET imaging or has low CSF $A\beta_{1-42}$, but there is no sign of neurodegeneration by MRI and CSF tau values are normal. In the second stage, the patient has evidence of elevated CSF tau, neuronal injury, and amyloid plaques on imaging. In the third stage, the patient begins to experience subtle cognitive deficits that are less severe than those seen in MCI (McKhann et al., 2011; Sperling et al., 2011). Both cross-sectional and longitudinal studies for preclinical familial AD and preclinical sporadic AD showed lower CSF $A\beta_{1-42}$ and higher tau and p-tau among AD progressors (Rosén et al., 2013).

The IWG includes two diagnostic criteria for AD: (1) clinical AD phenotype criterion manifested by an episodic memory profile and (2) the presence of biomarker evidence suggestive of AD. Biomarkers include (1) volumetric MRI;

Table 5.8	Crosscorrelation of CSF and Genetic Biomarkers of Alzheimer's Disease		
Study Design	**Biomarkers**	**Comments**	**References**
ADNI study GWAS AD (n = 96), MCI (n = 176), controls (n = 102)	CSF versus APOE4, TOMM40	CSF biomarkers were correlated with most important risk factors of sporadic AD. This study needs to be confirmed	Kim et al. (2011)
GWAS Washington University AD (n = 102), control (n = 305) ADNI AD (n = 154), control (n = 103)	CSF Aβ_{1-42} and tau phosphorylated at threonine 181 (ptau181) versus BIN1, CLU, CR1, and PICALM	No correlation	Kauwe et al. (2011)
AD (n = 45), AD with cerebrovascular disease (n = 5) All autopsy confirmed	CSF biomarkers versus APOE4	No association between CSF biomarkers with APOE4	Engelborghs et al. (2007)
ADNI study AD (n = 309); prodromal AD (n = 287); non-AD dementia (n = 99); Stable MCI (n = 399); control (n = 251)	CSF Aβ_{1-42} versus APOE4	No association between CSF biomarkers with APOE4	Lautner et al. (2014)
The Finnish-AD multicenter cohort GWAS AD (n = 890), control (n= 701)	CSF Aβ_{1-42} versus APOE4, CLU, and MS4A4A	APOE4, CLU, and MS4A4A associated with both AD risk and CSF Aβ_{1-42}	Elias-Sonnenschein et al. (2013)
ADNI study AD (n = 87), MCI (n = 166), control (n = 100)	CSF Aβ_{1-42} versus SORL1	Significant correlation between CSF Aβ_{1-42} levels and SORL1	Alexopoulos et al. (2011)

Aβ, amyloid beta; AD, Alzheimer's disease; ADNI, Alzheimer's Disease Neuroimaging Initiative; APOE, apolipoprotein E; BIN1, bridging integrator 1; CLU, clusterin; CR1, complement cell-surface receptor; CSF, cerebrospinal fluid; GWAS, genome-wide association study; MCI, mild cognitive impairment; MS4A4A, membrane-spanning 4-domains subfamily A (MS4A) gene cluster; PICALM, phosphatidylinositol binding clathrin assembly protein; SORL1, sortilin-related receptor 1; TOMM40, translocase of outer mitochondrial membrane 40 homolog.

- **Phase I (Asymptomatic preclinical phase)**
  (a) Amyloid deposition measured by PET imaging
  (b) Low CSF $A\beta_{1-42}$
  (c) No traces of neurodegeneration measured by structural MRI
  (d) No changes in CSF tau values
  (e) No cognitive impairment

- **Phase II (Symptomatic mild cognitive impairment phase)**
  (a) Elevated CSF tau and p-tau (ie, indication of neuronal injury)
  (b) Low CSF $A\beta_{1-42}$
  (c) Changes in volumetric MRI (ie, indication of brain atrophy)
  (d) Mild cognitive impairment (MCI)

- **Phase III (Symptomatic Alzheimer's disease)**
  (a) The patients would experience memory decline that is not less than that of MCI.
  (b) Brain atrophy by volumetric MRI
  (c) Low CSF $A\beta_{1-42}$
  (d) Elevated CSF tau and p-tau

FIGURE 5.2

Application of CSF biomarkers in the National Institute on Aging-Alzheimer's Association 2011 criteria for Alzheimer's disease diagnosis of prodromal phases *(McKhann et al., 2011; Sperling et al., 2011) (MRI, Magnetic resonance imaging; PET positron emission tomography).*

(2) PET imaging (FDG PET ^{11}C-PiB PET); or (3) CSF $A\beta_{1-42}$ or tau protein (t-tau and p-tau concentrations; Dubois et al., 2010). According to the IWG-2 criteria, which are similar to the NIA-AA criteria, a patient with preclinical AD has no clinical signs or symptoms but has one of the following: (1) decreased $A\beta_{1-42}$, together with increased tau or p-tau in CSF; or (2) increased fibrillary amyloid on PET (Dubois et al., 2014). According to both working groups, CSF biomarkers may provide valuable information when combined with neuroimaging biomarkers for identifying the preclinical stages of AD. MCI patients with lower CSF $A\beta_{1-42}$ values had a faster rate of progression to AD (Herukka et al., 2007), which is consistent with the predominant Aβ-hypothesis of AD, in which defective clearance of toxic Aβ from the brain and the resulting neurodegeneration leads to late-onset AD. Toxic Aβ accumulated over years to decades causes progressive neuronal injury and synaptic loss. Biomarkers of late-onset AD, such as CSF biomarkers may detect events downstream of early defects in Aβ clearance, when the disease has reached an advanced stage. Therefore, the search continues for early defects in the signaling pathways involved in Aβ-clearance as diagnostic markers and therapeutic targets in earlier stages of disease progression.

5.14 CONCLUSIONS

Compared with other AD biomarkers such as PET, CSF biomarkers are currently the most accurate reflection of underlying AD pathology. CSF biomarkers may be useful in predicting future AD dementia in patients with MCI. Since AD is a multifactorial neurodegenerative disease with a long prodromal state, and often presents with comorbid non-AD neurodegenerative disease, it has been difficult to achieve 100% diagnostic accuracy with any CSF biomarker to date. Despite decades of expensive research on CSF biomarkers for AD, the conclusion remains that CSF assays require invasive sample collection, the overall variability of the assays is too high to establish universal cut-off values to distinguish AD from non-AD dementia, and the assay protocols have yet to be standardized for use in a clinical setting. According to the NIA-AA working group, CSF biomarkers are appropriate for research purposes only and are not ready to be applied in the clinical setting. The current body of literature suggests that CSF biomarkers and neuroimaging techniques eventually may be useful for selecting patient populations for inclusion in AD clinical trials; however, the utility of these biomarkers as surrogate endpoints of drug efficacy needs more validation. Although the IWG-2 working group has proposed to integrate biomarkers into the diagnostic scheme as a biological complement to the current clinical assessment of AD, the existing CSF biomarkers are insufficiently accurate for diagnosing preclinical AD. Most of the CSF biomarker studies included patients mainly in memory disorder centers, and some bias may occur due to the high prevalence of AD in memory disorder clinics compared to primary care. Once validation and standardization efforts have proven that these biomarkers are sufficiently accurate, they can be applied in the clinical setting.

Bibliography

Aisen, P.S., Saumier, D., Briand, R., Laurin, J., Gervais, F., Tremblay, P., Garceau, D., 2006. A phase II study targeting amyloid-beta with 3APS in mild-to-moderate Alzheimer disease. Neurology 67, 1757–1763.

Alexopoulos, P., Guo, L.H., Kratzer, M., Westerteicher, C., Kurz, A., Perneczky, R., 2011. Impact of SORL1 single nucleotide polymorphisms on Alzheimer's disease cerebrospinal fluid markers. Dement. Geriatr. Cogn. Disord. 32, 164–170.

Anderson, J.J., Holtz, G., Baskin, P.P., Turner, M., Rowe, B., Wang, B., Kounnas, M.Z., Lamb, B.T., Barten, D., Felsenstein, K., McDonald, I., Srinivasan, K., Munoz, B., Wagner, S.L., 2005. Reductions in beta-amyloid concentrations in vivo by the gamma-secretase inhibitors BMS-289948 and BMS-299897. Biochem. Pharmacol. 69, 689–698.

Andreasen, N., Minthon, L., Davidsson, P., Vanmechelen, E., Vanderstichele, H., Winblad, B., Blennow, K., 2001. Evaluation of CSF-tau and CSF-Abeta42 as diagnostic markers for Alzheimer disease in clinical practice. Arch. Neurol. 58, 373–379.

Andreasson, U., Vanmechelen, E., Shaw, L.M., Zetterberg, H., Vanderstichele, H., 2012. Analytical aspects of molecular Alzheimer's disease biomarkers. Biomark. Med. 6, 377–389.

Anoop, A., Singh, P.K., Jacob, R.S., Maji, S.K., 2010. CSF biomarkers for Alzheimer's disease diagnosis. Int. J. Alzheimers Dis. 2010, 606802.

Bateman, R.J., Wen, G., Morris, J.C., Holtzman, D.M., 2007. Fluctuations of CSF amyloid-beta levels: implications for a diagnostic and therapeutic biomarker. Neurology 68, 666–669.

Beher, D., Wrigley, J.D., Owens, A.P., Shearman, M.S., 2002. Generation of C-terminally truncated amyloid-beta peptides is dependent on gamma-secretase activity. J. Neurochem. 82, 563–575.

Bjerke, M., Andreasson, U., Rolstad, S., Nordlund, A., Lind, K., Zetterberg, H., Edman, A., Blennow, K., Wallin, A., 2009. Subcortical vascular dementia biomarker pattern in mild cognitive impairment. Dement. Geriatr. Cogn. Disord. 28, 348–356.

Bjerke, M., Portelius, E., Minthon, L., Wallin, A., Anckarsäter, H., Anckarsäter, R., Andreasen, N., Zetterberg, H., Andreasson, U., Blennow, K., 2010. Confounding factors influencing amyloid Beta concentration in cerebrospinal fluid. Int. J. Alzheimers Dis., (Article ID: 986310), 1–11.

Bjerke, M., Zetterberg, H., Edman, Å., Blennow, K., Wallin, A., Andreasson, U., 2011. Cerebrospinal fluid matrix metalloproteinases and tissue inhibitor of metalloproteinases in combination with subcortical and cortical biomarkers in vascular dementia and Alzheimer's disease. J. Alzheimers Dis. 27, 665–676.

Blennow, K., Humpel, H., 2003. CSF markers for incipient Alzheimer's disease. Lancet Neurol. 2, 605–613.

Blennow, K., Humpel, H., Weiner, M., Zetterberg, H., 2010. Cerebrospinal fluid and plasma biomarkers in Alzheimer disease. Nat. Rev. Neurol. 6, 131–144.

Blennow, K., Humpel, H., Zetterberg H, 2014. Biomarkers in amyloid-β immunotherapy trials in Alzheimer's disease. Neuropsychopharmacology 39, 189–201.

Blennow, K., Zetterberg, H., Rinne, J.O., Salloway, S., Wei, J., Black, R., Grundman, M., Liu, E., AAB-001 201/202 Investigators, 2012. Effect of immunotherapy with bapineuzumab on cerebrospinal fluid biomarker levels in patients with mild to moderate Alzheimer disease. Arch. Neurol. 69, 1002–1010.

Bloudek, L.M., Spackman, D.E., Blankenburg, M., Sullivan, S.D., 2011. Review and meta-analysis of biomarkers and diagnostic imaging in Alzheimer's disease. J. Alzheimer Dis. 26, 627–645.

Bouwman, F.H., Schoonenboom, N.S., Verwey, N.A., van Elk, E.J., Kok, A., Blankenstein, M.A., Scheltens, P., van der Flier, W.M., 2009. CSF biomarker levels in early and late onset Alzheimer's disease. Neurobiol. Aging 30, 1895–1901.

Braak, H., Zetterberg, H., Del Tredici, K., Blennow, K., 2013. Intraneuronal tau aggregation precedes diffuse plaque deposition, but amyloid-β changes occur before increases of tau in cerebrospinal fluid. Acta. Neuropathol. 126, 631–641.

Brunnström, H., Rawshani, N., Zetterberg, H., Blennow, K., Minthon, L., Passant, U., Englund, E., 2010. Cerebrospinal fluid biomarker results in relation to neuropathological dementia diagnoses. Alzheimer's Dement. 6, 104–109.

Cairns, N.J., Ikonomovic, M.D., Benzinger, T., Storandt, M., Fagan, A.M., Shah, A.R., Reinwald, L.T., Carter, D., Felton, A., Holtzman, D.M., Mintun, M.A., Klunk, W.E., Morris, J.C., 2009. Absence of Pittsburgh compound B detection of cerebral amyloid beta in a patient with clinical, cognitive, and cerebrospinal fluid markers of Alzheimer disease: a case report. Arch. Neurol. 66, 1557–1562.

Coley, N., Andrieu, S., Delrieu, J., Voisin, T., Vellas, B., 2009. Biomarkers in Alzheimer's disease: not yet surrogate endpoints. Ann. NY Acad. Sci. 1180, 119–124.

Coric, V., van Dyck, C.H., Salloway, S., Andreasen, N., Brody, M., Richter, R.W., Soininen, H., Thein, S., Shiovitz, T., Pilcher, G., Colby, S., Rollin, L., Dockens, R., Pachai, C., Portelius, E., Andreasson, U., Blennow, K., Soares, H., Albright, C., Feldman, H.H., Berman, R.M., 2012. Safety and tolerability of the γ-secretase inhibitor avagacestat in a phase 2 study of mild to moderate Alzheimer disease. Arch. Neurol. 69, 1430–1440.

Craft, S., Baker, L.D., Montine, T.J., Minoshima, S., Watson, G.S., Claxton, A., Arbuckle, M., Callaghan, M., Tsai, E., Plymate, S.R., Green, P.S., Leverenz, J., Cross, D., Gerton, B., 2012. Intranasal insulin therapy for Alzheimer disease and amnestic mild cognitive impairment: a pilot clinical trial. Arch. Neurol. 69, 29–38.

Crary, J.F., Trojanowski, J.Q., Schneider, J.A., Abisambra, J.F., Abner, E.L., Alafuzoff, I., Arnold, S.E., Attems, J., Beach, T.G., Bigio, E.H., Cairns, N.J., Dickson, D.W., Gearing, M., Grinberg, L.T., Hof, P.R., Hyman, B.T., Jellinger, K., Jicha, G.A., Kovacs, G.G., Knopman, D.S., Kofler, J., Kukull, W.A., Mackenzie, I.R., Masliah, E., McKee, A., Montine, T.J., Murray, M.E., Neltner, J.H., Santa-Maria, I., Seeley, W.W., Serrano-Pozo, A., Shelanski, M.L., Stein, T., Takao, M., Thal, D.R., Toledo, J.B., Troncoso, J.C., Vonsattel, J.P., White, 3rd., C.L., Wisniewski, T., Woltjer, R.L., Yamada, M., Nelson, P.T., 2014. Primary age-related tauopathy (PART): a common pathology associated with human aging. Acta. Neuropathol. 128, 755–766.

Csernansky, J.G., Miller, J.P., Mckeel, D., Morris, J.C., 2002. Relationships among cerebrospinal fluid biomarkers in dementia of the Alzheimer type. Alzheimer Dis. Assoc. Disord. 16, 144–149.

de Almeida, S.M., Shumaker, S.D., LeBlanc, S.K., Delaney, P., Marquie-Beck, J., Ueland, S., Alexander, T., Ellis, R.J., 2011. Incidence of post-dural puncture headache in research volunteers. Headache 51, 1503–1510.

de Jong, D., Jansen, R.W., Kremer, B.P., Verbeek, M.M., 2006. Cerebrospinal fluid amyloid-beta42/phosphorylated tau ratio discriminates between Alzheimer's disease and vascular dementia. J. Gerontol. A Biol. Sci. Med. Sci. 61, 755–758.

de Rino, F., Martinelli-Boneschi, F., Caso, F., Zuffi, M., Zabeo, M., Passerini, G., Comi, G., Magnani, G., Franceschi, M., 2012. CSF metabolites in the differential diagnosis of Alzheimer's disease from frontal variant of frontotemporal dementia. Neurol. Sci. 33, 973–977.

Diniz, B.S., Pinto Júnior, J.A., Forlenza, O.V., 2008. Do CSF total tau, phosphorylated tau, and beta-amyloid 42 help to predict progression of mild cognitive impairment to Alzheimer's disease? A systematic review and meta-analysis of the literature. World J. Biol. Psychiatry 9, 172–182.

Doody, R.S., Raman, R., Farlow, M., Iwatsubo, T., Vellas, B., Joffe, S., Kieburtz, K., He, F., Sun, X., Thomas, R.G., Aisen, P.S., Alzheimer's Disease Cooperative Study Steering Committee, Siemers, E., Sethuraman, G., Mohs, R., Semagacestat Study Group, 2013. A phase 3 trial of semagacestat for treatment of Alzheimer's disease. N. Engl. J. Med. 369, 341–350.

Doody, R.S., Thomas, R.G., Farlow, M., Iwatsubo, T., Vellas, B., Joffe, S., Kieburtz, K., Raman, R., Sun, X., Aisen, P.S., Siemers, E., Liu-Seifert, H., Mohs, R., Alzheimer's Disease Cooperative Study Steering Committee; Solanezumab Study Group, 2014. Phase 3 trials of solanezumab for mild-to-moderate Alzheimer's disease. N. Engl. J. Med. 370, 311–321.

Dubois, B., Feldman, H.H., Jacova, C., Cummings, J.L., Dekosky, S.T., Barberger-Gateau, P., Delacourte, A., Frisoni, G., Fox, N.C., Galasko, D., Gauthier, S., Humpel, H., Jicha, G.A., Meguro, K., O'Brien, J., Pasquier, F., Robert, P., Rossor, M., Salloway, S., Sarazin, M., de Souza, L.C., Stern, Y., Visser, P.J., Scheltens, P., 2010. Revising the definition of Alzheimer's disease: a new lexicon. Lancet Neurol. 9, 1118–1127.

Dubois, B., Feldman, H.H., Jacova, C., Humpel, H., Molinuevo, J.L., Blennow, K., DeKosky, S.T., Gauthier, S., Selkoe, D., Bateman, R., Cappa, S., Crutch, S., Engelborghs, S., Frisoni, G.B., Fox, N.C., Galasko, D., Habert, M.O., Jicha, G.A., Nordberg, A., Pasquier, F., Rabinovici, G., Robert, P., Rowe, C., Salloway, S., Sarazin, M., Epelbaum, S., de Souza, L.C., Vellas, B., Visser, P.J., Schneider, L., Stern, Y., Scheltens, P., Cummings, J.L., 2014. Advancing research diagnostic criteria for Alzheimer's disease: the IWG-2 criteria. Lancet Neurol. 13, 614–629.

Elias-Sonnenschein, L.S., Helisalmi, S., Natunen, T., Hall, A., Paajanen, T., Herukka, S.K., Laitinen, M., Remes, A.M., Koivisto, A.M., Mattila, K.M., Lehtimäki, T., Verhey, F.R., Visser, P.J., Soininen, H., Hiltunen, M., 2013. Genetic loci associated with Alzheimer's disease and cerebrospinal fluid biomarkers in a Finnish case-control cohort. PLoS One. 8, e59676.

Engelborghs, S., Sleegers, K., Cras, P., Brouwers, N., Serneels, S., De Leenheir, E., Martin, J.J., Vanmechelen, E., Van Broeckhoven, C., De Deyn, P.P., 2007. No association of CSF biomarkers with APOEepsilon4, plaque and tangle burden in definite Alzheimer's disease. Brain 130, 2320–2326.

Fagan, A.M., Mintun, M.A., Shah, A.R., Aldea, P., Roe, C.M., Mach, R.H., Marcus, D., Morris, J.C., Holtzman, D.M., 2009. Cerebrospinal fluid tau and p-tau(181) increase with cortical amyloid deposition in cognitively normal individuals: implications for future clinical trials of Alzheimer's disease. EMBO Mol. Med. 1, 371–380.

Fagan, A.M., Xiong, C., Jasielec, M.S., Bateman, R.J., Goate, A.M., Benzinger, T.L., Ghetti, B., Martins, R.N., Masters, C.L., Mayeux, R., Ringman, J.M., Rossor, M.N., Salloway, S., Schofield, P.R., Sperling, R.A., Marcus, D., Cairns, N.J., Buckles, V.D., Ladenson, J.H., Morris, J.C., Holtzman, D.M., Dominantly Inherited Alzheimer Network, 2014. Longitudinal change in CSF biomarkers in autosomal-dominant Alzheimer's disease. Sci. Transl. Med. 6, (Article ID: 226ra30), 1–35.

Ferreira, D., Perestelo-Pérez, L., Westman, E., Wahlund, L.O., Sarría, A., Serrano-Aguilar, P., 2014. Meta-review of CSF core biomarkers in Alzheimer's disease: The State-of-the-Art after the New Revised Diagnostic Criteria. Front Aging Neurosci. 6, 47.

Fleisher, A.S., Raman, R., Siemers, E.R., Becerra, L., Clark, C.M., Dean, R.A., Farlow, M.R., Galvin, J.E., Peskind, E.R., Quinn, J.F., Sherzai, A., Sowell, B.B., Aisen, P.S., Thal, L.J., 2008. Phase 2 safety trial targeting amyloid beta production with a gamma-secretase inhibitor in Alzheimer disease. Arch. Neurol. 65, 1031–1038.

Fukuyama, R., Mizuno, T., Mori, S., Nakajima, K., Fushiki, S., Yanagisawa, K., 2000. Age-dependent change in the levels of Abeta40 and Abeta42 in cerebrospinal fluid from control subjects, and a decrease in the ratio of Abeta42 to Abeta40 level in cerebrospinal fluid from Alzheimer's disease patients. Eur. Neurol. 43, 155–160.

Gabelle, A., Dumurgier, J., Vercruysse, O., Paquet, C., Bombois, S., Laplanche, J.L., Peoc'h, K., Schraen, S., Buée, L., Pasquier, F., Hugon, J., Touchon, J., Lehmann, S., 2013. Impact of the 2008-2012 French Alzheimer Plan on the use of cerebrospinal fluid biomarkers in research memory center: the PLM Study. J. Alzheimers Dis. 34, 297–305.

Galasko, D., Chang, L., Motter, R., Clark, C.M., Kaye, J., Knopman, D., Thomas, R., Kholodenko, D., Schenk, D., Lieberburg, I., Miller, B., Green, R., Basherad, R., Kertiles, L., Boss, M.A., Seubert, P., 1998. High cerebrospinal fluid tau and low amyloid beta42 levels in the clinical diagnosis of Alzheimer disease and relation to apolipoprotein E genotype. Arch. Neurol. 55, 937–945.

Georganopoulou, D.G., Chang, L., Nam, J.M., Thaxton, C.S., Mufson, E.J., Klein, W.L., Mirkin, C.A., 2005. Nanoparticle-based detection in cerebral spinal fluid of a soluble pathogenic biomarker for Alzheimer's disease. Proc. Natl. Acad. Sci. USA 102, 2273–2276.

Green, A.J.E., Harvey, R.J., Thompson, E.J., Rossor, M.N., 1999. Increased tau in the cerebrospinal fluid of patients with frontotemporal dementia and Alzheimer's disease. Neurosci. Lett. 259, 133–135.

Hake, A., Trzepacz, P.T., Wang, S., Yu, P., Case, M., Hochstetler, H., Witte, M.M., Degenhardt, E.K., Dean, R.A., Alzheimer's Disease Neuroimaging Initiative, 2015. Florbetapir positron emission tomography and cerebrospinal fluid biomarkers. Alzheimers Dement. 11, 986–993.

Hansson, O., Zetterberg, H., Buchhave, P., Londos, E., Blennow, K., Minthon, L., 2006. Association between CSF biomarkers and incipient Alzheimer's disease in patients with mild cognitive impairment: a follow-up study. Lancet Neurol. 5, 228–234.

Hardy, J., Selkoe, D.J., 2002. The amyloid hypothesis of Alzheimer's disease: progress and problems on the road to therapeutics. Science 297, 353–355.

Hart, I.K., Bone, I., Hadley, D.M., 1988. Development of neurological problems after lumber puncture. Br. Med. J. 296, 51–52.

Herukka, S.K., Helisalmi, S., Hallikainen, M., Tervo, S., Soininen, H., Pirttilä, T., 2007. CSF Abeta42, Tau and phosphorylated Tau, APOE epsilon4 allele and MCI type in progressive MCI. Neurobiol. Aging 28, 507–514.

Hogervorst, E., Bandelow, S., Combrinck, M., Irani, S.R., Smith, A.D., 2003. The validity and reliability of 6 sets of clinical criteria to classify Alzheimer's disease and vascular dementia in cases confirmed post-mortem: added value of a decision tree approach. Dement. Geriatr. Cogn. Disord. 16, 170–180.

Holtzman, D.M., 2011. CSF biomarkers for Alzheimer's disease: current utility and potential future use. Neurobiol. Aging 32 (Suppl 1), S4–S9.

Hongpaisan, J., Sun, M.K., Alkon, D.L., 2011. PKCε activation prevents synaptic loss, Aβ elevation, and cognitive deficits in Alzheimer's disease transgenic mice. J. Neurosci. 31, 630–643.

Humpel, C., 2011. Identifying and validating biomarkers for Alzheimer's disease. Trends Biotechnol. 29, 26–32.

Humpel, H., Buerge, K., Zinkowski, R., Teipel, S.J., Goernitz, A., Andreasen, N., Sjoegren, M., DeBernardis, J., Kerkman, D., Ishiguro, K., Ohno, H., Vanmechelen, E., Vanderstichele, H., McCulloch, C., Mölle, H.-J., Davies, P., Blennow, K., 2004. Measurement of phosphorylated tau epitopes in the differential diagnosis of Alzheimer disease: A comparative cerebrospinal fluid study. Arch. Gen. Psychiatry. 61, 95–102.

Humpel, H., Bürger, K., Teipel, S.J., Bokde, A.L.W., Zetterberg, H., Blennow, K., 2008. Core candidate neurochemical and imaging biomarkers of Alzheimer's disease. Alzheimer Dement. 4, 38–48.

Humpel, H., Lista, S., Khachaturian, Z.S., 2012. Development of biomarkers to chart all Alzheimer's disease stages: the royal road to cutting the therapeutic Gordian Knot. Alzheimers Dement. 8, 312–336.

Iqbal, K., Grundke-Iqbal, I., 2008. Alzheimer neurofibrillary degeneration: significance, etiopathogenesis, therapeutics and prevention. J. Cell Mol. Med. 12, 38–55.

Jagust, W.J., Landau, S.M., Shaw, L.M., Trojanowski, J.Q., Koeppe, R.A., Reiman, E.M., Foster, N.L., Petersen, R.C., Weiner, M.W., Price, J.C., Mathis, C.A., 2009. Relationships between biomarkers in aging and dementia. Neurology 73, 1193–1199.

Kaerst, L., Kuhlmann, A., Wedekind, D., Stoeck, K., Lange, P., Zerr, I., 2013. Cerebrospinal fluid biomarkers in Alzheimer's disease, vascular dementia and ischemic stroke patients: a critical analysis. J. Neurol. 260, 2722–2727.

Kang, J.-E., Lim, M.M., Bateman, R.J., Lee, J.J., Smyth, L.P., John, R., Cirrito, J.R., Fujiki, N., Nishino, S., Holtzman, D.M., 2009. Amyloid-β dynamics are regulated by orexin and the sleep-wake cycle. Science 326, 1005–1007.

Kang, J.H., Korecka, M., Toledo, J.B., Trojanowski, J.Q., Shaw, L.M., 2013. Clinical utility and analytical challenges in measurement of cerebrospinal fluid amyloid-beta(1-42) and tau proteins as Alzheimer disease biomarkers. Clin. Chem. 59, 903–916.

Kauwe, J.S., Cruchaga, C., Karch, C.M., Sadler, B., Lee, M., Mayo, K., Latu, W., Su'a, M., Fagan, A.M., Holtzman, D.M., Morris, J.C., Alzheimer's Disease Neuroimaging Initiative, Goate, A.M., 2011. Fine mapping of genetic variants in BIN1, CLU, CR1 and PICALM for association with cerebrospinal fluid biomarkers for Alzheimer's disease. PLoS One 6, e15918.

Khan, T.K., Alkon, D.L., 2010. Early diagnostic accuracy and pathophysiologic relevance of an autopsy-confirmed Alzheimer's disease peripheral biomarker. Neurobiol. Aging 31, 889–900.

Khan, T.K., Alkon, D.L., 2015a. Peripheral biomarkers of Alzheimer's disease. J. Alzheimers Dis. 44, 729–744.

Khan, T.K., Alkon, D.L., 2015b. Alzheimer's disease cerebrospinal fluid and neuroimaging biomarkers: diagnostic accuracy and relationship to drug efficacy. J. Alzheimers Dis. 46, 817–836.

Kim, S., Swaminathan, S., Shen, L., Risacher, S.L., Nho, K., Foroud, T., Shaw, L.M., Trojanowski, J.Q., Potkin, S.G., Huentelman, M.J., Craig, D.W., DeChairo, B.M., Aisen, P.S., Petersen, R.C., Weiner, M.W., Saykin, A.J., Alzheimer's Disease Neuroimaging Initiative, 2011. Genome-wide association study of CSF biomarkers Aβ1-42, t-tau, and p-tau181p in the ADNI cohort. Neurology 76, 69–79.

Kuntz, K.M., Kokmen, E., Stevens, J.C., Miller, P., Offord, K.P., Ho, M.M., 1992. Post-lumbar puncture headaches: experience in 501 consecutive procedures. Neurology 42, 1884–1887.

Lammich, S., Kojro, E., Postina, R., Gilbert, S., Pfeiffer, R., Jasionowski, M., Haass, C., Fahrenholz, F., 1999. Constitutive and regulated alpha-secretase cleavage of Alzheimer's amyloid precursor protein by a disintegrin metalloprotease. Proc. Natl. Acad. Sci. USA 96, 3922–3927.

Landau, S.M., Lu, M., Joshi, A.D., Pontecorvo, M., Mintun, M.A., Trojanowski, J.Q., Shaw, L.M., Jagust, W.J., Alzheimer's Disease Neuroimaging Initiative, 2013. Comparing positron emission tomography imaging and cerebrospinal fluid measurements of β-amyloid. Ann. Neurol. 74, 826–886.

Lannfelt, L., Blennow, K., Zetterberg, H., Batsman, S., Ames, D., Harrison, J., Masters, C.L., Targum, S., Bush, A.I., Murdoch, R., Wilson, J., Ritchie, C.W., PBT2-201-EURO study group, 2008. Safety, efficacy, and biomarker findings of PBT2 in targeting Abeta as a modifying therapy for Alzheimer's disease: a phase IIa, double-blind, randomised, placebo-controlled trial. Lancet Neurol. 7, 779–786.

Lanz, T.A., Hosley, J.D., Adams, W.J., Merchant KM, 2004. Studies of Abeta pharmacodynamics in the brain, cerebrospinal fluid, and plasma in young (plaque-free) Tg2576 mice using the gamma-secretase inhibitor N2-[(2S)-2-(3,5-difluorophenyl)-2-hydroxyethanoyl]-N1-[(7S)-5-methyl-6-oxo-6,7-dihydro-5H-dibenzo[b,d]azepin-7-yl]-L-alaninamide (LY-411575). J. Pharmacol. Exp. Ther. 309, 49–55.

Lautner, R., Palmqvist, S., Mattsson, N., Andreasson, U., Wallin, A., Pålsson, E., Jakobsson, J., Herukka, S.K., Owenius, R., Olsson, B., Humpel, H., Rujescu, D., Ewers, M., Landén, M., Minthon, L., Blennow, K., Zetterberg, H., Hansson, O., Alzheimer's Disease Neuroimaging Initiative, 2014. Apolipoprotein E genotype and the diagnostic accuracy of cerebrospinal fluid biomarkers for Alzheimer disease. JAMA Psychiatry 71, 1183–1191.

Le Bastard, N., Coart, E., Vanderstichele, H., Vanmechelen, E., Martin, J.J., Engelborghs, S., 2013. Comparison of two analytical platforms for the clinical qualification of Alzheimer's disease biomarkers in pathologically-confirmed dementia. J. Alzheimers Dis. 33, 117–131.

Le Bastard, N., Martin, J.J., Vanmechelen, E., Vanderstichele, H., De Deyn, P.P., Engelborghs, S., 2010. Added diagnostic value of CSF biomarkers in differential dementia diagnosis. Neurobiol. Aging 31, 1867–1876.

Le Bastard, N., Van Buggenhout, M., De Leenheir, E., Martin, J.J., De Deyn, P.P., Engelborghs, S., 2007. Low specificity limits the use of the cerebrospinal fluid Aβ1-42/P-TAU181P ratio to discriminate Alzheimer's disease from vascular dementia. J. Gerontol. A Biol. Sci. Med. Sci. 62, 923–924.

Leuzy, A., Carter, S.F., Chiotis, K., Almkvist, O., Wall, A., Nordberg, A., 2015. Concordance and diagnostic accuracy of [11C]PIB PET and cerebrospinal fluid biomarkers in a sample of patients with mild cognitive impairment and Alzheimer's disease. J. Alzheimers Dis. 45, 1077–1088.

Lewczuk, P., Beck, G., Ganslandt, O., Esselmann, H., Deisenhammer, F., Regeniter, A., Petereit, H.F., Tumani, H., Gerritzen, A., Oschmann, P., Schröder, J., Schönknecht, P., Zimmermann, K., Humpel, H., Bürger, K., Otto, M., Haustein, S., Herzog, K., Dannenberg, R., Wurster, U., Bibl, M., Maler, J.M., Reubach, U., Kornhuber, J., Wiltfang, J., 2006. International quality control survey of neurochemical dementia diagnostics. Neurosci. Lett. 409, 1–4.

Lista, S., Garaci, F.G., Ewers, M., Teipel, S., Zetterberg, H., Blennow, K., Humpel, H., 2014. CSF Aβ1-42 combined with neuroimaging biomarkers in the early detection, diagnosis and prediction of Alzheimer's disease. Alzheimers Dement. 10, 381–392.

Lleó, A., Cavedo, E., Parnetti, L., Vanderstichele, H., Herukka, S.K., Andreasen, N., Ghidoni, R., Lewczuk, P., Jeromin, A., Winblad, B., Tsolaki, M., Mroczko, B., Visser, P.J., Santana, I., Svenningsson, P., Blennow, K., Aarsland, D., Molinuevo, J.L., Zetterberg, H., Mollenhauer, B., 2015. Cerebrospinal fluid biomarkers in trials for Alzheimer and Parkinson diseases. Nat. Rev. Neurol. 11, 41–55.

Maccecchini, M.L., Chang, M.Y., Pan, C., John, V., Zetterberg, H., Greig, N.H., 2012. Posiphen as a candidate drug to lower CSF amyloid precursor protein, amyloid-β peptide and τ levels: target engagement, tolerability and pharmacokinetics in humans. J. Neurol. Neurosurg. Psychiatry 83, 894–902.

Mattsson, N., Zetterberg, H., 2012. What is a certified reference material. Biomark. Med. 6, 369–370.

Mattsson, N., Andreasson, U., Persson, S., Carrillo, M.C., Collins, S., Chalbot, S., Cutler, N., Dufour-Rainfray, D., Fagan, A.M., Heegaard, N.H., Robin Hsiung, G.Y., Hyman, B., Iqbal, K., Kaeser, S.A., Lachno, D.R., Lleó, A., Lewczuk, P., Molinuevo, J.L., Parchi, P., Regeniter, A., Rissman, R.A., Rosenmann, H., Sancesario, G., Schröder, J., Shaw, L.M., Teunissen, C.E., Trojanowski, J.Q., Vanderstichele, H., Vandijck, M., Verbeek, M.M., Zetterberg, H., Blennow, K., Alzheimer's Association QC Program Work Group, 2013. CSF biomarker variability in the Alzheimer's Association quality control program. Alzheimers Dement. 9, 251–261.

Mattsson, N., Zegers, I., Andreasson, U., Bjerke, M., Blankenstein, M.A., Bowser, R., Carrillo, M.C., Gobom, J., Heath, T., Jenkins, R., Jeromin, A., Kaplow, J., Kidd, D., Laterza, O.F., Lockhart, A., Lunn, M.P., Martone, R.L., Mills, K., Pannee, J., Ratcliffe, M., Shaw, L.M., Simon, A.J., Soares, H., Teunissen, C.E., Verbeek, M.M., Umek, R.M., Vanderstichele, H., Zetterberg, H., Blennow, K.,

Portelius, E., 2012. Reference measurement procedures for Alzheimer's disease cerebrospinal fluid biomarkers: definitions and approaches with focus on amyloid β42. Biomark. Med. 6, 409–417.

Mattsson, N., Zetterberg, H., Hansson, O., Andreasen, N., Parnetti, L., Jonsson, M., Herukka, S.-K., van der Flier, W.M., Blankenstein, M.A., Ewers, M., Rich, K., Kaiser, E., Verbeek, M., Tsolaki, M., Mulugeta, E., Rosén, E., Aarsland, D., Visse, P.J., Schröder, J., Marcusson, J., de Leon, M., Humpel, H., Scheltens, P., Pirttilä, T., Wallin, A., Jönhagen, M.E., Minthon, L., Winblad, B., Blennow, K., 2009. CSF biomarkers and incipient Alzheimer disease in patients with mild cognitive impairment. JAMA 302, 385–393.

McKhann, G.M., Knopman, D.S., Chertkow, H., Hyman, B.T., Jack, Jr., C.R., Kawas, C.H., Klunk, W.E., Koroshetz, W.J., Manly, J.J., Mayeux, R., Mohs, R.C., Morris, J.C., Rossor, M.N., Scheltens, P., Carrillo, M.C., Thies, B., Weintraub, S., Phelps, C.H., 2011. The diagnosis of dementia due to Alzheimer's disease: recommendations from the National Institute on Aging-Alzheimer's Association workgroups on diagnostic guidelines for Alzheimer's disease. Alzheimers Dement. 7, 263–269.

McMillan, C.T., Avants, B., Irwin, D.J., Toledo, J.B., Wolk, D.A., Van Deerlin, V.M., Shaw, L.M., Trojanoswki, J.Q., Grossman, M., 2013. Can MRI screen for CSF biomarkers in neurodegenerative disease? Neurology 80, 132–138.

Mitchell, A.J., 2009. CSF phosphorylated tau in the diagnosis and prognosis of mild cognitive impairment and Alzheimer's disease: a meta-analysis of 51 studies. J. Neurol. Neurosurg. Psychiatry 80, 966–975.

Mitchell, A.J., Monge-Argilés, J.A., Sánchez-Paya, J., 2010. Do CSF biomarkers help clinicians predict the progression of mild cognitive impairment to dementia? Pract. Neurol. 10, 202–207.

Molinuevo, J.L., Gispert, J.D., Dubois, B., Heneka, M.T., Lleo, A., Engelborghs, S., Pujol, J., de Souza, L.C., Alcolea, D., Jessen, F., Sarazin, M., Lamari, F., Balasa, M., Antonell, A., Rami, L., 2013. The AD-CSF-index discriminates Alzheimer's disease patients from healthy controls: a validation study. J. Alzheimers Dis. 36, 67–77.

Motter, R., Vigo-Pelfrey, C., Kholodenko, D., Barbour, R., Johnson-Wood, K., Galasko, D., Chang, L., Miller, B., Clark, C., Green, R., Olson, D., Southwick, P., Wolfert, B., Munroe, B., Lieberburg, I., Seubert, P., Schenk, D., 1995. Reduction of β-amyloid peptide 42 in the cerebrospinal fluid of patients with Alzheimer's disease. Ann. Neurol. 38, 643–648.

Mucke, L., Selkoe, D.J., 2012. Neurotoxicity of amyloid β-protein: synaptic and network dysfunction. Cold Spring Harb. Perspect. Med. 2, a006338.

Muñoz-Ruiz, M.Á., Hartikainen, P., Hall, A., Mattila, J., Koikkalainen, J., Herukka, S.K., Julkunen, V., Vanninen, R., Liu, Y., Lötjönen, J., Soininen, H., 2013. Disease state fingerprint in frontotemporal degeneration with reference to Alzheimer's disease and mild cognitive impairment. J. Alzheimers Dis. 35, 727–739.

Oddo, S., Caccamo, A., Kitazawa, M., Tseng, B.P., LaFerla, F.M., 2003. Amyloid deposition precedes tangle formation in a triple transgenic model of Alzheimer's disease. Neurobiol. Aging 24, 1063–1070.

Otto, M., Esselmann, H., Schulz-Shaeffer, W., Neumann, M., Schröter, A., Ratzka, P., Cepek, L., Zerr, I., Steinacker, P., Windl, O., Kornhuber, J., Kretzschmar, H.A., Poser, S., Wiltfang, J., 2000. Decreased beta-amyloid 1-42 in cerebrospinal fluid of patients with Creutzfeldt–Jakob disease. Neurology 54, 1099–1102.

Otto, M., Wiltfang, J., Tumani, H., Zerr, I., Lantsch, M., Kornhuber, J., 1997. Elevated levels of tau protein in cerebrospinal fluid of patients with Creutzfeldt–Jakob disease. Neurosci. Lett. 225, 210–212.

Palmqvist, S., Zetterberg, H., Blennow, K., Vestberg, S., Andreasson, U., Brooks, D.J., Owenius, R., Hägerström, D., Wollmer, P., Minthon, L., Hansson, O., 2014. Accuracy of brain amyloid detection in clinical practice using cerebrospinal fluid β-amyloid 42: a cross-validation study against amyloid positron emission tomography. JAMA Neurol. 10, 1282–1289.

Parnetti, L., Amici, S., Lanari, A., Romani, C., Antognelli, C., Andreasen, N., Minthon, L., Davidsson, P., Pottel, H., Blennow, K., Gallai, V., 2002. Cerebrospinal fluid levels of biomarkers and activity of acetylcholinesterase (AChE) and butyrylcholinesterase in AD patients before and after treatment with different AChE inhibitors. Neurol. Sci. 23 (Suppl 2), S95–S96.

Peskind, E., Nordberg, A., Darreh-Shori, T., Soininen, H., 2009. Safety of lumbar puncture procedures in patients with Alzheimer's disease. Curr. Alzheimer Res. 6, 290–292.

Petzold, A., Sharpe, L.T., Keir, G., 2006. Spectrophotometry for cerebrospinal fluid pigment analysis. Neurocrit. Care 4, 153–162.

Portelius, E., Price, E., Brinkmalm, G., Stiteler, M., Olsson, M., Persson, R., Westman-Brinkmalm, A., Zetterberg, H., Simon, A.J., Blennow, K., 2011. A novel pathway for amyloid precursor protein processing. Neurobiol. Aging 32, 1090–1098.

Prince, J.A., Zetterberg, H., Andreasen, N., Marcusson, J., Blennow, K., 2004. APOE epsilon4 allele is associated with reduced cerebrospinal fluid levels of Abeta42 (2004). Neurology 62, 2116–2118.

Riekse, R.G., Li, G., Petrie, E.C., Leverenz, J.B., Vavrek, D., Vuletic, S., Albers, J.J., Montine, T.J., Lee, V.M., Lee, M., Seubert, P., Galasko, D., Schellenberg, G.D., Hazzard, W.R., Peskind, E.R., 2006. Effect of statins on Alzheimer's disease biomarkers in cerebrospinal fluid. J. Alzheimers Dis. 10, 399–406.

Riemenschneider, M., Lautenschlager, N., Wagenpfeil, S., Diehl, J., Drzezga, A., Kurz, A., 2002. Cerebrospinal fluid tau and beta-amyloid 42 proteins identify Alzheimer disease in subjects with mild cognitive impairment. Arch. Neurol. 59, 1729–1734.

Rissman, R.A., Trojanowski, J.Q., Shaw, L.M., Aisen, P.S., 2012. Longitudinal plasma amyloid beta as a biomarker of Alzheimer's disease. J. Neural Transm. 119, 843–850.

Rosén, C., Hansson, O., Blennow, K., Zetterberg, H., 2013. Fluid biomarkers in Alzheimer's disease—current concepts. Mol. Neurodegener. 8, 20.

Rosso, S.M., van Herpen, E., Pijnenburg, Y.A.L., Schoonenboom, N.S.M., Scheltens, P., Heutink, P., van Swieten, J.C., 2003. Total tau and phosphorylated tau 181 levels in the cerebrospinal fluid of patients with frontotemporal dementia due to P301L and G272V tau mutations. Arch. Neurol. 60, 1209–1213.

Salloway, S., Sperling, R., Fox, N.C., Blennow, K., Klunk, W., Raskind, M., Sabbagh, M., Honig, L.S., Porsteinsson, A.P., Ferris, S., Reichert, M., Ketter, N., Nejadnik, B., Guenzler, V., Miloslavsky, M., Wang, D., Lu, Y., Lull, J., Tudor, I.C., Liu, E., Grundman, M., Yuen, E., Black, R., Brashear, H.R., Bapineuzumab 301 and 302 Clinical Trial Investigators, 2014. Two phase 3 trials of bapineuzumab in mild-to-moderate Alzheimer's disease. N. Engl. J. Med. 370, 322–333.

Salloway, S., Sperling, R., Keren, R., Porsteinsson, A.P., van Dyck, C.H., Tariot, P.N., Gilman, S., Arnold, D., Abushakra, S., Hernandez, C., Crans, G., Liang, E., Quinn, G., Bairu, M., Pastrak, A., Cedarbaum, J.M., ELND005-AD201 Investigators, 2011. A phase 2 randomized trial of ELND005, scyllo-inositol, in mild to moderate Alzheimer disease. Neurology 77, 1253–1262.

Scheff, S.W., Price, D.A., Schmitt, F.A., Mufson, E.J., 2006. Hippocampal synaptic loss in early Alzheimer's disease and mild cognitive impairment. Neurobiol. Aging 27, 1372–1384.

Schmand, B., Huizenga, H.M., van Gool, W.A., 2010. Meta-analysis of CSF and MRI biomarkers for detecting preclinical Alzheimer's disease. Psychol. Med. 40, 135–145.

Schönknecht, P., Pantel, J., Hartmann, T., Werle, E., Volkmann, M., Essig, M., Amann, M., Zanabili, N., Bardenheuer, H., Hunt, A., Schröder, J., 2003. Cerebrospinal fluid tau levels in Alzheimer's disease are elevated when compared with vascular dementia but do not correlate with measures of cerebral atrophy. Psychiatry Res. 120, 231–238.

Seeburger, J.L., Holder, D.J., Combrinck, M., Joachim, C., Laterza, O., Tanen, M., Dallob, A., Chappell, D., Snyder, K., Flynn, M., Simon, A., Modur, V., Potter, W.Z., Wilcock, G., Savage, M.J., Smith, A.D., 2015. Cerebrospinal fluid biomarkers distinguish postmortem-confirmed Alzheimer's disease from other dementias and healthy controls in the OPTIMA cohort. J. Alzheimers Dis. 44, 525–539.

Selkoe, D.J., 2001. Alzheimer's disease: genes, proteins, and therapy. Physiol. Rev. 81, 741–766.

Selkoe, D.J., Wolfe, M.S., 2007. Presenilin: running with scissors in the membrane. Cell 131, 215–221.

Seppälä, T.T., Koivisto, A.M., Hartikainen, P., Helisalmi, S., Soininen, H., Herukka, S.K., 2011. Longitudinal changes of CSF biomarkers in Alzheimer's disease. J. Alzheimers Dis. 25, 583–594.

Seppälä, T.T., Nerg, O., Koivisto, A.M., Rummukainen, J., Puli, L., Zetterberg, H., Pyykkö, O.T., Helisalmi, S., Alafuzoff, I., Hiltunen, M., Jääskeläinen, J.E., Rinne, J., Soininen, H., Leinonen, V., Herukka, S.K., 2012. CSF biomarkers for Alzheimer disease correlate with cortical brain biopsy findings. Neurology 78, 1568–1575.

Shaw, L.M., Vanderstichele, H., Knapik-Czajka, M., Clark, C.M., Aisen, P.S., Petersen, R.C., Blennow, K., Soares, H., Simon, A., Lewczuk, P., Dean, R., Siemers, E., Potter, W., Lee, V.M., Trojanowski, J.Q., Alzheimer's Disease Neuroimaging Initiative, 2009. Cerebrospinal fluid biomarker signature in Alzheimer's disease neuroimaging initiative subjects. Ann. Neurol. 65, 403–413.

Shoji, M., Matsubara, E., Murakami, T., Manabe, Y., Abe, K., Kanai, M., Ikeda, M., Tomidokoro, Y., Shizuka, M., Watanabe, M., Amari, M., Ishiguro, K., Kawarabayashi, T., Harigaya, Y., Okamoto, K., Nishimura, T., Nakamura, Y., Takeda, M., Urakami, K., Adachi, Y., Nakashima, K., Arai, H., Sasaki, H., Kanemaru, K., Yamanouchi, H., Yoshida, Y., Ichise, K., Tanaka, K., Hamamoto, M., Yamamoto, H., Matsubayashi, T., Yoshida, H., Toji, H., Nakamura, S., Hirai, S., 2002. Cerebrospinal fluid tau in dementia disorders: a large scale multicenter study by a Japanese study group. Neurobiol. Aging 23, 363–370.

Sjögren, M., Davidsson, P., Tullberg, M., Minthon, L., Wallin, A., Wikkelso, C., Granérus, A.K., Vanderstichele, H., Vanmechelen, E., Blennow, K., 2001. Both total and phosphorylated tau are increased in Alzheimer's disease. J. Neurol. Neurosurg. Psychiatry 70, 624–630.

Sjögrena, M., Davidssona, P., Wallina, A., Granérusb, A.-K., Grundströmd, E., Askmarkd, H., Vanmechelenc, E., Blennowa, K., 2002. Decreased CSF-β-Amyloid 42 in Alzheimer's disease and Amyotrophic lateral sclerosis may reflect mismetabolism of β-Amyloid induced by disparate mechanisms. Dement. Geriatr. Cogn. Disord. 13, 112–118.

Slats D. Hour-to-hour variability CSF biomarkers in AD patients. Presented at the Alzheimer's Association International Conference on Alzheimer's Disease (AAICAD) July 10–15, 2010, Honolulu, Hawaii.

Sluimer, J.D., Bouwman, F.H., Vrenken, H., Blankenstein, M.A., Barkhof, F., van der Flier, W.M., Scheltens, P., 2010. Whole-brain atrophy rate and CSF biomarker levels in MCI and AD: a longitudinal study. Neurobiol. Aging 31, 758–764.

Sperling, R.A., Aisen, P.S., Beckett, L.A., Bennett, D.A., Craft, S., Fagan, A.M., Iwatsubo, T., Jack, Jr., C.R., Kaye, J., Montine, T.J., Park, D.C., Reiman, E.M., Rowe, C.C., Siemers, E., Stern, Y., Yaffe, K., Carrillo, M.C., Thies, B., Morrison-Bogorad, M., Wagster, M.V., Phelps, C.H., 2011. Toward defining the preclinical stages of Alzheimer's disease: recommendations from the National Institute on Aging-Alzheimer's Association workgroups on diagnostic guidelines for Alzheimer's disease. Alzheimers Dement. 7, 280–292.

Spies, P.E., Slats, D., Sjögren, J.M., Kremer, B.P., Verhey, F.R., Rikkert, M.G., Verbeek, M.M., 2009. The cerebrospinal fluid amyloid-beta(42/40) ratio in the differentiation of Alzheimer's disease from non-Alzheimer's dementia. Curr. Alzheimer Res. 16, 363–369.

Strozyk, D., Blennow, K., White, L.R., Launer, L.J., 2003. CSF Abeta 42 levels correlate with amyloid-neuropathology in a population-based autopsy study. Neurology 60, 652–656.

Strupp, M., Katsarava, Z., 2009. Post-lumbar puncture syndrome and spontaneous low CSF pressure syndrome. Nervenarzt 80, 1509–1519.

Struyfs, H., Molinuevo, J.L., Martin, J.J., De Deyn, P.P., Engelborghs, S., 2014. Validation of the AD-CSF-index in autopsy-confirmed Alzheimer's disease patients and healthy controls. J. Alzheimers Dis. 41, 903–909.

Sunderland, T., Linker, G., Mirza, N., Putnam, K.T., Friedman, D.L., Kimmel, L.H., Bergeson, J., Manetti, G.J., Zimmermann, M., Tang, B., Bartko, J.J., Cohen, R.M., 2003. Decreased beta-amyloid1-42 and increased tau levels in cerebrospinal fluid of patients with Alzheimer disease. JAMA 289, 2094–2103.

Sunderland, T., Mirza, N., Putnam, K.T., Linker, G., Bhupali, D., Durham, R., Soares, H., Kimmel, L., Friedman, D., Bergeson, J., Csako, G., Levy, J.A., Bartko, J.J., Cohen, R.M., 2004. Cerebrospinal fluid beta-amyloid1-42 and tau in control subjects at risk for Alzheimer's disease: the effect of APOE epsilon4 allele. Biol. Psychiatry 56, 670–676.

Süssmuth, S.D., Uttner, I., Landwehrmeyer, B., Pinkhardt, E.H., Brettschneider, J., Petzold, A., Kramer, B., Schulz, J.B., Palm, C., Otto, M., Ludolph, A.C., Kassubek, J., Tumani, H., 2010. Differential pattern of brain-specific CSF proteins tau and amyloid-β in Parkinsonian syndromes. Mov. Disord. 25, 1284–1288.

Teunissen, C.E., Verwey, N.A., Kester, M.I., van Uffelen, K., Blankenstein, M.A., 2010. Standardization of assay procedures for analysis of the CSF biomarkers amyloid β((1:42)) tau, and phosphorylated tau in Alzheimer's disease: report of an International Workshop. Int. J. Alzheimers Dis., (Article ID 635053), 1–6.

Tolboom, N., van der Flier, W.M., Yaqub, M., Boellaard, R., Verwey, N.A., Blankenstein, M.A., Windhorst, A.D., Scheltens, P., Lammertsma, A.A., van Berckel, B.N., 2009. Relationship of cerebrospinal fluid markers to ^{11}C-PIB and 18F-FDDNP binding. J. Nucl. Med. 50, 1464–1470.

Toledo, J.B., Xie, S.X., Trojanowski, J.Q., Shaw, L.M., 2013. Longitudinal change in CSF Tau and Aβ biomarkers for up to 48 months in ADNI. Acta Neuropathol. 126, 659–670.

van Der Flier, W.M., Scheltens, P., 2009. Knowing the natural course of biomarkers in AD: longitudinal MRI, CSF and PET data. J. Nutr. Health Aging 13, 353–355.

van Harten, A.C., Kester, M.I., Visser, P.-J., Blankenstein, M.A., Pijnenburg, Y.A.L., Van der Flier, W.M., et al., 2011. Tau and p-tau as CSF biomarkers in dementia: a meta-analysis. Clin. Chem. Lab. Med. 49, 353–366.

Vanderstichele, H., Bibl, M., Engelborghs, S., Le Bastard, N., Lewczuk, P., Molinuevo, J.L., Parnetti, L., Perret-Liaudet, A., Shaw, L.M., Teunissen, C., Wouters, D., Blennow, K., 2012. Standardization of preanalytical aspects of cerebrospinal fluid biomarker testing for Alzheimer's disease diagnosis: a consensus paper from the Alzheimer's Biomarkers Standardization Initiative. Alzheimers Dement. 8, 65–73.

Vassar, R., Bennett, B.D., Babu-Khan, S., Kahn, S., Mendiaz, E.A., Denis, P., Teplow, D.B., Ross, S., Amarante, P., Loeloff, R., Luo, Y., Fisher, S., Fuller, J., Edenson, S., Lile, J., Jarosinski, M.A., Biere, A.L., Curran, E., Burgess, T., Louis, J.C., Collins, F., Treanor, J., Rogers, G., Citron, M., 1999. Beta-secretase cleavage of Alzheimer's amyloid precursor protein by the transmembrane aspartic protease BACE. Science 286, 735–741.

Visser, P.J., Verhey, F., Knol, D.L., Scheltens, P., Wahlund, L.O., Freund-Levi, Y., Tsolaki, M., Minthon, L., Wallin, A.K., Humpel, H., Bürger, K., Pirttila, T., Soininen, H., Rikkert, M.O., Verbeek, M.M., Spiru, L., Blennow, K., 2009. Prevalence and prognostic value of CSF markers of Alzheimer's disease pathology in patients with subjective cognitive impairment or mild cognitive impairment in the DESCRIPA study: a prospective cohort study. Lancet Neurol. 8, 619–627.

Wahlund, L.-O., Blennow, K., 2003. Cerebrospinal fluid biomarkers for disease stage and intensity in cognitively impaired patients. Neurosci. Lett. 339, 99–102.

Wang, L.S., Leung, Y.Y., Chang, S.K., Leight, S., Knapik-Czajka, M., Baek, Y., Shaw, L.M., Lee, V.M., Trojanowski, J.Q., Clark, C.M., 2012. Comparison of xMAP and ELISA assays for detecting cerebrospinal fluid biomarkers of Alzheimer's disease. J. Alzheimers Dis. 31, 439–445.

Williams, J.H., Wilcock, G.K., Seeburger, J., Dallob, A., Laterza, O., Potter, W., Smith, A.D., 2011. Non-linear relationships of cerebrospinal fluid biomarker levels with cognitive function: an observational study. Alzheimer's Res. Ther. 3, 5–15.

Zetterberg, H., Tullhög, K., Hansson, O., Minthon, L., Londos, E., Blennow, K., 2010. Low incidence of post-lumbar puncture headache in 1,089 consecutive memory clinic patients. Eur. Neurol. 63, 326–330.

Zetterberg, H., Wahlund, L.O., Blennow, K., 2003. Cerebrospinal fluid markers for prediction of Alzheimer's disease. Neurosci. Lett. 352, 67–69.

CHAPTER 6

Peripheral Fluid-Based Biomarkers of Alzheimer's Disease

Chapter Outline

6.1 BACKGROUND

Discovery of therapeutics for Alzheimer's disease (AD) has been limited by the lack of simple and affordable tests for biomarkers of the disease. An ideal antemortem AD biomarker should satisfy the following criteria: (1) the ability to diagnose AD with high sensitivity and specificity as confirmed by the gold standard of autopsy validation; (2) the ability to detect early-stage disease and track the progression of AD; (3) utility in monitoring therapeutic efficacy; and most importantly (4) require tissue samples that can be collected easily, repeatedly, noninvasively, and inexpensively. Cerebrospinal fluid (CSF)-based biomarkers of AD (reduced $A\beta_{1-42}$ and elevated tau and p-tau-181) and neuroimaging AD biomarkers are moderately accurate and widely validated by cross-sectional and longitudinal studies, as well as by some autopsy cohort studies. In past two decades, considerable progress has been made in establishing the relationship of

Biomarkers in Alzheimer's Disease. http://dx.doi.org/10.1016/B978-0-12-804832-0.00006-7
Copyright © 2016 Elsevier Inc. All rights reserved.

these AD biomarkers to the pathophysiology of AD. However, use of CSF and neuroimaging biomarkers is limited in routine clinical care or in AD screening because of their invasiveness and cost. Repeated testing of these biomarkers to assess the efficacy of therapeutics in clinical trials has been difficult (see Chapters 3 and Chapter 5). Moreover, CSF and neuroimaging biomarkers are not able to fulfill the 1998 consensus criteria for AD biomarkers set forth by The Ronald and Nancy Reagan Research Institute of the Alzheimer's Association and the National Institute on Aging Working Group (1998). On the other hand genetic AD biomarkers (APP, PSEN1, and PSEN2 mutations) are only effective in diagnosing familial AD (>95% AD cases are sporadic in nature without mutation of those three geens). APOE4 mutation is a known risk factor of late-onset AD but has not proven to be an effective genetic biomarker for diagnosis (see Chapter 4). In contrast to CSF-based and neuroimaging biomarkers, peripheral fluid-based biomarkers utilize minimally invasive and more accessible tissue collection procedures that make them more attractive for use in clinical trials, in routine clinical work-up, and in screening for AD. Urine, saliva, and blood samples are easily collected, stored, and processed in a routine manner. The infrastructure for collection and processing of these samples is common across various clinical settings. Blood in particular contains a rich mixture of biomolecules including proteins, lipids, and other metabolites that can be easily measured using established technologies, such as ELISA, advanced mass spectrometry (MS), high throughput multiplex systems, and flow cytometry. In AD, certain proteins may be posttranslationally modified, produced as different isoforms, or splice variants and accumulated with time in peripheral system that can be detectable in peripheral fluids as an AD signature. In addition, DNA and RNA can be isolated from blood cells for transcriptomic study of genetic biomarkers of AD. Circulating microRNA is also easily isolated from blood samples.

Peripheral oxidative damage, mitochondrial dysfunction, and inflammation are common aspects of AD pathology. Oxidative damage occurs in membranes (lipid peroxidation), proteins (nitrosylation and other posttranslational changes), and nucleic acids. Given its multifactorial nature, the diagnosis of sporadic AD is challenging, and early diagnosis is most desirable for therapeutic intervention. There is an urgent need for the development of minimally invasive, simple, and inexpensive tests for the diagnosis of AD, ideally at the earliest stages, for predicting the likelihood of developing AD, and to monitor disease progression and therapeutic efficacy.

6.2 RATIONALE FOR PERIPHERAL FLUID-BASED ALZHEIMER'S DISEASE BIOMARKERS

Although AD is commonly regarded as a disease of the brain, it is now recognized that AD is a systemic disease that is manifested in peripheral tissues outside the central nervous system (CNS), starting at the earliest stages of the disease. In addition, there is a continuous exchange of constituent biomolecules between the circulation and CSF (Palmer, 2010). Factors that are important to the etiology of AD, including those involved in hypoxia, ischemia, and metabolic dysfunction, may be detected in peripheral fluids such as blood, saliva, and urine. Amyloid accumulation and tau metabolic pathways are not limited to the brain, but are ubiquitous in the human body and found in blood, saliva, skin, and other peripheral tissues (Perry et al., 1982; Joachim et al., 1989; Soininen et al., 1992; Eckert et al., 1998; Gasparini et al., 1998; Sevush et al., 1998; Baskin et al., 2000; Goldstein et al., 2003). Furthermore, Aβ generation by β- and γ-secretase metabolism of APP is not only limited to brain cells, but also occurs in skin cells, platelets, liver, kidney, intestine, vascular walls, and other glands. There may also be communication between peripheral and central pools of Aβ that is mediated via blood–brain barrier, receptor-based transport, or a passive mechanism. The amount of lipid peroxidation is indicated by increased isoprostane 8 concentration in blood plasma. Multicomponent analysis of blood sample is the ideal approach to detecting oxidative damage (Praticò et al., 2002).

In addition to neuropathological hallmarks of amyloid plaques and neurofibrillary tangles, neuro- inflammation is very common in AD pathology, and may alter immune system components detectable in blood. Neuro-inflammation can lead to increases in cytokines (TNF-α, IL-1β, and IL-6), chemokines (CCL, chemokine that contains a C-C motif; CXCL, chemokine that contains a C-X-C motif), and inflammatory growth factors (G-CSF, granulocyte-colony stimulating factor; GDNF, glial-derived neurotrophic factor; IGFBP-1, insulin-like growth factor–binding protein-6; PDGF-BB, platelet-derived growth factor BB) in the brain and well as in blood plasma (Ray et al., 2007).

Taken together, blood plasma, blood serum, saliva, and urine may hold considerable promise as peripheral fluid sample sources for AD biomarker assays, providing a less invasive and inexpensive sample source for AD diagnostic testing, particularly compared with CSF-based tests (Table 6.1).

6.3 METABOLOMICS TO IDENTIFY ALZHEIMER'S DISEASE BIOMARKERS

AD is a systemic and multifactorial disease; therefore, changes in the metabolome may provide us with opportunities to detect measurable differences between AD and non-AD cases. Identification of new biomarkers of AD using metabolomics has received enormous attention in recent years. Metabolomics is defined as global metabolic profiling using a combination of proteomic, lipidomic, and/ or genomic/transcriptomic approaches. Because metabolomic analyses detect end point perturbations in the proteome, genome, and lipid profile caused by disorders, they are much more relevant to the development of drug efficacy tests and pharmacodynamic analyses compared to other approaches. One of the reasons to use a metabolomic approach is that changes in the metabolome (the collective concentrations of various metabolites) are amplified relative to the transcriptome or the proteome. For example, an individual enzyme may be defective (a difference that may not be easily detectable at the transcriptome or proteome level), but the downstream consequences of this defect in activity will have a large and detectable effect on the metabolome.

Table 6.1	Peripheral Fluid-Based Biomarkers in Alzheimer's Disease
Blood serum/plasma Proteomics	*Aβ peptides*: $A\beta_{1-40}$, $A\beta_{1-42}$
	Tau proteins: tau
	Inflammatory proteins: CRP, antichymotrypsin, macroglobulin, interleukins, TNF-α, complement factors, homocysteine *Others*: pancreatic polypeptide, NT-proBNP, clusterin, APOE, SAP
Blood serum/plasma Lipidomics	Desmosterol/cholesterol, ceramide, isoprostane 8, DHA (22:6n-3), SM (39:1), PE(36:4), cholesteryl esters
Other peripheral fluids	Urinary AD7c_NTP
	Saliva p-tau/tau, $A\beta_{1-42}$

Aβ, beta-amyloid; AD7c-NTP, Alzheimer's disease associated neuronal thread protein; CRP, C-reactive protein; DHA, docosahexaenoic acid; N-terminal prohormone of brain natriuretic peptide; PE, phosphatidylethanolamine; SAP, serum amyloid P; SM, sphingomyelin; TNF-α, tumor necrosis factor alpha.

Two metabolomics approaches are commonly used for identifying new AD biomarkers: lipidomics and proteomics. While blood-based metabolic biomarkers are more attractive for use in AD diagnostic tests because sample collection is easy and the tests are relatively noninvasive and less time-consuming, metabolic biomarker-based tests have limited sensitivity and specificity. Using single component analysis of metabolites provides low accuracy. Profiling of a broad spectrum of chemically diverse metabolites using modern technologies such as ultraperformance liquid chromatography and mass spectroscopy (MS), multiplex chemiluminescence ELISA, surface-enhanced laser desorption/ionization (SELDI) MS, ultrasensitive laser ablation inductively coupled plasma MS are recent approaches in AD peripheral biomarkers.

6.3.1 Lipidomic Alzheimer's Disease Biomarkers

Lipidomics is the analysis of lipid and lipid derivatives in biological fluids, such as blood plasma and serum. There are several convincing reasons to take a lipidomic approach to identify AD biomarkers. First, AD results from abnormality in the brain, which is the most lipid-rich organ in the human body. Second, the lipid transporter protein APOE4 is a known risk factor of late-onset AD. Third, in the liver of AD patients, the expression level of peroxisomal D-bifunctional protein, which catalyzes the conversion of tetracosahexaenoic acid into DHA (docosahexaenoic acid) is selectively reduced (Kou et al., 2011). Abnormal lipid metabolism has been extensively implicated in the pathogenesis of AD in both genetic (Harold et al., 2009; Hollingworth et al., 2011; Lambert et al., 2009a, 2013) and cell biological studies (Reitz, 2013; Reitz and Mayeux, 2014). In addition, peroxisomal dysfunction in AD contributes to glycerophospholipid deficits (Wood, 2012). Results from studies of animal models of AD have also provided fundamental information on lipid dysregulation during various stages of AD. For example, the levels of docosahexaenoyl (22:6), cholesterol ester (ChE), ethanolamine plasmalogens (pPEs), and sphingomyelins (SMs) are markedly increased in APP/tau mice compared to age-matched controls (Tajima et al., 2013). Moreover, cholesterol, sphingomyelins, and ceramides are important components of lipid rafts that affect the sorting and recruiting of signaling proteins/receptors. Extensively studied lipidomic biomarkers of AD include abnormal glycerophospholipids (due to abnormality in integrity of cell membranes) (Mapstone et al., 2014); lower desmosterol (Sato et al., 2012); higher ceramide/sphingomyelin ratios (Han et al., 2011); abnormal lipid peroxidation (Praticò et al., 2002); lower nonesterified fatty acid (22:6n-3, DHA; AD versus age-matched control, $P < 0.01$); higher phophatidylethanolamine (PE[36:4]; AD versus age-matched control, $P < 0.01$); lower sphingomyelin (SM[39:1]; AD versus age-matched control, $P < 0.01$) (Olazarán et al., 2015); and reduced metabolic products of long-chain cholesteryl esters (ChE): ChE (32:0), ChE (34:0), ChE (34:6), ChE (32:4), ChE (33:6), ChE (40:4) mass spectroscopic entities Mass/Z 315, Mass/Z 367, Mass/Z 628, Mass/Z 906 (a combination of 10 metabolic products in AD vs age-matched control: accuracy = 79.2%, sensitivity = 81.8%, specificity 76.9%) (Proitsi et al., 2015) (Table 6.2).

Lipid peroxidation is a consequence of oxidative stress caused by AD molecular abnormalities. The isoprostane 8,12-iso-iPF(2α)-VI concentration in blood plasma is a specific marker of in vivo lipid peroxidation. The concentration of isoprostane 8,12-iso-iPF(2α)-VI was investigated as a lipidomics biomarker of AD in blood plasma (Praticò et al., 2002). Using gas chromatography mass spectrometry (GC-MS), this study measured isoprostane 8,12-iso-iPF(2α)-VI concentration in blood plasma of patients with AD ($n = 50$) or mild cognitive impairment (MCI; $n = 33$) and in age-matched control ($n = 40$) cases. The order of measured values of concentration of isoprostane 8,12-iso-iPF(2α)-VI order was AD > MCI > age-matched control (AD [0.61 ± 0.03 ng/mL] vs. MCI [0.44 ± 0.03 ng/mL]; $P < .03$ and MCI vs. age-matched control subjects [0.19 ± 0.01 ng/mL]; $P < .001$; Table 6.2). The isoprostane 8,12-iso-iPF(2α)-VI concentration levels in blood plasma of AD and MCI groups were significantly higher than that of age-matched controls ($P < 0.001$).

In a cohort study of community-dwelling participants ($n = 46$ AD [discovery $n = 35$; validation $n = 11$]); ($n = 56$ MCI [discovery $n = 36$; validation: $n = 20$]); ($n = 73$ age-matched control [discovery $n = 53$; validation $n = 20$]) liquid chromatography stable isotope dilution-multiple reaction monitoring (MRM) mass spectrometry (LS-SID-MRM-MS) was used to detect various lipid biomarkers in blood plasma, and found lower phophatidylinositol concentration, higher dioleoylphosphatidic acid concentration, and lower phosphatidylcholine C38:4 concentration in those with AD (Table 6.2). This study predicted conversion of MCI to AD with a high accuracy (90%) (Mapstone et al., 2014). Another study found that the ratio of desmosterol/cholesterol concentration in blood plasma for AD cases was significantly lower than age-matched control cases (Sato et al., 2012). Although sample size was low (10 AD and 10 age-matched control), the study found 80% accuracy by ROC analysis (Table 6.2). This lipidomic

Table 6.2   Lipid Biomarkers of Alzheimer's Disease in Blood Plasma

Cohort/Lipid Identified	Method of Detection/Patient Population	Biomarker Performance	References
University Alzheimer's Disease Center Memory Disorders Clinic (MDC), University of Pennsylvania Lipid peroxidation quantification: isoprostane 8,12-iso-iPF$_{2\alpha}$-VI ↑	GC-MS AD ($n = 50$), MCI ($n = 33$), age-matched control ($n = 40$)	AD (0.61 ± 0.03 ng/mL) versus MCI (0.44 ± 0.03 ng/mL) ($P < .03$); MCI versus age-matched control subjects (0.19 ± 0.01 ng/mL) ($P < .001$) Lipid peroxidation indicators were significantly increased in both AD and MCI groups compared with age-matched controls ($P < 0.001$)	Praticò et al. (2002)
Joseph and Kathleen Bryan Alzheimer's Disease Research Center (Bryan ADRC) and the Department of Psychiatry, Duke University Ratiometric quantification: Ceramide/sphingomyelin ratio for the same fatty acid chain ↑	Shotgun lipidomics MS AD ($n = 26$), age-matched control ($n = 26$)	Genotype-specific differences within AD group. Low specificity and sensitivity. N16:0 and N21:0 were significantly higher in AD than age-matched control cases ($P < 0.05$).	Han et al. (2011)
Samples purchased from Precision Med, Inc. (San Diego, CA) Ratiometric quantification: Desmosterol/cholesterol↓	LC-MS and GC-MS. AD ($n = 10$), age-matched control ($n = 10$); $P < 0.0001$. There were several overlaps. Sensitivity in males was much lower than for females	The difference between AD and age-matched control cases was statistically significant. Sensitivity: not reported; specificity: not reported; accuracy: 0.8 by ROC curve	Sato et al. (2012)

Community-dwelling participants Phospholipid quantification: Phosphatidyl inositol ↓; Dioleoylphosphatidic acid↑; Phosphatidyl choline C38:4↓	LC-stable isotope dilution–multiple reaction monitoring (MRM) mass spectrometry (SID-MRM-MS) AD (*n* = 46 (discovery: *n* = 35; validation: *n* = 11), MCI (*n* = 56 (discovery: *n* = 36; validation: *n* = 20), age-matched control (*n* = 73 (discovery: *n* = 53; validation: *n* = 20)	Conversion of MCI to AD with a 90% accuracy. Disturbed cell membrane integrity at preclinical AD may cause difference in phospholipid concentration in blood plasma	Mapstone et al. (2014)
Dementia Case Register (DCR) at King's College London and the AddNeuroMed study. Quantification of metabolic products of cholesteryl esters (ChE): ChE (32:0), ChE (34:0), ChE (34:6), ChE (32:4), ChE (33:6), ChE (40:4), Mass/Z 315, Mass/Z 367, Mass/Z 628, Mass/Z 906.	Lipidomic LC-MS-MS Long chain ChEs were reduced in AD. AD (*n* = 36), MCI (*n* = 48), age-matched control (*n* = 40)	A combination of 10 metabolic products: AD versus age-matched control Accuracy = 79.2%, sensitivity = 81.8%, specificity 76.9%	Proitsi et al. (2015)
Six Spanish university hospitals and one Spanish research institution Nonesterified fatty acid DHA (22:6n-3); phosphatidylethanolamine [PE(36:4)]; sphingomyelin [SM(39:1)]	Ultra-performance LC-MS. AD (*n* = 100), MCI (*n* = 58), age-matched control (*n* = 93). Nonesterified fatty acid DHA (22:6n-3) (↓); phosphatidyethanolamine [PE(36:4)] (↑); sphingomyelin [SM(39:1)] (↓)	AD versus age-matched control [NEFA22:6n-3]: *P* < 0.01, [PE(36:4)]: *P* < 0.01, [SM(39:1)]: *P* < 0.01	Olazarán et al. (2015)

AD, Alzheimer's disease; DHA, docosahexaenoic acid, GC, gas chromatography; LC, liquid chromatography; MS, mass spectroscopy; MCI, mild cognitive impairment.

study was conducted by shotgun lipidomics mass spectroscopy. A study at the Joseph and Kathleen Bryan Alzheimer's Disease Research Center (Bryan ADRC) that examined ratios of specific ceramide/sphingomyelin with the same fatty acid chain found the ratios were higher for AD cases compared to age-matched control cases (Han et al., 2011). This study found low specificity and sensitivity and genotype-specific differences within the AD group (Table 6.2). The Dementia Case Register (DCR) at King's College London and the AddNeuroMed study (n = 36 AD; n = 48 MCI; n = 40 age-matched control) found that the metabolic products of long-chain ChE (32:0), ChE (34:0), ChE (34:6), ChE (32:4), ChE (33:6), ChE (40:4), entity Mass/Z 315, entity Mass/Z 367, entity Mass/Z 628, entity Mass/Z 906 in mass spectroscopy in blood plasma were reduced in AD cases (Proitsi et al., 2015). A combination of all 10 metabolic products in blood plasma showed accuracy of 79.2%, sensitivity of 81.8%, and specificity of 76.9% (Table 6.2). Using ultraperformance liquid chromatography and mass spectroscopy, a study by six Spanish University hospitals and one research institute (n = 100 AD; n = 58 MCI; n = 93; age-matched control) found lower nonesterified fatty acid (22:6n-3, DHA), higher phosphatidylethanolamine (PE[36:4]), and lower sphingomyelin (SM[39:1]; Table 6.2) (Olazarán et al., 2015). Blood plasma-based lipidomics will continue to reveal relevant biomarkers of AD for early-stage disease detection, risk assessment, and monitoring of drug efficacy.

6.3.2 Proteomic Alzheimer's Disease Biomarkers

6.3.2.1 *PLASMA Aβ ALZHEIMER'S DISEASE BIOMARKERS*

Several biomarkers have been investigated that are based in the dominating amyloid hypothesis of AD pathogenesis. Plasma $A\beta_{1-42}$ has long been proposed as a potential diagnostic biomarker for AD, with changes in $A\beta_{1-42}$ as a marker of disease progression, and it is the most widely researched peripheral fluid biomarker for AD (Table 6.3). The large Australian Imaging, Biomarkers and Lifestyle (AIBL) study of aging (n = 186 AD; n = 122 MCI; and n = 724 age-matched control) found consistently lower $A\beta_{1-42}$ levels in AD cases (Lui et al., 2010). A systematic meta-analysis of 13 studies consisting of more than 10,000 cases confirmed that $A\beta_{1-42}$ and a ratio $A\beta_{1-42}/A\beta_{1-40}$ are predictors of AD (Koyama et al., 2012). This widely acknowledged study found a statistically significant association between a decline in the $A\beta_{1-42}/A\beta_{1-40}$ ratio and a decline in cognitive measures. However, an Alzheimer's Disease Neuroimaging Initiative study showed no correlation between plasma Aβ concentration and clinical symptoms of AD (Toledo et al., 2011). Unfortunately, the majority of cross-sectional studies of plasma $A\beta_{1-42}$ concentration in humans have not revealed any differences between individuals with or without AD (Fukumoto et al., 2003; Le Bastard et al., 2009; Figurski et al., 2012). While a decrease in CSF $A\beta_{1-42}$ levels correlates well with AD and disease progression (Jensen et al., 1999; Sunderland et al., 2003), the link between AD and changes in human plasma $A\beta_{1-42}$ remains inconsistent (Irizarry, 2004), particularly for sporadic AD. The peripheral blood has no direct contact with brain, and hence the widely researched CSF biomarkers $A\beta_{1-42}$, tau, and phosphor-tau-181 may not be present in the blood at detect-

Table 6.3	Peripheral Blood-Based Aβ and tau as Alzheimer's Disease Biomarkers		
Sample/Patient Population	**Methods**	**Comments**	**References**
Blood plasma AD ($n = 78$), age-matched control ($n = 61$)	ELISA using 6E10 antibody for Aβ	High $Aβ_{1-40}$, no change in $Aβ_{1-42}$ for AD.	Mehta et al. (2000)
Blood plasma sporadic AD ($n = 146$), MCI ($n = 37$), PD ($n = 96$), age-matched control ($n = 92$)	Sandwich ELISA	Neither sensitive nor specific for the diagnosis of sporadic AD	Fukumoto et al. (2003)
Blood plasma AD ($n = 79$), age-matched control ($n = 365$).	Sandwich ELISA using 6E10 antibody for Aβ	No change in $Aβ_{1-40}$, and higher $Aβ_{1-42}$ for AD	Mayeux et al. (2003)
Blood plasma AD ($n = 127$), MCI (stable) ($n = 62$), MCI (progress to AD) ($n = 137$). Other dementia ($n = 25$)	Multiplexing assay platform	AD or MCI type AD (MCI-AD) had significantly lower $Aβ_{1-42}$ ($P < 0.007$).	Lewczuk et al. (2010)
Blood minus plasma AD ($n = 43$), age-matched control ($n = 52$), MCI ($n = 23$)	$Aβ_{1-42}$ Dimers by surface enhanced laser desorption ionization time of flight (SELDI-TOF) mass spectrometry	Significant difference AD and age-matched control group	Villemagne et al. (2010)
Blood minus plasma AD ($n = 186$), age-matched control ($n = 724$), MCI ($n = 122$)	ELISA and multiplex assay	No change in $Aβ_{1-40}$, and lower $Aβ_{1-42}$ for AD	Lui et al. (2010)
Blood plasma AD ($n = 162$), age-matched control ($n = 187$), age-matched control (progress to AD) ($n = 10$), MCI (stable) ($n = 162$), MCI (progress to AD) ($n = 145$)	Luminex immunoassay platform (INNO-BIA) (ADNI study)	No difference between in AD versus age-matched controls No difference between in AD versus non-AD dementia cases	Toledo et al. (2011), Figurski et al. (2012)

(Continued)

Table 6.3	Peripheral Blood-Based Aβ and tau as Alzheimer's Disease Biomarkers (*cont.*)		
Sample/Patient Population	**Methods**	**Comments**	**References**
Blood plasma components AD ($n = 47$), age-matched control ($n = 47$), non-AD dementia ($n = 50$)	Multiparameter fluorimetric bead-based immunoassay using xMAP® technology	No difference between in AD versus age-matched controls No difference between in AD versus non-AD dementia cases	Le Bastard et al. (2009)
Blood serum AD ($n = 21$)	ELISA assay designed to quantify fragmented tau protein in serum	Fragmented tau concentration in blood serum of AD cases was inversely correlated to MMSE and Mattis Dementia Rating Scale	Henriksen et al. (2013)

Aβ, beta-amyloid; AD, Alzheimer's disease; ADNI, Alzheimer's disease Neuroimaging Initiative; ELISA, enzyme-linked immunosorbent assay; GSK-3, glycogen synthase kinase 3; MAPK, mitogen-activated protein kinase; MCI, mild cognitive impairment; MMSE, minimental state examination; PD, Parkinson's disease.

able levels. Furthermore, the source of Aβ in blood plasma is not limited to that which crosses the blood–brain barrier and blood–CSF barrier. APP is expressed in a wide variety in organs such as skin, intestine, lung, vascular walls, liver, skeletal muscles, and platelets, and each of these tissues is a potential source of Aβ in blood plasma. Platelets are important source of $Aβ_{1-40}$, $Aβ_{1-42}$, and APP. Activated platelets have been shown to produce a considerable amount of Aβ and APP, and activation of platelets may occur as a consequence of a variety of physiological responses. The liver and kidney are two major organs responsible for clearance of peripheral Aβ, and the efficiency of Aβ clearance may vary person to person, specifically at the older age. Another issue is that some studies show an increase or no change in the plasma $Aβ_{1-42}$ level in normal aging humans without dementia (Tamaoka et al., 1996; Fukumoto et al., 2003; van Oijen et al., 2006; Sundelöf et al., 2008). Mouse studies have also shown inconsistent trends in $Aβ_{1-42}$ blood plasma levels across AD models and age-matched controls. In addition, plasma Aβ levels show circadian fluctuation, similar to CSF Aβ (Huang et al., 2012). Taken together, plasma Aβ measurements are prone to more variability compared to CSF Aβ. As a result, peripheral fluid-based Aβ biomarkers for AD have been difficult to validate and with adequate accuracy, sensitivity, and specificity.

Considerably less attention has been paid to the metabolic product of tau as a peripheral fluid biomarker for AD. One study quantified the fragmented tau concentration in blood serum by ELISA (Henriksen et al., 2013), and found it to be inversely correlated with cognitive decline (measured by mini mental score examination (MMSE) and Mattis Dementia Rating Scale; Table 6.3).

6.3.2.2 *MULTICOMPONENT PLASMA ALZHEIMER'S DISEASE BIOMARKERS*

AD is a multifactorial disease and several interrelated cellular/molecular abnormalities may occur during the long prodromal period of pathoprogression. Several multicomponent AD biomarker panels are currently under investigation as a way to assess multiple biomarkers across various cellular and molecular pathways simultaneously. A number of high-throughput proteomics technologies have been explored to evaluate the usefulness of these multicomponent biomarker panels for the diagnosis of AD (Table 6.4), including proteomics antibody microarrays (Ray et al., 2007; Marksteiner et al., 2011), flow-cytometry-based immunoassays (Hye et al., 2014), multiplex immune assays (O'Bryant et al., 2010; Doecke et al., 2012), and liquid chromatography-mass spectrometry (LC-MS) (Thambisetty et al., 2010; Olazarán et al., 2015).

Inflammation occurs in the brain of AD patients at both preclinical and clinical stages of the disease, possibly even before Aβ and tau changes (Akiyama et al., 2000; Veerhuis et al., 2003; Schuitemaker et al., 2009). Positron emission tomography (PET) imaging with compounds such as 11C-PK11195 has provided in vivo evidence of inflammation in AD and used in the early diagnosis of AD or MCI (Kropholler et al., 2007) (for further detail, see Chapter 3, section "*3.4 Positron Emission Tomography (PET)*" subsection "*3.4.3 AD brain inflammation imaging by PET*"). There is some evidence that activation of microglia and astrocytes produces cytokines, chemokines, and inflammatory growth factors in the brain and blood plasma. Using a protein antibody-based microarray, one study reported that a set of 18 inflammatory biomarkers was able to distinguish patients with AD from those with MCI with an accuracy of 90% (Ray et al., 2007). This study was the first to show the potential of blood plasma-based multicomponent profiling. This study also found a correlation between the 18-biomarker panel with CSF biomarkers obtained from the same patients (Britschgi et al., 2011). However, attempts to validate these findings by other laboratories found a diagnostic accuracy of only 60%–70% (Soares et al., 2009; Marksteiner et al., 2011). Other important serum inflammatory markers being investigated as potential AD biomarkers include C-reactive protein (CRP), antichymotrypsin, macroglobulin, interleukins, and homocystine. Using a more advanced multiplex technology, a recent study found 10 plasma proteins that are strongly associated with disease severity and disease progression (Hye et al., 2014). To improve the accuracy of the study, some unusually stringent conditions were applied for data analysis. The most important blood-based AD biomarkers identified by proteomics are summarized in Table 6.4. It is noteworthy that a few proteins have been found to be consistently up- or downregulated across all studies (Ray et al., 2007; Thambisetty et al., 2010; O'Bryant et al., 2010; Doecke et al., 2012; Soares et al., 2012; Hye et al., 2014). In a ADNI study using Human Discovery Multi-Analyte Profile 1.0 panel 16 (detecting c-peptide, fibrinogen, alpha-1-antitrypsin, pancreatic

Table 6.4   Multicomponent Blood-Based Alzheimer's Disease Biomarkers

Cohort/Protein Identified	Biochemical Method of Detection	Biomarker Performance	References
Ray *All ↓ in AD:* TNF-α; PDGF-BB; M-CSF; G-CSF; CCL5; CCL7; CCL15; EGF; GDNF; IL-1α; IL-3 *All ↑ in AD:* Ang-2; ICAM-1; CCL18; CXCL8; IGFBP-6; IL-11; Trail-R4	Proteomics antibody array. A panel of 18 protein antibodies used to detect from blood leukocytes in arrayed sandwich ELISA Patient groups: non-AD dementia (*n* = 11); AD (*n* = 86); MCI (*n* = 47); age-matched control (*n* = 21); rheumatoid arthritis (*n* = 16)	Immune-responsive analytes and cytokines. High accuracy for detecting AD. However, failed to produce same level of accuracy during validation studies	Ray et al. (2007)
AddNeuroMed (ANM) and Kings Health Partners–Dementia Case Register (KHP-DCR) *All ↓ in AD:* APOC3; TTR; ICAM-1; RANTES; cystatin *All ↑ in AD:* PEDF; CC4; A1AcidG; clusterin	Multiplex bead assays (Luminex xMAP) incorporated in 7 MILLIPLEX MAP panels run on the Luminex 200 instrument; patient groups: AD (*n* = 476); age-matched control (*n* = 452); MCI (*n* = 220)	This study showed conversion of MCI to AD with an accuracy =87%, sensitivity = 85%, and specificity =88%	Hye et al. (2014)
Australian Imaging Biomarker and Lifestyle study (AIBL) validated by ADNI *All ↓ in AD:* IL-17; EGFPR *All ↑ in AD:* insulin-like growth factor binding protein 2; pp; Ang-2; cortisol; beta-2-microglobulin	Multiplex immunoassay. *Study group:* AD (*n* = 207); age-matched control (*n* = 754) *Validation study:* AD (*n* = 112); age-matched control (*n* = 58)	*Study group:* Sensitivity= 85% and Specificity= 93% *Validation study:* Sensitivity = 80% Specificity = 85%	Doecke et al. (2012)

ADNI + Washington University AD Cohort + University of Pennsylvania AD Cohort Four most significant components associated with from proteomics study *All ↑ in AD* APOE, BNP, CRP, PP [CDR: clinical dementia rating scale developed by the Washington University; CDR= 0 indicates no dementia, CDR = 0.5 indicates very mild dementia, and CDR = 1 indicates mild dementia; CDR = 0.5 participants in the study were considered to be same as MCI in other two cohorts for comparison]	Myrad RBM – Luminex X MAP Patient groups: *ADNI:* AD (n = 112); age-matched control (n = 58); MCI (n = 396). *University of Pennsylvania AD cohort:* AD (n = 88); age-matched control (n = 126); MCI (n = 16). *Washington University AD cohort:* CDR1 (n = 28); age-matched control (n = 242); CDR0.5 (n = 63)	Four blood plasma analytes were consistently associated with the diagnosis of very mild dementia/MCI/AD in all three independent clinical cohorts *AD versus age-matched control:* APOE $P < 0.001$ BNP $P < 0.001$ CRP $P < 0.001$ PP $P < 0.001$	Hu et al. (2012)
Alzheimer Research Trust-funded cohort at King's College London (KCL-ART) and the AddNeuroMed study Clusterin ↑ in AD	Two-dimensional gel electrophoresis and liquid chromatography coupled to tandem mass spectrometry (LC-MS-MS) Total subjects (n = 744). This study found high levels of plasma clusterin in AD cases compared to age-matched control cases	AD pathogenesis. Significantly ($P < 0.001$) associated with the rate of progression of AD. Clusterin has a role in atrophy in AD brain.	Thambisetty et al. (2010)
AddNeuroMed cohort study Complement components C3 ↑ in AD, complement components C3a ↓ in AD and MCI, complement factor-I ↓ in AD and MCI, gamma-fibrinogen ↓ in AD and ↑MCI	Two-dimensional gel electrophoresis LC-MS-MS AD (n = 79); age-matched control (n = 95); MCI (n = 88)	Correlated with AD brain atrophy measured by MRI.	Thambisetty et al. (2011)

(Continued)

Table 6.4 Multicomponent Blood-Based Alzheimer's Disease Biomarkers (cont.)

Cohort/Protein Identified	Biochemical Method of Detection	Biomarker Performance	References
The Texas Alzheimer's Research Consortium *All ↓ in AD:* creatine MB; G-CSF; S-100B; IL-10; IL-1ra; prostatic acid phosphatase; C-reactive protein; TNF-α; stem cell factor; MIP1α *All ↑ in AD:* thromboprotein; alpha-2-macroglobulin; tenascin; TNF-β; beta-2-microglobulin; eotaxin; PP; von Willebrand factor; IL-15; VCAM-1; IL-8; IGFBP2; Fas ligand; prolactin resistin	Myriad RBM-Multiplex Luminex XMAP Immunoassay: a panel of 25 proteins. Patient groups: AD (n = 197); age-matched control (n = 203)	Specific algorithm in data analysis provided high specificity and sensitivity. Sensitivity = 80% Specificity = 91% Accuracy = 91%	O'Bryant et al. (2010)
Biomarkers Consortium Alzheimer's Disease Plasma Proteomics Project by ADNI *All ↑ in AD:* eotaxin 3, pancreatic polypeptide, tenascin C, and NT-proBNP *All ↓ in AD:* IgM and APOE	Myriad RBM-Multiplex Luminex XMAP Immunoassay AD (n = 97), MCI (n = 345), age-matched control (n = 58)	Moderate sensitivity and specificity; similar changes reported in CSF biomarkers patients with AD and MCI	Soares et al. (2012)
Six Spanish university hospitals and one Spanish research institution Plasma metabolites: glutamic acid (↓), alanine (↓), and aspartic acid (↓)	Ultraperformance LC-MS. AD (n = 100), MCI (n = 58), age-matched control (n = 93)	AD versus age-matched control P < 0.01	Olazarán et al. (2015)

ADNI c-peptide, fibrinogen, alpha-1-antitrypsin, PP, complement C3, vitronectin, cortisol, AXL receptor kinase, interleukin-3, interleukin-13, matrix metalloproteinase-9 total, apolipoprotein E, leptin, von Willebrand factor, serum amyloid p-component, and immunoglobulin E.	Human Discovery Multi-Analyte Profile 1.0 panel AD (n = 16), MCI (n = 52), age-matched control (n = 3)	Sixteen of the analytes of the Rules Based Medicine Human Discovery Multi-Analyte Profile 1.0 panel were found to associate with [11C]-PiB PET imaging biomarkers. Sensitivity = 0.918, specificity = 0.545.	Kiddle et al. (2012)
AddNeuroMed (ANM) *Alzheimer's Research UK/Maudsley BRC Dementia Case Registry at King's Health Partners (ARUK/DCR)* *All ↓ in AD:* α-1-antitrypsin, G-CSF, clusterin, complement C6, inter-α-trypsin inhibitor heavy chain H4, C-C chemokine 18 *All ↑ in AD:* complement C3, Pancreatic prohormone, IGFBP2	SomaLogic's SOMAscan proteomics technology (SOMAscan, SomaLogic, Inc., Boulder, Colorado) AD (n = 319), MCI-AD (n = 43), MCI-MCI (n = 106), age-matched control (n = 209)	This validation study found 9 out of 94 previously discovered plasma AD biomarkers were replicated.	Kiddle et al. (2014).
AclarusDx 170 probesets that map 136 annotated genes	Transcriptomic signature using human Genome-Wide Splice Array *Test study cohort:* AD (n = 90), age-matched control (n = 87) *Validation study cohort:* AD (n = 111), age-matched control (n = 98)	In validation study: Sensitivity = 81.3% Specificity = 67.1%	Fehlbaum-Beurdeley et al. (2012)

(Continued)

Table 6.4 Multicomponent Blood-Based Alzheimer's Disease Biomarkers (cont.)

Cohort/Protein Identified	Biochemical Method of Detection	Biomarker Performance	References
Memory Clinics at the Department of Psychiatry in Innsbruck or the Landeskrankenhaus Hall/Tirol in Austria *All ↑ in AD:* α2 Macroglobulin, apolipoprotein A1, NT-proBNP, trombospondin-2, serum amyloid A	*Screening study:* age-matched control (*n* = 23), MCI (*n* = 24), AD (*n* = 33); *Follow up study* Age-matched controls converted to MCI (*n* = 7), MCI patients converted to AD (*n* = 7), stable MCI (*n* = 5), AD (*n* = 3) *Validation study* *focused NT-proBNP* Young control (*n* = 15), age-matched control (*n* = 40), MCI (*n* = 27), and AD (*n* = 43)	NT-proBNP significantly enhanced both in MCI and AD patients compared to age-matched as well as young control cases Plasma NT-proBNP can be a stable candidate for both diagnosis and AD disease progression	Marksteiner et al. (2014)

Aβ, beta-amyloid; AD, Alzheimer's disease; ADNI, Alzheimer Disease Neuroimaging Initiative; Ang-2, angiopoietin-2; APOC3, apolipoprotein C3; APOE, apolipoprotein E; BNP, brain natriuretic peptide; CCL, chemokine containing a C-C motif; CDR, clinical dementia rating scale; CRP, C reactive protein; CXCL, chemokine containing a C-X-C motif; EGF, epidermal growth factor; G-CSF, granulocyte-colony stimulating factor; GDNF, glial-derived neurotrophic factor; ICAM-1, intercellular adhesion molecule-1; IGFBP2, insulin-like growth factor binding protein 2; IL, interleukin; IL-3, interleukin 3; IL-1ra, interleukin 1 receptor antagonist; MCI, mild cognitive impairment; M-CSF, macrophage-colony stimulating factor; MIP1α, macrophage inflammatory protein 1-α; MRI, magnetic resonance imaging; NT-proBNP, N-terminal prohormone of brain natriuretic peptide; PDGF-BB, platelet-derived growth factor BB; PEDF, pigment epithelium-derived factor; 11C-PIB, [11C]-Pittsburgh Compound B; PP, pancreatic polypeptide; RANTES, regulated on activation, normal T cell expressed and secreted; TNF-α, tumor necrosis factor-α; TNF-β, tumor necrosis factor-β; TRAIL-R4, TNF-related apoptosis-inducing ligand receptor-4; TTR, transthyretin type receptor; VCAM-1, vascular cell adhesion molecule 1.

polypeptide, complement C3, vitronectin, cortisol, AXL receptor kinase, inter-leukin-3, interleukin-13, matrix metalloproteinase-9 total, apolipoprotein E, Leptin, von Willebrand factor, serum amyloid p-component, and immunoglob-ulin E) the component analytes were found to correlate with the neuroimaging AD biomarker [11C]-PIB PET with high sensitivity (91.8%) but low specificity (54.5%) (Kiddle et al., 2012). To validate previously published most promis-ing multicomponent blood-based AD proteomics biomarkers a very interest-ing large-scale validation study was conducted using SomaLogic's SOMAscan proteomics technology. This study selected 94 out of 163 candidate biomarkers from 21 most interesting published studies in plasma samples. This validation study selected patients from of the AddNeuroMed (ANM) and the Alzheimer's Research UK/Maudsley BRC Dementia Case Registry at King's Health Partners (ARUK/DCR) research cohorts and found 9 out of 94 selected AD plasma bio-markers are strongly associated with AD (Table 6.4; Kiddle et al., 2014). Using Myrad RBM – Luminex X MAP technology, four blood plasma analytes (apoli-poprotein E [APOE], B-type natriuretic peptide [BNP], C-reactive protein [CRP], and pancreatic polypeptide [PP]) were consistently associated with the diagno-sis of very mild dementia/MCI/AD in three independent clinical cohorts (APOE, $P < 0.001$; BNP, $P < 0.001$; CRP, $P < 0.001$, PP, $P < 0.001$) (Hu et al., 2012). This multicentered study used subjects from three different cohorts in US (ADNI cohorts: AD ($n = 112$); age-matched control ($n = 58$); MCI ($n = 396$); University of Pennsylvania AD cohort: AD ($n = 88$); age-matched control ($n = 126$); MCI ($n = 16$) Washington University AD cohort: CDR = 1 ($n = 28$); age-matched control ($n = 242$); CDR = 0.5 ($n = 63$); CDR: clinical dementia rating scale devel-oped by the Washington University; CDR= 0 indicates no dementia, CDR = 0.5 indicates very mild dementia, and CDR = 1 indicates mild dementia; CDR = 0.5 participants in the study were considered to be same as MCI in other two cohorts for comparison) (Table 6.4).

The most promising blood-based biomarkers emerging from multicomponent plasma AD biomarker studies that have been reproducible are N-terminal prohor-mone of brain natriuretic peptide (NT-proBNP) (significantly elevated in AD pa-tients: Soares et al., 2012; Marksteiner et al., 2014), CRP (decreased significantly in AD patients: O'Bryant et al., 2010; Doecke et al., 2012), and pancreatic polypeptide (significantly elevated in AD patients: O'Bryant et al., 2010; Doecke et al., 2012; Soares et al., 2012) (Table 6.4). NT-proBNP was 1.9 times higher in AD and MCI cases compared to age-matched control cases (Marksteiner et al., 2014). This study also found that NT-proBNP has the potential to diagnose AD disease progression and was significantly ($P < 0.01$) higher in MCI and AD patients compared to age-matched as well as young control cases (Marksteiner et al., 2014). A combina-tion of these three proteins with APOE4 genotyping might constitute a good bio-signature of LOAD (multicomponent blood-based AD biomarkers using genomic technologies are discussed in detail in Chapter 4).

Bioinformatics approaches have been applied to handle the highly heteroge-neous data emerging from studies of multicomponent AD biomarkers. High-throughput results from proteomics and genomics studies should be integrated through a multidisciplinary approach that combines computer science, mathematics, statistics, and graphic arts (Merrick et al., 2011). Well-validated computer simulation models can be used to identify the optimal multicomponent blood-based biomarkers for AD and develop a "fingerprint" of AD that can be easily interpreted for individual patients.

Poor reproducibility of study findings using blood-based AD biomarkers has posed a significant challenge to AD biomarker research. No blood-based biomarkers have been established so far by multisite validation study that have the level of accuracy, sensitivity, and specificity (sensitivity >80% for detecting AD and a specificity of >80% for distinguishing other dementias) recommended by The Ronald and Nancy Reagan Research Institute of the Alzheimer's Association and the National Institute on Aging Working Group (1998) Consensus Report of the Working Group on Molecular and Biochemical Markers of Alzheimer's Disease (published by Neurobiology of Aging 1998 Mar–Apr;19(2):109–116).

6.4 OTHER BODY FLUIDS AS A SAMPLE SOURCES OF ALZHEIMER'S DISEASE BIOMARKERS

In addition to blood, saliva and urine have been explored as sample sources for AD biomarkers. Several studies have reported changes in pathological brain proteins detected in human saliva, including $A\beta_{1-42}$ and p-tau, and AD-associated neuronal thread protein (AD7c-NTP [MW ~41 kDa]; Table 6.5). Bermejo-Pareja et al. (2010) found a small but statistically significant increase in saliva $A\beta_{1-42}$ levels in patients with mild AD compared to age-matched controls, but found no differences in saliva concentration of $A\beta_{1-42}$ between patients with Parkinson's disease (PD) and age-matched control cases. However, another study showed that the p-tau to tau ratio (p-tau/tau), but not $A\beta_{1-42}$, was significantly higher in AD compared to age-matched control cases (Shi et al., 2011). Almost all studies have found no change in plasma $A\beta_{1-40}$ between AD and non-AD cases.

Neuronal thread proteins (NTPs) can be altered in AD. AD7c-NTP was found to be selectively elevated in the AD brain (De La Monte et al., 1996). AD7c-NTP was also increased in cortical neurons, CSF, and urine in the early stages of AD, and its level was proportional to the degree of dementia. Selectively increased AD7c-NTP in urine of AD cases has been reported by several laboratories (Ghanbari et al., 1998; Munzar et al., 2002; Goodman et al., 2007). Urinary AD7C-NTP has the same molecular weight as AD7C-NTP obtained from brain and CSF. Urinary AD7c-NTP is currently the most promising non–blood-based peripheral fluid biomarker of AD (Table 6.5).

6.5 LONGITUDINAL ASSESSMENT OF PERIPHERAL BLOOD-BASED ALZHEIMER'S DISEASE BIOMARKERS

One of the main characteristics of AD is its long prodromal stages. A great deal of evidence suggests that neuropathological changes of the disease begin before any clinical manifestation occurs. Plasma $A\beta$ levels begin to increase

Table 6.5	Other Peripheral Fluid-Based Biomarkers of Alzheimer's Disease		
Study Designed	**Biomarkers/ Method**	**Comments**	**References**
Saliva tau AD ($n = 21$); age-matched control ($n = 38$)	▪ p-tau/tau ; $P < 0.05$ ▪ Luminex assay ▪ No change in Aβ levels	▪ p-tau/tau ratio was significant difference between AD and age-matched control cases	Shi et al. (2011)
Saliva Aβ$_{1-42}$ AD ($n = 70$), age-matched control ($n = 56$), PD ($n = 51$)	▪ Change in Aβ$_{1-42}$ levels ▪ Aβ-sensitive ELISA kit	▪ Small but significant increase in Aβ$_{1-42}$ levels in AD cases	Bermejo-Pareja et al. (2010)
Urinary AD7c-NTP AD ($n = 66$), age-matched control ($n = 134$)	Sandwich ELISA specific to AD7c-NTP Sensitivity = 82% Specificity = 91%	Cut-off values were low (1.5 µg/L); clinical confirmation	Ghanbari et al. (1998)
AD ($n = 82$), age-matched control ($n = 27$), non-AD dementia ($n = 13$)	Competitive ELISA format of AD7c-NTP Sensitivity = 89% Specificity = 90%	Significant difference between AD and non-AD cases (P <0.001); clinical confirmation	Munzar et al. (2002)
AD ($n = 88$), definite non-AD ($n = 43$)	ELISA format assay Sensitivity = 91.4% Specificity = 90.7%	Blinded multicentered study; clinical confirmation	Goodman et al. (2007)

AD, Alzheimer's disease; AD7c-NTP, Alzheimer's disease associated neuronal thread protein; ELISA, enzyme-linked immunosorbent assay; PD, Parkinson's disease.

in familial AD and in Down syndrome cases before dementia is seen. Longitudinal assessment of plasma Aβ with a repeated sampling could help correlate biomarker values with risk and progression of dementia related to AD. Longitudinal studies of changes in plasma Aβ$_{1-42}$ levels over time in humans have produced some promising results (Mayeux and Sano 1999; Mayeux et al., 2003; Oh et al., 2008). Data from studies including longitudinal monitoring of plasma Aβ suggest that Aβ$_{1-40}$, Aβ$_{1-42}$, and their ratio may be useful for detecting AD prodromal stages and preclinical AD risk assessment (Table 6.6). There have been conflicting data emerging from different longitudinal studies of plasma Aβ. Some studies showed that progressive decreases in the Aβ$_{1-42}$/Aβ$_{1-40}$ ratio represent the conversion from a cognitively normal individual to AD (van Oijen et al., 2006; Schupf et al., 2008; Lambert et al., 2009b), though similar trends were not observed in other studies (Toledo et al., 2011). While longitudinal cohort studies may help predict future onset of disease and test

Table 6.6	Longitudinal Studies of Blood Plasma Aβ Biomarkers of Alzheimer's Disease		
Study Designed	**Results**	**Comments**	**References**
Cognitively normal ($n = 105$); cognitively normal progression to AD ($n = 64$)	Follow-up: 3.5 years High baseline $A\beta_{1-42}$	Higher baseline $A\beta_{1-42}$ long before clinical manifestation of AD	Mayeux and Sano (1999)
Rotterdam study Cognitively normal ($n = 1364$); cognitively normal progression to AD ($n = 289$); cognitively normal progression to non-AD dementia ($n = 103$)	Follow-up: 8.5 years High baseline $A\beta_{1-40}$ Low baseline $A\beta_{1-42}/A\beta_{1-40}$	Higher baseline $A\beta_{1-42}/A\beta_{1-40}$ corresponds to lower risk of AD	van Oijen et al. (2006)
Cognitively normal ($n = 105$); MCI progress ($n = 1021$)	Follow-up: 4.5 years High baseline $A\beta_{1-42}$ Decrease of $A\beta_{1-42}$ Decrease of $A\beta_{1-42}/A\beta_{1-40}$	Relates both baseline and change of Aβ with the risk of AD	Schupf et al. (2008)
Uppsala Longitudinal Study on Adult Men (ULSAM) cognitively normal ($n = 608$); cognitively normal progression to AD ($n = 74$)	Follow-up: 11.2 years Low baseline $A\beta_{1-40}$	Plasma $A\beta_{1-42}$ levels were not significantly associated with AD incidence, whereas $A\beta_{1-40}$ levels were significantly associated with AD incidence	Sundelöf et al. (2008)
Three-City Study Cognitively normal ($n = 985$); cognitively normal progression to AD ($n = 233$)	Follow-up: 4 years Low baseline $A\beta_{1-42}/A\beta_{1-40}$	Higher baseline ratio ($A\beta_{1-42}/A\beta_{1-40}$) showed significantly lower risk of developing AD	Lambert et al. (2009b)
ADNI study MCI stable ($n = 162$) MCI progress ($n = 145$)	Follow-up: 3 years Not significant change in baseline $A\beta_{1-42}/A\beta_{1-40}$	Plasma $A\beta_{1-42}$ and $A\beta_{1-40}$ levels showed modest prognostic factor	Toledo et al. (2011)

Table 6.6	Longitudinal Studies of Blood Plasma Aβ Biomarkers of Alzheimer's Disease (cont.)		
Study Designed	**Results**	**Comments**	**References**
Cognitively normal (n = 677); MCI progress (n = 37); MCI progress-VaD (n = 11); MCI progress-OD (n = 5)	Follow-up: 5 years High baseline $Aβ_{1-40}$	Plasma Aβ should not be used clinically to predict dementia due to AD but may be regarded as a moderate risk marker comparable to other risk markers for AD such as family history of dementia	Hansson et al. (2012)
Honolulu Asia Aging Study Cognitively normal (n = 590); MCI progress (n = 53); MCI progress-VaD (n = 24)	Follow-up: 15.8 years Low baseline $Aβ_{1-40}$ Low baseline $Aβ_{1-42}$	Cerebral amyloid angiopathy and impaired Aβ clearance from the brain	Shah et al. (2012)

Aβ, beta-amyloid; AD, Alzheimer's disease; ADNI, Alzheimer's disease neuroimaging initiative; MCI, mild cognitive impairment; OD, other dementia; VaD, vascular dementia.

the usefulness of a particular biomarker in prognostic assessment, because of the long prodromal stages of AD, data from cross-sectional cohort studies of blood plasma Aβ measurement may be less useful for evaluating the prognostic value of AD biomarkers (Table 6.6).

Studies of multicomponent blood-based AD biomarkers for predicting the conversion of MCI to AD have produced some encouraging results. A study of a panel of 10 proteins detected on a multiplex array platform claimed strong predictability for progression from MCI to AD (accuracy = 87%, sensitivity = 85%, and specificity = 88%) (Hye et al., 2014). In a large multicenter cohort (AddNeuroMed (ANM), a multicenter European study; Kings Health Partners-Dementia Case Register (KHP-DCR), a UK clinic and population based study; Genetics AD Association (GenADA), a multisite case–control longitudinal study based in Canada) consisting of 1148 subjects, this same panel of 10 plasma proteins was also found to be strongly associated with disease severity and disease progression

((Hye et al., 2014). However, another large cohort study using a different panel of blood plasma proteins (all low in AD: α-1-antitrypsin, G-CSF, clusterin, complement C6, Inter-α-trypsin inhibitor heavy chain H4, C-C chemokine 18; all elevated in AD: complement C3 Pancreatic prohormone IGFBP2) (SOMAscan, SomaLogic, Inc., Boulder, Colorado) found no association with conversion of MCI to AD (Kiddle et al., 2014). Three consecutive studies (screening, follow up, and validation) conducted by the Memory Clinics at the Department of Psychiatry in Innsbruck, Laboratory for Experimental Alzheimer's disease in Austria found that NT-proBNP in blood plasma may be useful for detecting disease progression (Marksteiner et al., 2014).

6.6 INTERRELATIONSHIP OF CSF, NEUROIMAGING, AND PERIPHERAL BLOOD-BASED ALZHEIMER'S DISEASE BIOMARKERS

Multicomponent blood plasma-based AD biomarkers show moderate to significant correlation with CSF and neuroimaging biomarkers (Table 6.7). Blood plasma clusterin/apolipoprotein J concentration was associated with atrophy of the entorhinal cortex measured by magnetic resonance imaging (MRI; Thambisetty et al., 2010). CSF $A\beta_{1-42}$ levels and t-tau/$A\beta_{1-42}$ ratios were shown to be well correlated with the number of APOE4 alleles, and plasma levels of BNP and PP by regression analysis of all three independent cohorts [ADNI: AD ($n = 112$); age-matched control ($n = 58$); MCI ($n = 396$); University of Pennsylvania AD cohort: AD ($n = 88$); age-matched control ($n = 126$); MCI ($n = 16$) Washington University AD cohort: CDR1 ($n = 28$); age-matched control ($n = 242$); CDR0.5 ($n = 63$)] (Table 6.7) (Hu et al., 2012). All 16 of the analytes on the Rules Based Medicine Human Discovery Multi-Analyte Profile 1.0 panel (c-peptide, fibrinogen, alpha-1-antitrypsin, pancreatic polypeptide, complement C3, vitronectin, cortisol, AXL receptor kinase, interleukin-3, interleukin-13, matrix metalloproteinase-9 total, apolipoprotein E, Leptin, von Willebrand factor, serum amyloid p-component, and immunoglobulin E) were found to be well correlated with the neuroimaging more established AD biomarker such as [11C]-PIB PET (Kiddle et al., 2012).

There is no or very low correlation between plasma Aβ and CSF biomarkers (Table 6.7). The possible reasons for this are as follows: (1) sources of Aβ species in blood and CSF are different. CSF Aβ species are the result of APP metabolism in the CNS and diffuse from interstitial fluid to the CSF. Aβ species in blood originate from a wide variety of peripheral organs such as platelets, skin cells, vascular walls, liver, kidney, skeletal muscles, intestine, and some other glands. (2) Degradation/clearance mechanisms of plasma Aβ species in blood are different. The liver and kidney primarily remove blood Aβ species. (3) Blood plasma Aβ species interact with other protein components in blood such as serum amyloid P component, immunoglobulin, α2-macroglobulin, apolipoproteins (A-I, A-IV, E, and J), transthyretin, apoferritin and complement factors, which may affect their levels detectable in blood-based assays.

Table 6.7   Crosscorrelation of Blood Plasma-Based Biomarkers and Other Biomarkers of Alzheimer's Disease

Study Designed	Alzheimer's Disease Biomarker Modalities	Comments	References
Probable AD ($n = 50$)	CSF Aβ_{1-40} and Aβ_{1-42} versus blood plasma Aβ_{1-40} and Aβ_{1-42}	No relation between CSF Aβ_{1-40} and Aβ_{1-42} levels with those of plasma. There was no correlation with MMSE.	Mehta et al. (2001)
Alzheimer Research Trust-funded cohort at King's College London (KCL-ART) and the AddNeuroMed study	Clusterin was high in AD Blood plasma clusterin measured by two-dimensional gel electrophoresis and liquid chromatography coupled to tandem mass spectrometry (LC-MS-MS) versus MRI	Total subjects ($n = 744$). This study found high levels of plasma clusterin in AD cases compared to age-matched control cases. Clusterin concentration was proportional to atrophy in AD brain	Thambisetty et al. (2010)
ADNI study AD and non-AD ($n = 368$)	CSF Aβ_{1-42} versus blood plasma Aβ_{1-42} CSF tau and p-tau versus blood plasma Aβ_{1-42}	Low relation between CSF Aβ_{1-42} levels with those of plasma. Mild inverse correlation of CSF tau and p-tau with plasma Aβ_{1-42}	Toledo et al. (2011)
ADNI study AD and non-AD ($n = 95$)	[11C]PIB PET versus blood plasma Aβ_{1-42}	Mild inverse correlation of [11C]PIB PET and p-tau with plasma Aβ_{1-42}	Toledo et al. (2011)
MCI ($n = 20$), age-matched control ($n = 19$)	(Aβ_{1-42}/ Aβ_{1-40}) versus [11C]PIB PET	(Aβ_{1-42}/ Aβ_{1-40}) inversely related to [11C]PIB PET in MCI cases	Devanand et al. (2011)
Biomarkers Consortium Alzheimer's Disease Plasma Proteomics Project AD ($n = 112$), MCI ($n = 396$), age-matched control ($n = 58$)	Blood plasma biomarkers versus CSF biomarkers and APOE genotype	Plasma biomarkers are correlated with CSF biomarkers and APOE genotype	Soares et al. (2012)

(Continued)

Table 6.7	Crosscorrelation of Blood Plasma-Based Biomarkers and Other Biomarkers of Alzheimer's Disease *(cont.)*		
Study Designed	**Alzheimer's Disease Biomarker Modalities**	**Comments**	**References**
ADNI + Washington University AD Cohort + University of Pennsylvania AD Cohort *ADNI:* AD (*n* = 112); age-matched control (*n* = 58); MCI (*n* = 396). *University of Pennsylvania AD cohort:* AD (*n* = 88); age-matched control (*n* = 126); MCI (*n* = 16). *Washington University AD cohort:* CDR1 (*n* = 28); age-matched control (*n* = 242); CDR0.5 (*n* = 63).	Blood plasma biomarkers versus CSF biomarkers and APOE genotype	CSF Aβ_{1-42} levels and t-tau/Aβ_{1-42} ratios were well correlated with the number of APOE4 alleles and plasma levels of B-type natriuretic peptide and pancreatic polypeptide	Hu et al. (2012)
ADNI AD (*n* = 16), MCI (*n* = 52), age-matched control (*n* = 3)	Blood plasma biomarkers versus [11C]-PIB PET Imaging	16 of the analytes of the Rules Based Medicine Human Discovery Multi-Analyte Profile 1.0 panel were found to associate with [11C]-PiB PET imaging biomarkers. Sensitivity = 0.918, specificity = 0.545	Kiddle et al. (2012)

Aβ, beta-amyloid; AD, Alzheimer's disease; ADNI, Alzheimer's disease neuroimaging initiative; APOE, apolipoprotein E; CDR, clinical dementia rating; CSF, cerebrospinal fluid; LC-MS, liquid chromatography-mass spectrometry; MCI, mild cognitive impairment; MMSE, mini mental score examination; MRI, magnetic resonance imaging; PET, position emission tomography; [11C]PIB, [11C] Pittsburgh compound B.

6.7 POTENTIAL CHALLENGES IN DEVELOPING BLOOD-BASED ALZHEIMER'S DISEASE BIOMARKERS

To evaluate the future of blood-based AD biomarkers, including standardization of preanalytical variables and harmonization of technological methodologies, the Blood-Based Biomarker Interest Group was established consisting of leading AD scientists from industry and academics (Henriksen et al., 2014). This group identified three clear advantages of peripheral fluid-based AD biomarkers: cost-effectiveness, capacity for repeated testing, and tissue collection is easy, less invasive, and can be performed in primary care settings (Table 6.8). According to The Ronald and Nancy Reagan Research Institute of the Alzheimer's Association and the National Institute on Aging Working Group (1998) Consensus Report of the Working Group on Molecular and Biochemical Markers of Alzheimer's Disease CSF biomarkers may not be qualified as an ideal AD biomarker for several reasons.

Cost and time: Blood-based biomarkers of AD provide a cost- and time-effective measure of analytes.

Easy and common protocol for sample collection: Noninvasive and very common way of sample collection procedure would be plus point for blood-based AD biomarkers easily applicable to the general care setting compare to other AD biomarker modalities.

Repeated measurement: Repeated measurements are necessary for disease progression and therapeutic efficacy tests. Blood-based fluid biomarkers are the best method with respect to the requirement of repeated sample collection.

Yet the greatest limitation to blood-based AD biomarkers is that their accuracy, sensitivity, and specificity are lower than those of CSF biomarkers. The reasons are as follows:

1. *Interference of blood plasma proteins*: Blood is a complex bio-fluid. There are numerous plasma proteins that can interfere with the detection of minute changes in blood-based biomarkers related to AD pathology. Proteins in blood may be present in many isoforms with different biophysical states, and complexed with other analytes. Altered activities and variability in posttranslational modification may affect biomarker detection. Interfering protein components include: serum amyloid P components, transthyretin, apoferritin, immunoglobulin, α2-macroglobulin, complement factors, apolipoprotein A-I, A-IV, E and J, etc. Erythrocytes can sequester plasma $A\beta_{1-42}$ peptides to a greater extent than other $A\beta$ forms and human serum albumin can bound $A\beta$ peptides rapidly with a 1:1 stoichiometry (Kuo et al., 2000).

2. *Degradation of analytes in blood*: AD-pathologic proteins can be degraded by protease activity in blood, phagocytosis by macrophages, hepatic metabolism, and renal excretion. Degradation of blood-based biomarkers by proteases in blood cells is common problem. The activity of proteases is dependent on a variety of factors, which may introduce error in quantification of blood analytes. Proteins and lipids may be metabolized to a different extent. Storage of blood samples at $-70°C$ has been shown to

Table 6.8 Cost Comparison of Alzheimer's Disease Biomarker Testing Modalities

Modality	Invasiveness/Side Effects	Estimated Cost per Test[a]	Other Considerations	References
CSF biomarkers	Highly invasive; significant number (~40%) of patients have headache; other side effects are: back pain. Very rare case found subdural hematoma, meningitis	~$450–$1000	Highly skilled personal are required for LP. LP procedure is yet to be included as AD diagnostic evaluations in the US. LP is routinely performed in AD diagnostic procedure in Europe and approved by European Medicines Agency (EMA). Three different kind of ELISA are required for thee CSF biomarkers (of $A\beta_{1-42}$, total tau (t-tau), and phosphorylated tau-181 (p-tau)).	Valcárcel-Nazco et al. (2014), Fiandaca et al. (2014)
sMRI	Mild invasive	$1694–$3624	Specialized staff, expensive equipment, highly qualified bioinformatics personal, variability center to center. Claustrophobia for elderly person is a minor problem	http://www.comparemricost.com/. Study of ten cities in USA (Orlando, FL Dallas, TX—MRI Testing Facility A MRI and Dallas, TX— MRI Testing Facility B, San Diego, CA, Salt Lake City, UT, Detroit, MI, New York, NY—MRI Testing Facility A, New York, NY—MRI Testing Facility B, Raleigh, NC, Omaha, NE)
fMRI	Mild invasive	Generally more than sMRI	More specialized staff, more expensive equipment, extremely high qualified bioinformatics personal are necessary compared to sMRI. Variability center to center is a big problem. Extremely sophisticated and standardized data analysis system are required.	

MRS	Mild invasive	Similar to MRI	Not yet standardized like MRI for AD diagnosis	
PET	Moderately invasive	$825–$6800	Radioactive isotopes are required to work	http://www.newchoicehealth.com/procedures/pet-scan-brain: National PET Scan Brain Procedure Pricing Summary
SPECT	Moderately invasive	~ $1100	Radioactive isotopes are required to work	AmenClinics Inc.
Blood-based biomarkers	Minimally invasive	<$100	Common to all diagnostic center	

AD, Alzheimer's disease; Aβ, beta-amyloid; CSF, cerebrospinal fluid; ELISA, enzyme-linked immunosorbent assay; fMRI, functional magnetic resonance imaging; ELISA, enzyme-linked immunosorbent assay; LP, lumbar puncture; sMRI, structural magnetic resonance imaging; MRS, magnetic resonance spectroscopy; PET, positron emission tomography; SPECT, single-photon emission computed tomography.
aCost calculation was an estimate of different published data.

result in no reduction of Aβ concentration and were found to be stable for 1 year; however, longer duration storage at −20°C resulted in decreases in Aβ concentration. At room temperature, Aβ concentration decreases ~20% every 24 h (Toledo et al., 2013)

3. *High dilution in blood plasma*: Brain-specific AD pathological proteins are significantly diluted in circulating blood plasma and extracellular fluids of peripheral tissues. As a result, plasma Aβ measurements are found to be more variable than CSF Aβ measurements (Rissman et al., 2012).

4. *Different extent of loss of blood–brain barrier (BBB):* The extent of integrity of the BBB is not known for a slowly progressing disease like AD, but differences in BBB integrity are expected to affect the amount of Aβ detected in the blood.

All of these issues pose considerable challenges to developing blood-based AD biomarkers. The most important challenges are variability in sampling, unresolved issues in preanalytical procedures, standardization of analytical techniques used in different laboratories, and fixing universal cut-off values of analytes that distinguish between AD versus non-AD patients. Variabilities in sampling include choice of anticoagulant, protease inhibitor, material used in collection tubes, and needle size. Variabilities in preanalytical procedures include storage temperature, dilution factor, number of freeze/thaw cycles, and protocols for separating blood plasma/serum. Analytical procedures such as ELISA, multiplex chemiluminescence, antibody/gene micro assay system need to be standardized and calibrated. Consensus on universal cut-off values to differentiate between AD versus non-AD cases is urgently needed.

To accelerate the development of blood-based AD biomarkers, the Alzheimer's Association and Alzheimer's Drug Discovery Foundation convened a meeting in 2013. This working group outlined the most important challenges in terms of research and clinical regulatory perspectives and standardization of analytical procedures (Snyder et al., 2014). Despite the challenges, considerable progress has been made in the research and development of blood-based AD biomarkers. For example, a wide variety of new technological platforms are now available in proteomics, transcriptomics, and micro RNA detection. Those analytical platforms are now ready for next step validation and standardization.

6.8 PERFORMANCE OF BLOOD-BASED BIOMARKERS IN ASSESSING DRUG EFFICACY IN ALZHEIMER'S DISEASE CLINICAL TRIALS

The purpose of incorporation of blood-based biomarkers into AD clinical trials is to easily assess drug response, provide surrogate endpoints for drug efficacy, and give insights into the mechanisms of drug action. Repeated sampling and measurement of plasma Aβ levels to monitor drug efficacy make this an attractive biomarker for clinical trials; however, the utility and accuracy of plasma Aβ measurement in clinical trials of AD is not supported by clear data. Several clinical trials that have used blood-based AD biomarkers to assess drug efficacy produced inconclusive results (Table 6.9). Recently, plasma Aβ concentration

Table 6.9	**Use of Blood-Based Biomarkers of Alzheimer's Disease in Assessing Drug Efficacy in Clinical Trials**			
Modality	**Clinical Efficacy**	**Biomarkers**	**Comments**	**References**
Phase II trial: γ-secretase inhibitor LY450139	After oral administration of drug no significant improvement in cognitive function	■ Dose dependence reduction of $A\beta_{1-42}$ ■ Slight decrease in $A\beta$ plaques by	No clinical improvement with treatment	Fleisher et al. (2008)
Donepezil	No significant improvement in cognitive or functional ability	No change $A\beta$ levels	No significant clinical improvement with treatment	Roher et al. (2009)
Multicenter Tow clinical trials: Donepezil + vitamin E in MCI Simvastatin	Baseline plasma $A\beta$ was not related to cognitive or clinical progression	Longitudinal measurement of plasma $A\beta$	No support for the utility of plasma $A\beta$ biomarker as a prognostic factor or correlate of cognitive change	Donohue et al. (2015)
Single doseγ-secretase inhibitor LY450139	Reduction in plasma $A\beta_{1-40}$ was demonstrated that did not return to baseline for more than 12 h	Noncompetitive sandwich ELISA	Plasma $A\beta$ response were not observed in CSF $A\beta$ response	Siemers et al. (2007)
$A\beta$ monoclonal antibody Solanezumab	Marked increase in plasma total $A\beta$ was observed	Total (bound + unbound) $A\beta_{1-40}$ and $A\beta_{1-42}$ using validated solanezumab-tolerant ELISA methods	Passive immunization to slow the progression of AD	Uenaka et al. (2012)

AD, Alzheimer's disease; $A\beta$, beta-amyloid; CSF, cerebrospinal fluid; ELISA, enzyme-linked immunosorbent assay; MCI, mild cognitive impairment.

changes have been monitored as a marker of treatment response for donepezil and vitamin E in MCI ($n = 405$, for 24 months) and simvastatin in mild to moderate AD ($n = 225$, for 18 months (Donohue et al., 2015). Baseline plasma Aβ was not related to cognitive or clinical progression, although simvastatin was associated with a significant increase in plasma Aβ compared to placebo. Plasma Aβ measurement may not be a good prognostic biomarker or a good indicator of cognitive decline, but plasma Aβ values correlated with antiamyloid immunotherapies and amyloid removing drugs (Donohue et al., 2015).

6.9 CONCLUSIONS

Amyloid pathogenesis and tau metabolic pathways are not limited to the brain, but are ubiquitous in the human body and found in blood, skin, saliva, and other peripheral tissues. Therefore, measurement of AD biomarkers in easily accessible peripheral fluids is a highly promising area of research. Unfortunately, the most widely researched peripheral AD biomarker, blood plasma Aβ has produced conflicting results. As a result, the initial focus on blood-based Aβ as an AD biomarker has shifted to blood-based multicomponent plasma AD biomarkers. There are several reasons for inconsistent results from studies of blood serum/plasma-based AD biomarkers. First, the integrity of the BBB in AD has not been extensively studied. The degree of crossing analytes (proteins/peptides) is related to the degree of loss of BBB integrity. AD is a slow heterogeneous progressive disease and that may affect the BBB integrity differently. Second, brain proteins/peptides that cross the BBB may be degraded or metabolized in blood. Third, the levels of fluctuation of proteins/peptides concentration depend on physical state of the patients (sleep cycle, food intake, etc.). Finally, and most importantly, interference of other conditions such as blood pressure, blood glucose levels, and concentration of inflammatory molecules, may hamper the diagnosis.

Some circulating proteins and lipids have been identified as potential biomarkers for AD, though their relationship to AD or MCI pathology is still unknown and merits further research. Those studies would provide further support for the systemic manifestation of AD pathophysiology. Discovery and development of candidate AD biomarkers may also lead to the identification of new therapeutic targets and approaches. Assessment of peripheral fluid biomarkers in longitudinal cohorts provides more information than cross-sectional studies. Use of neuroimaging and CSF biomarkers hold promise, but continue to face challenges related to invasiveness of sample collection, cost, interlaboratory variation, and the ability to distinguish AD from non-AD dementias. AD blood-based biomarkers also require validation and standardization in terms of sample collection procedures, methodologies, and data interpretation. Combinations of AD biomarkers into a molecular signature or index may prove to be more accurate than any single biomarker. Peripheral fluid-based biomarkers using multicomponent proteomics and lipidomics assays are likely to produce a more promising AD "signature." The future focus for peripheral fluid-based AD biomarker research will include (1) improvements in diagnostic specificity sensitivity; (2)

improved ability to differentiate AD from non-AD dementias and MCI; (3) improved identification of different AD phenotypes; (4) the capacity to monitor prodromal stages of AD; and (5) development of peripheral AD biomarker to detect preclinical AD.

Bibliography

Akiyama, H., Barger, S., Barnum, S., Bradt, B., Bauer, J., Cole, G.M., Cooper, N.R., Eikelenboom, P., Emmerling, M., Fiebich, B.L., Finch, C.E., Frautschy, S., Griffin, W.S., Hampel, H., Hull, M., Landreth, G., Lue, L., Mrak, R., Mackenzie, I.R., McGeer, P.L., O'Banion, M.K., Pachter, J., Pasinetti, G., Plata-Salaman, C., Rogers, J., Rydel, R., Shen, Y., Streit, W., Strohmeyer, R., Tooyoma, I., Van Muiswinkel, F.L., Veerhuis, R., Walker, D., Webster, S., Wegrzyniak, B., Wenk, G., Wyss-Coray, T, 2000. Inflammation and Alzheimer's disease. Neurobiol. Aging 21, 383–421.

Baskin, F., Rosenberg, R.N., Iyer, L., Hynan, L., Cullum, C.M., 2000. Platelet APP isoform ratios correlate with declining cognition in AD. Neurology 54, 1907–1909.

Bermejo-Pareja, F., Antequera, D., Vargas, T., Molina, J.A., Carro, E., 2010. Saliva levels of Abeta1-42 as potential biomarker of Alzheimer's disease: a pilot study. BMC Neurol. 10, 108.

Britschgi, M., Rufibach, K., Huang, S.L., Clark, C.M., Kaye, J.A., Li, G., Peskind, E.R., Quinn, J.F., Galasko, D.R., Wyss-Coray, T., 2011. Modeling of pathological traits in Alzheimer's disease based on systemic extracellular signaling proteome. Mol. Cell Proteomics 10, M111.

De La Monte, S.M., Carlson, R.I., Brown, N.V., Wands, J.R., 1996. Profiles of neuronal thread protein expression in Alzheimer's disease. J. Neuropathol. Exp. Neurol. 55, 1038–1050.

Devanand, D.P., Schupf, N., Stern, Y., Parsey, R., Pelton, G.H., Mehta, P., Mayeux, R., 2011. Plasma Aβ and PET PiB binding are inversely related in mild cognitive impairment. Neurology 77, 125–131.

Doecke, J.D., Laws, S.M., Faux, N.G., Wilson, W., Burnham, S.C., Lam, C.P., Mondal, A., Bedo, J., Bush, A.I., Brown, B., De Ruyck, K., Ellis, K.A., Fowler, C., Gupta, V.B., Head, R., Macaulay, S.L., Pertile, K., Rowe, C.C., Rembach, A., Rodrigues, M., Rumble, R., Szoeke, C., Taddei, K., Taddei, T., Trounson, B., Ames, D., Masters, C.L., Martins, R.N., Alzheimer's Disease Neuroimaging Initiative; Australian Imaging Biomarker and Lifestyle Research Group, 2012. Blood-based protein biomarkers for diagnosis of Alzheimer disease. Arch. Neurol. 69, 1318–1325.

Donohue, M.C., Moghadam, S.H., Roe, A.D., Sun, C.K., Edland, S.D., Thomas, R.G., Petersen, R.C., Sano, M., Galasko, D., Aisen, P.S., Rissman, R.A., 2015. Longitudinal plasma amyloid beta in Alzheimer's disease clinical trials. Alzheimers Dement. 11, 1069–1079.

Eckert, A., Cotman, C.W., Zerfass, R., Hennerici, M., Müller, W.E., 1998. Lymphocytes as cell model to study apoptosis in Alzheimer's disease: vulnerability to programmed cell death appears to be altered. J. Neural. Transm. Suppl. 54, 259–267.

Fehlbaum-Beurdeley, P.1., Sol, O., Désiré, L., Touchon, J., Dantoine, T., Vercelletto, M., Gabelle, A., Jarrige, A.C., Haddad, R., Lemarié, J.C., Zhou, W., Hampel, H., Einstein, R., Vellas, B., EH-TAD/002 study group, 2012. Validation of AclarusDx™, a blood-based transcriptomic signature for the diagnosis of Alzheimer's disease. J. Alzheimers Dis. 32, 169–181.

Fiandaca, M.S., Mapstone, M.E., Cheema, A.K., Federoff, H.J., 2014. The critical need for defining preclinical biomarkers in Alzheimer's disease. Alzheimers Dement. 10 (3 Suppl), S196–S212.

Figurski, M.J., Waligorska, T., Toledo, J., Vander stichele, H., Korecka, M., Lee, V.M., Trojanowski, J.Q., Shaw, L.M., 2012. Improved protocol for measurement of plasma beta-amyloid in longitudinal evaluation of Alzheimer's Disease Neuroimaging Initiative study patients. Alzheimers Dement. 8, 250–260.

Fleisher, A.S., Raman, R., Siemers, E.R., Becerra, L., Clark, C.M., Dean, R.A., Farlow, M.R., Galvin, J.E., Peskind, E.R., Quinn, J.F., Sherzai, A., Sowell, B.B., Aisen, P.S., Thal, L.J., 2008. Phase 2 safety trial targeting amyloid beta production with a gamma-secretase inhibitor in Alzheimer disease. Arch. Neurol. 65, 1031–1038.

Fukumoto, H., Tennis, M., Locascio, J.J., Hyman, B.T., Growdon, J.H., Irizarry, M.C., 2003. Age but not diagnosis is the main predictor of plasma amyloid beta-protein levels. Arch. Neurol. 60, 958–964.

Gasparini, L., Racchi, M., Binetti, G., Trabucchi, M., Solerte, S.B., Alkon, D.L., Etcheberrigaray, R., Gibson, G., Blass, J., Paoletti, R., Govoni, S., 1998. Peripheral markers in testing pathophysiological hypotheses and diagnosing Alzheimer's disease. FASEB J. 12, 17–34.

Ghanbari, H., Ghanbari, K., Beheshti, I., Munzar, M., Vasauskas, A., Averback, P., 1998. Biochemical assay for AD7C-NTP in urine as an Alzheimer's disease marker. J. Clin. Lab. Anal. 12, 285–288.

Goldstein, L.E., Muffat, J.A., Cherny, R.A., Moir, R.D., Ericsson, M.H., Huang, X., Mavros, C., Coccia, J.A., Faget, K.Y., Fitch, K.A., Masters, C.L., Tanzi, R.E., Chylack, Jr., L.T., Bush, A.I., 2003. Cytosolic beta-amyloid deposition and supranuclear cataracts in lenses from people with Alzheimer's disease. Lancet 361, 1258–1265.

Goodman, I., Golden, G., Flitman, S., Xie, K., McConville, M., Levy, S., Zimmerman, E., Lebedeva, Z., Richter, R., Minagar, A., Averback, P., 2007. A multi-center blinded prospective study of urine neural thread protein measurements in patients with suspected Alzheimer's disease. J. Am. Med. Dir. Assoc. 8, 21–30.

Han, X., Rozen, S., Boyle, S.H., Hellegers, C., Cheng, H., Burke, J.R., Welsh-Bohmer, K.A., Doraiswamy, P.M., Kaddurah-Daouk, R., 2011. Metabolomics in early Alzheimer's disease: identification of altered plasma sphingolipidome using shotgun lipidomics. PLoS One 6, e21643.

Hansson, O., Stomrud, E., Vanmechelen, E., Östling, S., Gustafson, D.R., Zetterberg, H., Blennow, K., Skoog, I., 2012. Evaluation of plasma Aβ as predictor of Alzheimer's disease in older individuals without dementia: a population-based study. J. Alzheimers Dis. 28, 231–238.

Harold, D., Abraham, R., Hollingworth, P., Sims, R., Gerrish, A., Hamshere, M.L., et al., 2009. Genome-wide association study identifies variants at CLU and PICALM associated with Alzheimer's disease. Nat. Genet. 41, 1088–1093.

Henriksen, K., O'Bryant, S.E., Hampel, H., Trojanowski, J.Q., Montine, T.J., Jeromin, A., Blennow, K., Lönneborg, A., Wyss-Coray, T., Soares, H., Bazenet, C., Sjögren, M., Hu, W., Lovestone, S., Karsdal, M.A., Weiner, M.W., Blood-Based Biomarker Interest Group, 2014. The future of blood-based biomarkers for Alzheimer's disease. Alzheimers Dement. 10, 115–131.

Henriksen, K., Wang, Y., Sørensen, M.G., Barascuk, N., Suhy, J., Pedersen, J.T., Duffin, K.L., Dean, R.A., Pajak, M., Christiansen, C., Zheng, Q., Karsdal, M.A., 2013. An enzyme-generated fragment of tau measured in serum shows an inverse correlation to cognitive function. PLoS One 8, e64990.

Hollingworth, P., Harold, D., Sims, R., Gerrish, A., Lambert, J.C., Carrasquillo, M.M., et al., 2011. Common variants at ABCA7, MS4A6A/MS4A4E, EPHA1, CD33 and CD2AP are associated with Alzheimer's disease. Nat. Genet. 43, 429–435.

Hu, W.T., Holtzman, D.M., Fagan, A.M., Shaw, L.M., Perrin, R., Arnold, S.E., Grossman, M., Xiong, C., Craig-Schapiro, R., Clark, C.M., Pickering, E., Kuhn, M., Chen, Y., Van Deerlin, V.M., McCluskey, L., Elman, L., Karlawish, J., Chen-Plotkin, A., Hurtig, H.I., Siderowf, A., Swenson, F., Lee, V.M., Morris, J.C., Trojanowski, J.Q., Soares, H., Alzheimer's Disease Neuroimaging Initiative, 2012. Plasma multianalyte profiling in mild cognitive impairment and Alzheimer disease. Neurology 79, 897–905.

Huang, Y., Potter, R., Sigurdson, W., Kasten, T., Connors, R., Morris, J.C., Benzinger, T., Mintun, M., Ashwood, T., Ferm, M., Budd, S.L., Bateman, R.J., 2012. β-amyloid dynamics in human plasma. Arch. Neurol. 69, 1591–1597.

Hye, A., Riddoch-Contreras, J., Baird, A.L., Ashton, N.J., Bazenet, C., Leung, R., Westman, E., Simmons, A., Dobson, R., Sattlecker, M., Lupton, M., Lunnon, K., Keohane, A., Ward, M., Pike, I., Zucht, H.D., Pepin, D., Zheng, W., Tunnicliffe, A., Richardson, J., Gauthier, S., Soininen, H., Kłoszewska, I., Mecocci, P., Tsolaki, M., Vellas, B., Lovestone, S., 2014. Plasma proteins predict conversion to dementia from prodromal disease. Alzheimers Dement. 10, 799–807.

Irizarry, M.C., 2004. Biomarker of Alzheimer's disease in plasma. NeuroRx 1, 226–234.

Jensen, M., Schröder, J., Blomberg, M., Engvall, B., Pantel, J., Ida, N., Basun, H., Wahlund, L.O., Werle, E., Jauss, M., Beyreuther, K., Lannfelt, L., Hartmann, T., 1999. Cerebrospinal fluid A beta42 is increased early in sporadic Alzheimer's disease and declines with disease progression. Ann. Neurol. 45, 504–511.

Joachim, C.L., Mori, H., Selkoe, D.J., 1989. Amyloid beta-protein deposition in tissues other than brain in Alzheimer's disease. Nature 341, 226–230.

Kiddle, S.J., Sattlecker, M., Proitsi, P., Simmons, A., Westman, E., Bazenet, C., Nelson, S.K., Williams, S., Hodges, A., Johnston, C., Soininen, H., Kłoszewska, I., Mecocci, P., Tsolaki, M., Vellas, B., Newhouse, S., Lovestone, S., Dobson, R.J., 2014. Candidate blood proteome markers of Alzheimer's disease onset and progression: a systematic review and replication study. J. Alzheimers Dis. 38, 515–531.

Kiddle, S.J., Thambisetty, M., Simmons, A., Riddoch-Contreras, J., Hye, A., Westman, E., Pike, I., Ward, M., Johnston, C., Lupton, M.K., Lunnon, K., Soininen, H., Kloszewska, I., Tsolaki, M., Vellas, B., Mecocci, P., Lovestone, S., Newhouse, S., Dobson, R., Alzheimers Disease Neuroimaging Initiative, 2012. Plasma based markers of [11C] PiB-PET brain amyloid burden. PLoS One 7, e44260.

Kou, J., Kovacs, G.G., Höftberger, R., Kulik, W., Brodde, A., Forss-Petter, S., Hönigschnabl, S., Gleiss, A., Brügger, B., Wanders, R., Just, W., Budka, H., Jungwirth, S., Fischer, P., Berger, J., 2011. Peroxisomal alterations in Alzheimer's disease. Acta Neuropathol. 122, 271–283.

Koyama, A., Okereke, O.I., Yang, T., Blacker, D., Selkoe, D.J., Grodstein, F., 2012. Plasma amyloid-beta as a predictor of dementia and cognitive decline:a systematic review and meta-analysis. Arch. Neurol. 69, 824–831.

Kropholler, M.A., Boellaard, R., van Berckel, B.N., Schuitemaker, A., Kloet, R.W., Lubberink, M.J., Jonker, C., Scheltens, P., Lammertsma, A.A., 2007. Evaluation of reference regions for (R)-[(11) C]PK11195 studies in Alzheimer's disease and mild cognitive impairment. J. Cereb. Blood Flow Metab. 27, 1965–1974.

Kuo, Y.M., Kokjohn, T.A., Kalback, W., Luehrs, D., Galasko, D.R., Chevallier, N., Koo, E.H., Emmerling, M.R., Roher, A.E., 2000. Amyloid-beta peptides interact with plasma proteins and erythrocytes: implications for their quantitation in plasma. Biochem. Biophys. Res. Commun. 268, 750–756.

Lambert, J.C., Heath, S., Even, G., Campion, D., Sleegers, K., Hiltunen, M., et al., 2009a. Genome-wide association study identifies variants at CLU and CR1 associated with Alzheimer's disease. Nat. Genet. 41, 1094–1099.

Lambert, J.C., Ibrahim-Verbaas, C.A., Harold, D., Naj, A.C., Sims, R., Bellenguez, C., et al., 2013. Meta-analysis of 74,046 individuals identifies 11 new susceptibility loci for Alzheimer's disease. Nat. Genet. 45, 1452–1458.

Lambert, J.C., Schraen-Maschke, S., Richard, F., Fievet, N., Rouaud, O., Berr, C., Dartigues, J.F., Tzourio, C., Alpérovitch, A., Buée, L., Amouyel, P., 2009b. Association of plasma amyloid beta with risk of dementia: the prospective Three-City Study. Neurology 73, 847–853.

Le Bastard, N., Aerts, L., Leurs, J., Blomme, W., De Deyn, P.P., Engelborghs, S., 2009. No correlation between time-linked plasma and CSF Abeta levels. Neurochem. Int. 55, 820–825.

Lewczuk, P., Kornhuber, J., Vanmechelen, E., Peters, O., Heuser, I., Maier, W., Jessen, F., Bürger, K., Hampel, H., Frölich, L., Henn, F., Falkai, P., Rüther, E., Jahn, H., Luckhaus, Ch., Perneczky, R., Schmidtke, K., Schröder, J., Kessler, H., Pantel, J., Gertz, H.J., Vanderstichele, H., de Meyer, G., Shapiro, F., Wolf, S., Bibl, M., Wiltfang, J., 2010. Amyloid beta peptides in plasma in early diagnosis of Alzheimer's disease: a multicenter study with multiplexing. Exp. Neurol. 223, 366–370.

Lui, J.K., Laws, S.M., Li, Q.X., Villemagne, V.L., Ames, D., Brown, B., Bush, A.I., De Ruyck, K., Dromey, J., Ellis, K.A., Faux, N.G., Foster, J., Fowler, C., Gupta, V., Hudson, P., Laughton, K., Masters, C.L., Pertile, K., Rembach, A., Rimajova, M., Rodrigues, M., Rowe, C.C., Rumble, R., Szoeke, C., Taddei, K., Taddei, T., Trounson, B., Ward, V., Martins, R.N., AIBL Research Group, 2010. Plasma amyloid-beta as a biomarker in Alzheimer's disease: the AIBL study of aging. J. Alzheimers Dis. 20, 1233–1242.

Mapstone, M., Cheema, A.K., Fiandaca, M.S., Zhong, X., Mhyre, T.R., MacArthur, L.H., Hall, W.J., Fisher, S.G., Peterson, D.R., Haley, J.M., Nazar, M.D., Rich, S.A., Berlau, D.J., Peltz, C.B., Tan, M.T., Kawas, C.H., Federoff, H.J., 2014. Plasma phospholipids identify antecedent memory impairment in older adults. Nat. Med. 20, 415–418.

Marksteiner, J., Imarhiagbe, D., Defrancesco, M., Deisenhammer, E.A., Kemmler, G., Humpel, C., 2014. Analysis of 27 vascular-related proteins reveals that NT-proBNP is a potential biomarker for Alzheimer's disease and mild cognitive impairment: a pilot-study. Exp. Gerontol. 50, 114–121.

Marksteiner, J., Kemmler, G., Weiss, E.M., Knaus, G., Ullrich, C., Mechtcheriakov, S., Oberbauer, H., Auffinger, S., Hinterhölzl, J., Hinterhuber, H., Humpel, C., 2011. Five out of 16 plasma signaling proteins are enhanced in plasma of patients with mild cognitive impairment and Alzheimer's disease. Neurobiol. Aging 32, 539–555.

Mayeux, R., Honig, L.S., Tang, M.X., Manly, J., Stern, Y., Schupf, N., Mehta, P.D., 2003. Plasma Abeta40 and Abeta42 and Alzheimer's disease: relation to age, mortality, and risk. Neurology 61, 1185–1190.

Mayeux, R., Sano, M., 1999. Treatment of Alzheimer's disease. New Engl J Med 341, 1670–1679.

Mehta, P.D., Pirttila, T., Mehta, S.P., Sersen, E.A., Aisen, P.S., Wisniewski, H.M., 2000. Plasma and cerebrospinal fluid levels of amyloid beta proteins 1-40 and 1-42 in Alzheimer disease. Arch. Neurol. 57, 100–105.

Mehta, P.D., Pirttila, T., Patrick, B.A., Barshatzky, M., Mehta, S.P., 2001. Amyloid beta protein 1-40 and 1-42 levels in matched cerebrospinal fluid and plasma from patients with Alzheimer disease. Neurosci. Lett. 304, 102–106.

Merrick, B.A., London, R.E., Bushel, P.R., Grissom, S.F., Paules, R.S., 2011. Platforms for biomarker analysis using high-throughput approaches in genomics, transcriptomics, proteomics, metabolomics, and bioinformatics. IARC Sci. Publ. 163, 121–142.

Munzar, M., Levy, S., Rush, R., Averback, P., 2002. Clinical study of a urinary competitive ELISA for neural thread protein in Alzheimer disease. Neurol. Clin. Neurophysiol. 2002, 2–8.

O'Bryant, S.E., Xiao, G., Barber, R., Reisch, J., Doody, R., Fairchild, T., Adams, P., Waring, S., Diaz-Arrastia, R., Texas Alzheimer's Research Consortium, 2010. A serum protein-based algorithm for the detection of Alzheimer disease. Arch. Neurol. 67, 1077–1081.

Oh, E.S., Troncoso, J.C., Tucker, S.M.F., 2008. Maximizing the potential of plasma amyloid-beta as a diagnostic biomarker for Alzheimer's disease. Neuromol. Med. 10, 195–207.

Olazarán, J., Gil-de-Gómez, L., Rodríguez-Martín, A., Valentí-Soler, M., Frades-Payo, B., Marín-Muñoz, J., Antúnez, C., Frank-García, A., Jiménez, C.A., Gracia, L.M., Torregrossa, R.P., Guisasola, M.C., Bermejo-Pareja, F., Sánchez-Ferro, Á., Pérez-Martínez, D.A., Palomo, S.M., Farquhar, R., Rábano, A., Calero, M., 2015. A blood-based, 7-metabolite signature for the early diagnosis of Alzheimer's disease. J. Alzheimers Dis. 45, 1157–1173.

Palmer, A.M., 2010. The role of the blood-CNS barrier in CNS disorders and their treatment. Neurobiol. Dis. 37, 3–12.

Perry, R.H., Wilson, I.D., Bober, M.J., Atack, J., Blessed, G., Tomlinson, B.E., Perry, E.K., 1982. Plasma and erythrocyte acetylcholinesterase in senile dementia of Alzheimer type. Lancet 1, 174–175.

Praticò, D., Clark, C.M., Liun, F., Rokach, J., Lee, V.Y., Trojanowski, J.Q., 2002. Increase of brain oxidative stress in mild cognitive impairment: a possible predictor of Alzheimer disease. Arch. Neurol. 59, 972–976.

Proitsi, P., Kim, M., Whiley, L., Pritchard, M., Leung, R., Soininen, H., Kloszewska, I., Mecocci, P., Tsolaki, M., Vellas, B., Sham, P., Lovestone, S., Powell, J.F., Dobson, R.J., Legido-Quigley, C., 2015. Plasma lipidomics analysis finds long chain cholesteryl esters to be associated with Alzheimer's disease. Transl. Psychiatry 5, e494.

Ray, S., Britschgi, M., Herbert, C., Takeda-Uchimura, Y., Boxer, A., Blennow, K., Friedman, L.F., Galasko, D.R., Jutel, M., Karydas, A., Kaye, J.A., Leszek, J., Miller, B.L., Minthon, L., Quinn, J.F., Rabinovici, G.D., Robinson, W.H., Sabbagh, M.N., So, Y.T., Sparks, D.L., Tabaton, M., Tinklenberg,

J., Yesavage, J.A., Tibshirani, R., Wyss-Coray, T., 2007. Classification and prediction of clinical Alzheimer's diagnosis based on plasma signaling proteins. Nat. Med. 13, 1359–1362.

Reitz, C., 2013. Dyslipidemia and the risk of Alzheimer's disease. Curr. Atheroscler. Rep. 15, 307.

Reitz, C., Mayeux, R., 2014. Alzheimer disease: epidemiology, diagnostic criteria, risk factors and biomarkers. Biochem. Pharmacol. 88, 640–651.

Rissman, R.A., Trojanowski, J.Q., Shaw, L.M., Aisen, P.S., 2012. Longitudinal plasma amyloid beta as a biomarker of Alzheimer's disease. J. Neural. Transm. 119, 843–850.

Roher, A.E., Esh, C.L., Kokjohn, T.A., Castaño, E.M., Van Vickle, G.D., Kalback, W.M., Patton, R.L., Luehrs, D.C., Daugs, I.D., Kuo, Y.M., Emmerling, M.R., Soares, H., Quinn, J.F., Kaye, J., Connor, D.J., Silverberg, N.B., Adler, C.H., Seward, J.D., Beach, T.G., Sabbagh, M.N., 2009. Amyloid beta peptides in human plasma and tissues and their significance for Alzheimer's disease. Alzheimers Dement. 5, 18–29.

Sato, Y., Suzuki, I., Nakamura, T., Bernier, F., Aoshima, K., Oda, Y., 2012. Identification of a new plasma biomarker of Alzheimer's disease using metabolomics technology. J. Lipid Res. 53, 567–576.

Schuitemaker, A., Dik, M.G., Veerhuis, R., Scheltens, P., Schoonenboom, N.S., Hack, C.E., Blankenstein, M.A., Jonker, C., 2009. Inflammatory markers in AD and MCI patients with different biomarker profiles. Neurobiol. Aging 30, 1885–1889.

Schupf, N., Tang, M.X., Fukuyama, H., Manly, J., Andrews, H., Mehta, P., Ravetch, J., Mayeux, R., 2008. Peripheral Abeta subspecies as risk biomarkers of Alzheimer's disease. Proc. Natl. Acad. Sci. USA 105, 14052–14057.

Sevush, S., Jy, W., Horstman, L.L., Mao, W.W., Kolodny, L., Ahn, Y.S., 1998. Platelet activation in Alzheimer disease. Arch. Neurol. 55, 530–536.

Shah, N.S., Vidal, J.S., Masaki, K., Petrovitch, H., Ross, G.W., Tilley, C., DeMattos, R.B., Tracy, R.P., White, L.R., Launer, L.J., 2012. Midlife blood pressure, plasma β-amyloid, and the risk for Alzheimer disease: the Honolulu Asia Aging Study. Hypertension 59, 780–786.

Shi, M., Sui, Y.T., Peskind, E.R., Li, G., Hwang, H., Devic, I., Ginghina, C., Edgar, J.S., Pan, C., Goodlett, D.R., Furay, A.R., Gonzalez-Cuyar, L.F., Zhang, J., 2011. Salivary tau species are potential biomarkers of Alzheimer's disease. J. Alzheimers Dis. 27, 299–305.

Siemers, E.R., Dean, R.A., Friedrich, S., Ferguson-Sells, L., Gonzales, C., Farlow, M.R., May, P.C., 2007. Safety, tolerability, and effects on plasma and cerebrospinal fluid amyloid-beta after inhibition of gamma-secretase. Clin. Neuropharmacol. 30, 317–325.

Snyder, H.M., Carrillo, M.C., Grodstein, F., Henriksen, K., Jeromin, A., Lovestone, S., Mielke, M.M., O'Bryant, S., Sarasa, M., Sjøgren, M., Soares, H., Teeling, J., Trushina, E., Ward, M., West, T., Bain, L.J., Shineman, D.W., Weiner, M., Fillit, H.M., 2014. Developing novel blood-based biomarkers for Alzheimer's disease. Alzheimers Dement. 10, 109–114.

Soares, H.D., Chen, Y., Sabbagh, M., Roher, A., Schrijvers, E., Breteler, M., 2009. Identifying early markers of Alzheimer's disease using quantitative multiplex proteomic immunoassay panels. Ann. NY Acad. Sci. 1180, 56–67.

Soares, H.D., Potter, W.Z., Pickering, E., Kuhn, M., Immermann, F.W., Shera, D.M., Ferm, M., Dean, R.A., Simon, A.J., Swenson, F., Siuciak, J.A., Kaplow, J., Thambisetty, M., Zagouras, P., Koroshetz, W.J., Wan, H.I., Trojanowski, J.Q., Shaw, L.M., Biomarkers Consortium Alzheimer's Disease Plasma Proteomics Project, 2012. Plasma biomarkers associated with the apolipoprotein E genotype and Alzheimer disease. Arch. Neurol. 69, 1310–1317.

Soininen, H., Syrjänen, S., Heinonen, O., Neittaanmäki, H., Miettinen, R., Paljärvi, L., Syrjänen, K., Beyreuther, K., Riekkinen, P., 1992. Amyloid beta-protein deposition in skin of patients with dementia. Lancet 339, 245–1245.

Sundelöf, J., Giedraitis, V., Irizarry, M.C., Sundström, J., Ingelsson, E., Rönnemaa, E., Arnlöv, J., Gunnarsson, M.D., Hyman, B.T., Basun, H., Ingelsson, M., Lannfelt, L., Kilander, L., 2008. Plasma beta amyloid and the risk of Alzheimer disease and dementia in elderly men: a prospective, population-based cohort study. Arch. Neurol. 65, 256–263.

Sunderland, T., Linker, G., Mirza, N., Putnam, K.T., Friedman, D.L., Kimmel, L.H., Bergeson, J., Manetti, G.J., Zimmermann, M., Tang, B., Bartko, J.J., Cohen, R.M., 2003. Decreased beta-amyloid1-42 and increased tau levels in cerebrospinal fluid of patients with Alzheimer disease. JAMA 289, 2094–2103.

Tajima, Y., Ishikawa, M., Maekawa, K., Murayama, M., Senoo, Y., Nishimaki-Mogami, T., Nakanishi, H., Ikeda, K., Arita, M., Taguchi, R., Okuno, A., Mikawa, R., Niida, S., Takikawa, O., Saito, Y., 2013. Lipidomic analysis of brain tissues and plasma in a mouse model expressing mutated human amyloid precursor protein/tau for Alzheimer's disease. Lipids Health Dis. 12, 68–82.

Tamaoka, A., Fukushima, T., Sawamura, N., Ishikawa, K., Oguni, E., Komatsuzaki, Y., Shoji, S., 1996. Amyloid beta protein in plasma from patients with sporadic Alzheimer's disease. J. Neurol. Sci. 141, 65–68.

Thambisetty, M., Simmons, A., Hye, A., Campbell, J., Westman, E., Zhang, Y., Wahlund, L.O., Kinsey, A., Causevic, M., Killick, R., Kloszewska, I., Mecocci, P., Soininen, H., Tsolaki, M., Vellas, B., Spenger, C., Lovestone, S., AddNeuroMed Consortium, 2011. Plasma biomarkers of brain atrophy in Alzheimer's disease. PLoS One 6, e28527.

Thambisetty, M., Simmons, A., Velayudhan, L., Hye, A., Campbell, J., Zhang, Y., Wahlund, L.O., Westman, E., Kinsey, A., Güntert, A., Proitsi, P., Powell, J., Causevic, M., Killick, R., Lunnon, K., Lynham, S., Broadstock, M., Choudhry, F., Howlett, D.R., Williams, R.J., Sharp, S.I., Mitchelmore, C., Tunnard, C., Leung, R., Foy, C., O'Brien, D., Breen, G., Furney, S.J., Ward, M., Kloszewska, I., Mecocci, P., Soininen, H., Tsolaki, M., Vellas, B., Hodges, A., Murphy, D.G., Parkins, S., Richardson, J.C., Resnick, S.M., Ferrucci, L., Wong, D.F., Zhou, Y., Muehlboeck, S., Evans, A., Francis, P.T., Spenger, C., Lovestone, S., 2010. Association of plasma clusterin concentration with severity, pathology, and progression in Alzheimer disease. Arch. Gen. Psychiatry 67, 739–748.

The Ronald and Nancy Reagan Research Institute of the Alzheimer's Association and the National Institute on Aging Working Group, 1998. Consensus report of the Working Group on: "Molecular and Biochemical Markers of Alzheimer's Disease". Neurobiol. Aging 19, 109–116.

Toledo, J.B., Shaw, L.M., Trojanowski, J.Q., 2013. Plasma amyloid beta measurements—a desired but elusive Alzheimer's disease biomarker. Alzheimers Res. Ther. 5, 8.

Toledo, J.B., Vanderstichele, H., Figurski, M., Aisen, P.S., Petersen, R.C., Weiner, M.W., Jack, Jr., C.R., Jagust, W., Decarli, C., Toga, A.W., Toledo, E., Xie, S.X., Lee, V.M., Trojanowski, J.Q., Shaw, L.M., Alzheimer's Disease Neuroimaging Initiative, 2011. Factors affecting Abeta plasma levels and their utility as biomarkers in ADNI. Acta Neuropathol. 122, 401–413.

Uenaka, K., Nakano, M., Willis, B.A., Friedrich, S., Ferguson-Sells, L., Dean, R.A., Ieiri, I., Siemers, E.R., 2012. Comparison of pharmacokinetics, pharmacodynamics, safety, and tolerability of the amyloid beta monoclonal antibody solanezumab in Japanese and white patients with mild to moderate Alzheimer disease. Clin. Neuropharmacol. 35, 25–29.

Valcárcel-Nazco, C., Perestelo-Pérez, L., Molinuevo, J.L., Mar, J., Castilla, I., Serrano-Aguilar, P., 2014. Cost-effectiveness of the use of biomarkers in cerebrospinal fluid for Alzheimer's disease. J. Alzheimers Dis. 42, 777–788.

van Oijen, M., Hofman, A., Soares, H.D., Koudstaal, P.J., Breteler, M.M., 2006. Plasma Abeta(1-40) and Abeta(1-42) and the risk of dementia: a prospective case-cohort study. Lancet Neurol. 5, 655–660.

Veerhuis, R., Van Breemen, M.J., Hoozemans, J.M., Morbin, M., Ouladhadj, J., Tagliavini, F., Eikelenboom, P., 2003. Amyloid beta plaque-associated proteins C1q and SAP enhance the Abeta1-42 peptide-induced cytokine secretion by adult human microglia in vitro. Acta Neuropathol. 105, 135–144.

Villemagne, V.L., Perez, K.A., Pike, K.E., Kok, W.M., Rowe, C.C., White, A.R., Bourgeat, P., Salvado, O., Bedo, J., Hutton, C.A., Faux, N.G., Masters, C.L., Barnham, K.J., 2010. Blood-borne amyloid-beta dimer correlates with clinical markers of Alzheimer's disease. J. Neurosci. 30, 6315–6322.

Wood, P.L., 2012. Lipidomics of Alzheimer's disease: current status. Alzheimer's Res. Ther. 4, 5.

CHAPTER 7

Cell-Based Alzheimer's Disease Biomarkers

7.1 BACKGROUND

A simple, inexpensive, minimally invasive procedure for detecting biomarkers of Alzheimer's disease (AD), preferably in its earliest stages, is needed. Although AD is considered to be a disease of the central nervous system, it is associated with several systemic manifestations, including dysfunction in metabolic,

Biomarkers in Alzheimer's Disease. http://dx.doi.org/10.1016/B978-0-12-804832-0.00007-9
Copyright © 2016 Elsevier Inc. All rights reserved.

oxidative, inflammatory, and biochemical pathways in peripheral cells. AD pathophysiological changes, therefore, are not restricted to the brain, raising the possibility of detecting changes in cells as biomarkers of AD. During development, the ectoderm differentiates into skin, the sense organs, and components of the early nervous system. A great deal of evidence supports the notion of a "brain–skin axis" in which biochemical changes in the brain are mirrored in ectoderm-derived peripheral tissues such as the skin (Zoumakis et al., 2007; Schreml et al., 2010). Cerebrospinal fluid (CSF), neuroimaging, genetic, and peripheral fluid-based AD biomarkers have received increasing attention in the last two decades; however, their capacity for early diagnosis of AD and application for widespread screening remains questionable because of the degree of test invasiveness (eg, CSF collection requires lumbar puncture), expense (neuroimaging biomarkers), and variability in measurements (eg, genetic and peripheral fluid-based AD biomarkers). The CSF is in direct contact with the brain, thus providing a promising source for measuring biomarkers of AD, but the CSF does not contain any cellular components and biomarkers are limited to RNA and DNA and preclinical changes in AD molecular signaling. Although plasma-based AD biomarkers are under investigation, protein components in the blood, such as serum amyloid P components, transthyretin, apoferritin, immunoglobulin, α2-macroglobulin, complement factors, and apolipoprotein A-I, A-IV, E and J, can interfere with the analysis of AD biomarkers in plasma. Detection of AD biomarkers with neuroimaging is very expensive and inadequate to some extent to differentially diagnose AD and non-AD dementias.

AD is an irreversible progressive dementia with long prodromal stages. Minimally invasive diagnostic assays that measure biomarkers reflecting systemic manifestations of AD pathology could be the best alternative to CSF-based assays for early diagnosis of AD. Some of the best-known peripheral cells that are affected by AD pathology are skin fibroblasts, blood cells (blood platelets, lymphocytes, leukocytes, white blood cells,), buccal cells, nasal cells, and peripheral tissues such as eye lenses (deposited Aβ). Each of these cell types contains measurable components that are different in individuals with AD versus non-AD dementias or no dementia, and thus are candidate biomarkers for early diagnosis of AD.

7.2   RATIONALE FOR CELL-BASED ALZHEIMER'S DISEASE BIOMARKERS

There are several advantages to studying alterations in cellular systems as potential biomarkers for AD. First, AD pathology is reflected systemically. It is now well recognized that AD pathology is not restricted to the central nervous system (CNS) but is also manifested systemically in peripheral tissues/cells, and these cells can be used as sample sources for AD diagnostic assays. Consistent with the amyloidal hypothesis of AD pathogenesis, it has been shown that Aβ secretion is elevated in the skin fibroblasts of patients with familial AD compared with unaffected patients (Soininen et al., 1992; Citron et al., 1994; Johnston et al., 1994; Scheuner et al., 1996).

First, beyond changes in molecular signaling pathways, several studies have reported an effect of AD on telomere lengths; in certain peripheral cells, telomere lengths have been shown to shorten with aging and to be even shorter in patients with AD. For example, telomere lengths were significantly shorter in T cells obtained from AD patients compared to age-matched control cases (Panossian et al., 2003). Telomere lengths were also found to be shortened in subjects with Down syndrome (Jenkins et al., 2006) and in peripheral blood leukocytes (PBLs) isolated from AD patients compared to age-matched control cases (Lukens et al., 2009). PBL-telomere length was found to be related to both dementia and mortality in subjects with the AD risk factor gene APOE4 (Honig et al., 2006). Telomere length has been examined as an example of cell-based markers of AD compared to age-matched controls.

Second, in the pathogenesis of AD, alterations in molecular signaling may occur in the early stages of the disease, long before synaptic loss, loss of dendritic spines, and neuronal degeneration, and far earlier than the development of clinical symptoms. For example, the gene expression signature of familial AD (FAD) genes was found to be present in cultured skin fibroblasts before the appearance of clinical symptoms of AD (Nagasaka et al., 2005). The same study also found evidence that the disease process starts several decades before the onset of cognitive decline, suggesting that presymptomatic diagnosis of AD may be feasible using cultured skin fibroblasts isolated from patients with familial AD. Another study found excess production of Aβ in the peripheral cells from presymptomatic and symptomatic individuals carrying the Swedish familial Alzheimer's mutation (Citron et al., 1994). Inflammation in AD brains and peripheral systems also occur at clinical as well as preclinical stages, possibly even before Aβ plaque deposition and tau concentration changes (Akiyama et al., 2000; Veerhuis et al., 2003; Schuitemaker et al., 2009). Inflammatory markers in peripheral cells may reflect inflammation in the brain and permit early diagnosis of AD. For example, abnormal expression levels of the bradykinin receptor and its ligand bradykinin have been seen in the AD brain (Viel and Buck, 2011) and in peripheral systems (Jong et al., 2003).

Third, some specific deficiencies in molecular signaling can distinguish between AD and non-AD dementias. Alterations in AD-specific molecular signaling signatures may better distinguish between AD and non-AD dementias. For example, skin fibroblast signaling (Erk1/2 phosphorylation ratios stimulated by bradykinin) is different in cultured skin fibroblasts obtained from patients with non-AD dementia compared to those obtained from AD patients (Khan and Alkon, 2006, 2010).

Fourth, a growing body of literature has reported cell cycle dysfunction in peripheral cells, specifically blood lymphocytes, from AD patients (Nagy et al., 2002; Stieler et al., 2012; Song et al., 2012a). Neurons lack proteins that control the cell cycle; as a result, neurons do not undergo cell division. Unlike other peripheral cells, the cell cycle of neurons is kept in check by genetic programming. Neurons isolated from postmortem brains of AD patients express some of the proteins that control cell cycle (Nagy et al., 1997a,b, 1998, 2000), suggesting that neurons in AD brains may be unable to stop the cell cycle program and enter into lethal neuronal cell cycle events. Inappropriate cellular proliferation signaling in neurons of AD brains may lead to abnormal cellular integrity (Vincent et al., 1997; Arendt et al., 1998; Yang et al., 2001), neuronal dysfunction, and death, a major feature of early AD (Nagy, 2000; Herrup and Arendt, 2002; Herrup et al., 2004; Herrup, 2012; Keeney et al., 2012; Arendt, 2012). Therefore, cell cycle abnormalities in the brain that are reflected in peripheral blood cells provide a unique opportunity to probe the disease and potential new therapies (Webber et al., 2005; Lopes et al., 2009).

Finally, cellular biomarkers may identify new drug targets. Identification of very early markers of AD may reveal new targets for drug discovery and development. Alterations in several AD-specific cellular systems in blood cells, cultured skin fibroblasts, buccal cells, and nasal cells have been intensely studied as potential biomarkers of AD, and each of these systems may reflect pathophysiology in the brain that could be targeted by novel AD therapeutics.

7.3 BLOOD CELL-BASED ALZHEIMER'S DISEASE BIOMARKERS

Blood plasma/serum-based AD biomarkers have not provided an acceptable level of diagnostic sensitivity and specificity, which has led many researchers to focus their attention on blood-cell-based biomarkers. A large body of work has established that the hemotopoietic system is affected in AD (Hye et al., 2005; Tang et al., 2006; Janoshazi et al., 2006; Esteras et al., 2013; Vignini et al., 2013), with abnormalities described in platelets, red blood cells, and white blood cells from patients with AD compared to age-matched controls (Table 7.1). Each of these abnormalities is a potential biomarker of AD that could be detected in a diagnostic assay. For example, a panel of 12 microRNAs detected in whole blood cells was recently found to distinguish AD patients from age-matched controls with high sensitivity (91.5%) and specificity (95%) (Leidinger et al., 2013). An obvious advantage of using blood cells as samples for AD diagnostic assays is their accessibility through established, fast, and noninvasive collection procedures that are already in place across various clinical settings. Cell-based analytes can also be studied in a static condition or after cells are conditioned in culture systems, or can be stimulated with AD-signaling specific stimuli to measure the response. Stimulus-elicited measurements are a unique approach to evaluating diagnostic biomarkers, because the result is offset against a reference of no stimulation. Therefore, the variability in results from interpatient heterogeneity normally seen in diseases such as AD would be minimized. Results of blood-cell-based AD biomarker studies are summarized in Table 7.1.

7.3.1 Platelets

Blood platelets share several biochemical properties with neurons. Platelets store and release neurotransmitters such as dopamine, glutamate, and serotonin

Table 7.1	Blood Cell-Based Alzheimer's Disease Biomarkers		
Cohort Study/ Blood Cell Type	**Biomarker/Measurement Techniques**	**Patient Population**	**References**
Hospital Doce de Octubre, Madrid, Spain; and Hospital Donostia, San Sebastián, Spain Lymphocytes	↑ Calmodulin Sensitivity = 89% Specificity = 82%	Patient groups: Total ($n = 165$); AD ($n = 56$); MCI ($n = 15$); control ($n = 48$); FTD ($n = 7$); LBD ($n = 4$); PD ($n = 20$); ALS ($n = 10$); progressive supranuclear palsy ($n = 5$). All clinically confirmed by NINCDS-ADRDA criteria and MMSE score	Esteras et al. (2013)
Xuan Wu Hospital of Capital Medical University, Beijing Geriatric Hospital, and Huairou Community Lymphocytes	G1/S check point protein concentration measured by ELISA and flow cytometry ↑ CDK2: sensitivity = 59%, specificity = 85%, accuracy = 75% (by ROC curve) ↑ Cyclin E: sensitivity = 52%, specificity = 84%, accuracy = 74.9% (by ROC curve) ↑ E2F-1: sensitivity = 59%, specificity = 80%, accuracy = 76.4% (by ROC curve) ↑ RB: sensitivity = 58%, specificity = 74%, accuracy = 74.7% (by ROC curve)	Patient groups: Total ($n = 176$); AD ($n = 74$); PD ($n = 11$); VaD ($n = 11$); age-matched control ($n = 80$). AD: all clinically confirmed by NINCDS-ADRDA criteria and MMSE score PD and VaD: UK Parkinson's Disease Society Brain Bank Criteria and NINDS-AIREN criteria	Song et al. (2012a)
Xuan Wu Hospital of Capital Medical University Lymphocytes	Dysfunctional p53 fails to arrest G1 phase in AD cases. ↓ P53 ($P < 0.01$)	Severe AD ($n = 15$), mild AD ($n = 15$), age-matched control ($n = 15$) All clinically confirmed by NINCDS-ADRDA criteria and MMSE score	Zhou and Jia (2010)
All subjects were residents of the UK. White blood cells	Immunoblot ↑ GSK-3	Patient groups: AD (60); MCI ($n = 33$); control (20) Overlap with control, MCI, and AD.	Hye et al. (2005)
Neuropsychological Unit of the Strasbourg University Hospital, France Red blood cells	Fluorescence spectroscopy Alteration of PKC conformation	Patient groups: AD ($n = 33$); control ($n = 25$) Distinguished between AD and PD. All clinically confirmed by NINCDS-ADRDA criteria	Janoshazi et al. (2006)

(Continued)

Table 7.1 Blood Cell-Based Alzheimer's Disease Biomarkers (*cont.*)

Cohort Study/ Blood Cell Type	Biomarker/Measurement Techniques	Patient Population	References
University Medical Centers: University of Milan, University of Brescia, University of Rome, Italy Platelets	Immunoblots Ratio between the band intensity of the 130-kDa and 106- to 110-kDa APP isoforms. AD: 0.31 ± 0.15 Age-matched control: 0.84 ± 0.20 Non-AD dementia: 0.97 ± 0.40	Patient groups: AD (n = 32), age-matched control (n = 25), non-AD dementia (n = 16). lllSignificant difference between AD versus age-matched control and non-AD dementia groups (P < 0.01). All clinically confirmed by NINCDS-ADRDA criteria	Di Luca et al. (1998)
Alzheimer's disease Center University of Texas Southwestern Medical Center at Dallas Platelets	Immunoblot analysis: β-Secretase activity ↑; α-secretase activities ↓; Aβ ↑; APP isoform ratios (120–130 kDa to 110 kDa) ↓	Patient groups: AD (n = 31); control (n = 10) Some overlap between AD and controls lllAll clinically confirmed by NINCDS-ADRDA criteria and MMSE score	Tang et al. (2006)
School of Medicine, Polytechnic University of Marche, Ancona, Italy Platelets (F, female; M, male)	*NO production assay* (means ± SD) AD (F): 15.38 ± 1.29 Control (F): 9.71 ± 1.22 AD (M): 20.11 ± 2.71 Control (M): 5.55 ± 0.97	Patient groups: AD (F): 60; control (F): 25; P < 0.001 AD (M): 40; control (M): 25; P < 0.001. All clinically confirmed by NINCDS-ADRDA criteria and MMSE score	Vignini et al. (2013)
Platelets (F, female; M, male)	*ONOO⁻ production assay* (means ± SD) AD (F): 18.529 ± 1.29 Control (F): 8.673 ± 0.795 AD (M): 29.047 ± 2.357 Control (M): 12.835 ± 1.146	Patient groups: AD (F): 60; control (F): 25; P < 0.001 AD (M): 40; control (M): 25; P < 0.001	Vignini et al. (2013)

Platelets (F, female; M, male)	Intracellular Ca^{2+} concentration assay (means ± SD) AD (F): 203.55 ± 16.21 Control (F): 148.97 ± 8.00 IIIAD (M): 222.56 ± 10.14 Control (M): 169.07 ± 14.10	Patient groups: AD (F): 60; control (F): 25; $P < 0.001$ AD (M): 40; control (M): 25; $P < 0.001$	Vignini et al. (2013)
Platelets (F, female; M, male)	Na^{+}/K^{+}-ATPase activity assay (means ± SD) AD (F): 7.697 ± 0.681 Control (F): 16.080 ± 1.476 IIIAD (M): 3.326 ± 0.249 Control (M): 6.331 ± 0.583	Patient groups: AD (F): 60; control (F): 25; $P < 0.001$ AD (M): 40; control (M): 25; $P < 0.001$	Vignini et al. (2013)
Washington Heights-Inwood Columbia Aging Project Leukocytes	Telomere length Real-time PCR to measure telomere length AD: 0.458 ± 0.207, age-matched control: 0.516 ± 0.229	DNA samples were available for $n = 257$ (demented and nondemented) AD vs age-matched control: $P < 0.03$	Honig et al. (2006)

AD, Alzheimer's disease; Aβ, beta-amyloid; APP, amyloid precursor protein; ALS, Amyotrophic lateral sclerosis; CDK2, cyclin-dependent kinase 2; E2F-1, E2F transcription factor 1; ELISA, enzyme-linked immunosorbent assay; FTD, frontotemporal dementia; GSK-3, glycogen synthase kinase 3; LBD, Lewy body dementia; MCI, mild cognitive impairment; MMSE, mini mental score examination; NINCDS-ADRDA, National Institute of Neurological Communicative Disorders and Stroke/Alzheimer's Disease and Related Disorders Association Work Group criteria of AD; NINDS-AIREN, National Institute of Neurological Disorders and Stroke (NINDS) and the Association Internationale pour la Recherche et l'Enseignement en Neurosciences (AIREN) criteria for VaD; NO, nitric oxide; PD, Parkinson's disease; $ONOO^{-}$, peroxynitrite; RB, retinoblastoma protein; VaD, vascular dementia.

(Cupello et al., 2005; Rainesalo et al., 2005), and produce some neuronal proteins such as N-methyl-D-aspartate (NMDA) receptors. Platelets use α-granules for storage of small molecules, for example, serotonin and adenosine diphosphate (ADP), similar to synaptic vesicles in neurons. More than 90% of circulating amyloid precursor protein (APP) in the human body outside of the brain is found in platelets, making them an attractive tissue source for detecting Aβ as a blood-cell-based AD biomarker.

Expression of APP in neurons and platelets is isoform specific. The isoform APP695 is predominantly expressed in neurons, whereas isoforms APP770 and APP751 are highly expressed in blood platelets. The majority of Aβ in whole blood originates from platelet APP. Cleavage of APP (by α-, β-, and γ-secretase) occurs at the platelet surface as well as within intracellular secretory pathway organelles (Smith, 1997; Evin et al., 2003). Several important AD-related Aβ-processing abnormalities have been described in platelets from AD patients compared to those from age-matched control cases (Table 7.1). One of the earliest efforts to develop a peripheral blood cell-based biomarker assay for AD was the measurement of APP isoforms in blood platelets (Di Luca et al., 1998). The study used immunoblot analysis to quantify three APP isoforms (106, 110, and 130 kDa) and found that the ratio between the intensities of the 130 and 106–110 kDa bands was significantly lower in AD patients compared to age-matched control and non-AD dementia patients ($P < 0.01$). A reduced ratio of 130–110 kDa APP isoforms was also found in platelets of AD patients compared to age-matched control cases in a later study ($P = 0.0003$) (Tang et al., 2006). Additionally, this study reported increased levels of $A\beta_{1-42}$ ($P = 0.0023$), increased activity of β-secretase (BACE1) ($P = 0.0005$), and decreased activity of α-secretase (ADAM10; $P = 0.0014$) in AD platelets versus age-matched control platelets (all P values were calculated by the Kruskal–Wallis test; Dumm multiple comparison post hoc test also found statistically significant differences) (Tang et al., 2006). However, scatter diagrams of all parameters in AD versus age-matched control showed considerable overlaps.

Oxidative stress-related signaling molecules in platelets, such as NO, intracellular Ca^{2+}, and ATPase activity, were found to be different between AD patients and age-matched control cases. This study reported significantly higher NO ($P < 0.001$) and $ONOO^-$ ($P < 0.001$) production, higher intracellular Ca^{2+} levels ($P < 0.001$), and low Na^+/K^+-ATPase activity ($P < 0.001$) in platelets isolated from AD patients compared to age-matched control cases (Vignini et al., 2013) (Table 7.1). Several reports have found alterations in membrane fluidity, cytosolic calcium effluxes, the cytoskeleton, cytochrome oxidase activity, phospholipase C activity, and cytosolic protein kinase C (PKC) in AD platelets (Matsushima et al., 1995; Casoli et al., 2010)

7.3.2 Red Blood Cells

Abnormal PKC activity (Janoshazi et al., 2006) and reduced folate concentration (Faux et al., 2011) have been observed in red blood cells from AD patients. PKC has been implicated in memory and synapse formation, and PKC signaling pathways are dysfunctional in patients with AD and in transgenic mouse models of AD (Alkon et al., 2007). Decreased PKC levels, activity, and cellular localization of PKC have been noted in the brains and skin fibroblasts of AD patients (van Huynh et al., 1989; Khan et al., 2015). Conformational changes in PKC in red blood cells, measured by a specialized fluorescence spectrum, are different in samples from patients with or without AD (Janoshazi et al., 2006; de Barry et al., 2010). In this assay, PKC in isolated red blood cells is activated by 12-o-Tetradecanoylphorbol-13-acetate (TPA) and Fim-1 fluorescence is measured at a specific excitation wavelength. Fim-1 (a derivative compound of bis-indolyl-maleimide) is a specific fluorescent probe that binds to the ATP-binding site on the catalytic domain of PKC and detects conformational changes (Janoshazi and de Barry, 1999). This study supports the use of alterations in PKC conformation in red blood cells as an early predictive peripheral cell-based biomarker for AD (Table 7.1). The same group showed the ability of the PKC conformation biomarker to distinguish between AD and Parkinson's disease (PD). However, they did not examine other non-AD dementias such as vascular dementia (VaD), Lewy body dementia (LBD), or frontotemporal dementia (FTD).

7.3.3 Lymphocytes

Lymphocytes are one of the most widely explored sources for peripheral AD biomarkers. Decreased telomere length (Thomas et al., 2008; Panossian et al., 2003), accumulation of neutral lipids (Pani et al., 2009a), increased total tau (Armentero et al., 2011), increased chromosomal micronuclear frequency (Migliore et al., 1999), and increased G1/S cellular checkpoint arrest (Song et al., 2012b) are observed in AD lymphocytes. In AD patients, failure of cell cycle regulation is systemic, affecting not only neurons but also peripheral blood cells. Several groups have hypothesized that loss of cellular proliferation control may be reflected in easily accessible blood cells (lymphocytes). Dysregulation of the cell cycle in AD lymphocytes has been demonstrated by multiple laboratories (Nagy et al., 2002; Nagy, 2007; Stieler et al., 2012; Song et al., 2012a, b). Abnormalities observed in the cell cycle in AD lymphocytes are presented in Fig. 7.1. Most studies have focused on two specific abnormal molecular/cellular abnormalities: G1/S check point failure (Zhou and Jia, 2010; Song et al., 2012b) and cell surface marker (CD4, CD45, CD69) expression levels (Tan et al., 2002; Stieler et al., 2012) that reflect cell cycle initiation and proliferation (Fig. 7.1). P53, a tumor suppressor protein, has a profound role in regulating the G1/S stage of the cell cycle and was found to be dysfunctional in AD lymphocytes (Zhou and Jia, 2010). The expression levels of p53 are significantly lower in AD lymphocytes (Table 7.1). Cell cycle abnormalities in AD lymphocytes appear to be one of the earliest neuropathological signatures reflected in peripheral systems (Yang et al., 2003). A summary of the method used to measure lymphocyte cell cycle abnormalities as AD biomarkers, and an example of a typical assay result, is presented in Fig. 7.2. The degree of cell cycle abnormality may not be sufficiently large to detect a difference between AD and non-AD lymphocytes; therefore, researchers first amplify the difference by triggering and/or blocking mitogenic activity and measuring differences in the lymphocyte response by

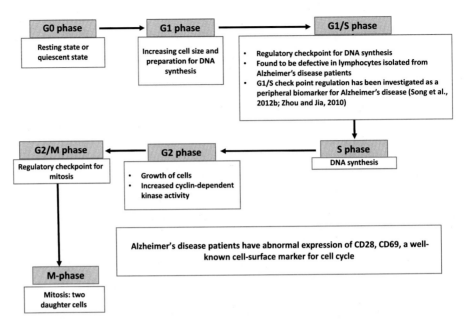

FIGURE 7.1

Schematic diagram of cell cycle abnormality in lymphocytes from Alzheimer's disease patients. Abnormalities in the G1/S phase as the regulatory checkpoint and differential expression levels of cell cycle markers have been investigated as Alzheimer's disease biomarkers.

flow cytometry. For example, phytohemagglutinin (PHA) and pokeweed mitogen (PWM) have been used to stimulate mitogenic activity in lymphocytes. In some studies, rapamycin, a well-known G1/S check point regulator has been used to amplify the difference in cell cycle between AD and non-AD lymphocytes (Zhou and Jia, 2010; Stieler et al., 2012; Song et al., 2012b).

Others have observed that the functional relationship between Ca^{2+}/calmodulin and the signaling pathways that control cell survival and death are altered in AD neurons. Calmodulin levels in AD lymphoblasts and peripheral blood mononuclear cells (PBMCs) were found to be higher in AD compared to age-matched controls or non-AD dementias (Esteras et al., 2013). In early-stage AD, glycogen synthase kinase-3 (GSK-3) expression levels have been found to be high in white blood cells (Hye et al., 2005). Other abnormalities include increased β-secretase and decreased α-secretase activities (Tang et al., 2006), increased Aβ levels (Tang et al., 2006), and low APP isoform ratios (120–130 kDa to 110 kDa) in AD compared to controls (Tang et al., 2006; Baskin et al., 2000).

7.3.4   Leukocytes

Neutrophils are major component of leukocytes. Telomere lengths in peripheral blood leukocytes have been found to be shorter in AD patients compared to

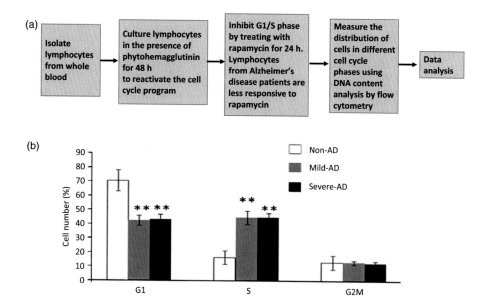

FIGURE 7.2

Cultured lymphocyte-based Alzheimer's disease (AD) biomarker assay. (a) Schematic flow diagram of lymphocyte-based AD cell cycle biomarker assay. After culturing lymphocytes in presence of phytohemagglutinin (PHA) for 48 h to reactivate cell cycle and inhibiting the G1/S phase by treatment with rapamycin, cell cycle distributions are measured by flow cytometry. (b) Typical results from a lymphocyte-based AD cell cycle biomarker assay. The number of cells in G1 phase was significantly higher (**$P < 0.001$) in non-AD ($n = 15$) cases compared to mild AD ($n = 15$) and severe AD ($n = 15$) patients. In contrast, the number of cells in S phase was significantly lower (**$P < 0.001$) in non-AD cases compared to mild AD and severe AD patients. There were no differences in the number of cells in G2 and M phases in non-AD cases compared to mild AD and severe AD patients. Lymphocytes from Alzheimer's disease patients are less responsive to rapamycin. *(Figure 7.2b is adapted from Zhou and Jia (2010); Neurosci. Lett. 468, 320–325; with copyright permission from Elsevier press).*

age-matched control cases (Honig et al., 2006; Lukens et al., 2009). Results of studies of telomere length in leukocytes as an AD biomarker are summarized in Table 7.1. Other abnormalities in AD leukocytes include greater frequency of single- and double-strand DNA breaks and greater amounts of DNA-oxidized pyrimidines and purines (Migliore et al., 2005).

7.4 SYSTEMIC PATHOPHYSIOLOGY OF ALZHEIMER'S DISEASE IN SKIN FIBROBLASTS

The pathophysiology of AD is manifested in skin fibroblasts in a variety of ways: abnormal gene expression (Nagasaka et al., 2005), overproduction of Aβ peptides (Soininen et al., 1992; Citron et al., 1994; Johnston et al., 1994; Scheuner et al., 1996), overproduction of Aβ from neurons transformed from fibroblasts of AD patients (Vierbuchen et al., 2010; Pang et al., 2011; Pfisterer et al., 2011; Caiazzo et al., 2011; Yoo et al., 2011; Ambasudhan et al., 2011), and dysfunctional

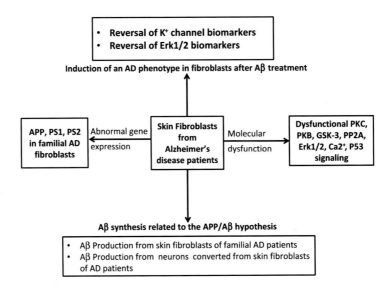

- Reversal of K⁺ channel biomarkers
- Reversal of Erk1/2 biomarkers

Induction of an AD phenotype in fibroblasts after Aβ treatment

| APP, PS1, PS2 in familial AD fibroblasts | Abnormal gene expression | Skin Fibroblasts from Alzheimer's disease patients | Molecular dysfunction | Dysfunctional PKC, PKB, GSK-3, PP2A, Erk1/2, Ca2⁺, P53 signaling |

Aβ synthesis related to the APP/Aβ hypothesis

- Aβ Production from skin fibroblasts of familial AD patients
- Aβ Production from neurons converted from skin fibroblasts of AD patients

FIGURE 7.3

Schematic diagram of skin fibroblasts as a systemic cellular model of Alzheimer's disease.

molecular pathways related to AD. Treatment of cultured normal skin fibroblasts with Aβ stimulates an AD phenotype (Etcheberrigaray et al., 1994; Khan et al., 2009). Abnormalities of cultured AD skin fibroblasts are summarized in Fig. 7.3. Several other studies have described that the basic pathogenic mechanism of amyloidogenesis is similar in the brain and in skin fibroblasts (Schreml et al., 2010). Differences in gene expression signatures in fibroblasts from familial AD patients as well as unaffected subjects indicate that the systemic manifestation of AD starts several decades before the onset of cognitive decline (Nagasaka et al., 2005).

Manipulation of human fibroblasts to produce functional neurons by targeted means provides further evidence for the relationship between skin and brain physiology (Vierbuchen et al., 2010; Pang et al., 2011; Pfisterer et al., 2011; Caiazzo et al., 2011; Yoo et al., 2011; Ambasudhan et al., 2011). Similar to skin fibroblasts, functional neurons generated from skin fibroblasts from AD patients produce excess $A\beta_{1-40}$ and $A\beta_{1-42}$ toxic oligomers, a signature of AD pathology.

As previously reviewed (Gasparini et al., 1998), peripheral tissues may reflect some of the systemic abnormalities of AD prior to development of any symptomatic effects. Specific systemic abnormalities of AD involve human fibroblast K⁺ channels (Etcheberrigaray et al., 1993, 1994), PKC isozymes (Govoni et al., 1993; Favit et al., 1998), Ca²⁺ signaling (Ito et al., 1994), phosphorylation of the Erk1/2 MAP kinases (Zhao et al., 2002; Khan and Alkon, 2006, 2010), altered folate binding (Cazzaniga et al., 2008), and abnormal cholesterol processing (Pani et al., 2009b) (Table 7.2).

Table 7.2	Altered Cellular Signaling Pathways in Skin Fibroblasts From Patients With Alzheimer's Disease	
Altered Pathway	**Cellular and Molecular Dysfunction in AD**	**References**
Altered Ca^{2+} homeostasis	Disturbed Ca^{2+} homeostasis by oxidative damage and $A\beta$ triggering	Peterson et al. (1985) Ito et al. (1994)
Dysfunctional PKC isozyme activity	PKC signaling deficits in relation to AD-specific memory loss. Defective PKC isozymes in familial and sporadic AD cases	van Huynh et al. (1989) Bruel et al. (1991) Favit et al. (1998) Govoni et al. (1993) Khan et al. (2015)
Enhanced folate binding	Higher expression levels of folate receptor in AD fibroblasts	Cazzaniga et al. (2008)
Bradykinin signaling pathway	Defective tau protein serine phosphorylation	Jong et al. (2003)
Abnormal Erk1/2 phosphorylation	Stimulus-activated differential Erk1/2 signaling	Zhao et al. (2002) Khan and Alkon (2006)
Apoptotic protein p53 conformation change	Unfolded conformation of p53 is high in AD cases and folded conformation is high in non-AD cases	Lanni et al. (2007) Lanni et al. (2008)
Altered cholesterol ester cycle	Like AD brain higher levels of cholesterol ester in cultured skin fibroblasts from AD patients	Pani et al. (2009b)

AD, Alzheimer's disease; Erk1/2, extracellular regulated kinase 1 and 2; MAP, mitogen-activated protein kinase; p53, a tumor suppressor protein; PKC, protein kinase C.

7.5 SKIN FIBROBLAST-BASED ALZHEIMER'S DISEASE BIOMARKERS

Numerous peripheral biomarkers of AD have been investigated in cultured skin fibroblasts (Table 7.3), including changes in Ca^{2+} distribution (Ito et al., 1994), abnormal bradykinin-induced Erk1/2 phosphorylation (Zhao et al., 2002), bradykinin-induced differential phosphorylation of Erk1 and Erk2 (Khan and Alkon, 2006, 2010), deficits in PKCε (Khan et al., 2015), changes in skin fibroblast network morphology (Chirila et al., 2013), apoptotic protein p53 conformational change (Uberti et al., 2006; Lanni et al., 2007), and electrophysiological K^+ channel dysfunction induced by soluble $A\beta$ (Etcheberrigaray et al., 1994). Some of these skin fibroblast-based AD biomarkers have been autopsy confirmed and show similar levels of sensitivity and specificity as established CSF

Table 7.3	Skin Fibroblast-Based Alzheimer's Disease Biomarkers	
Biomarker/Measurement Techniques	**Assay Performance and Comments**	**References**
Electrophysiology: Electrophysiological K$^+$ channel dysfunction induced by soluble Aβ	Need to be validated by other laboratory	Etcheberrigaray et al. (1994)
Fluorescence imaging—Altered Ca^{2+} homeostasis: Ratio of differential response of TAE challenge and bradykinin challenge in terms of Ca^{2+} concentration measured by calcium imaging technique	Low specificity and sensitivity	Ito et al. (1994)
Immunoblot: Increased Erk1/2 phosphorylation in response to bradykinin stimulation	Need to be validated by other laboratory Some of the AD cases are autopsy confirmed	Zhao et al. (2002)
Immunoblot: Differential stimulus-elicited phosphorylation of Erk1/2 (relatively higher Erk1 compared with Erk2) by bradykinin	High specificity and sensitivity. Most of the AD cases are autopsy confirmed. Most effective for early ($\leq$4 years of disease duration) diagnosis (confirmed by autopsy). Need to be validated by MCI cases	Khan and Alkon (2006) Khan and Alkon (2010)
Fibroblasts network formation in culture system: Spatiotemporal complexity of fibroblasts network morphology assay	High specificity and sensitivity; need to be validated by other laboratory. Some of the AD cases are autopsy confirmed.	Chirila et al. (2013)
High-throughput ELISA assay of PKCε: Reduced basal levels of PKCε in AD cases. After toxic oligomeric Aβ treatment levels of PKCε reduced in non-AD cases, but increased for AD cases	High specificity and sensitivity; need to be validated by other laboratory. Some of the AD cases are autopsy confirmed	Khan et al. (2015)
Protein confirmation analysis: Unfolded conformation of p53 is high in AD cases and folded conformation is high in non-AD cases	Moderate specificity and sensitivity. No non-AD dementia cases were included	Uberti et al. (2006) Lanni et al. (2007)

AD, Alzheimer's disease; Aβ, beta-amyloid; ELISA, enzyme-linked immunosorbent assay; Erk, extracellular regulated kinase; MAPK, mitogen-activated protein kinase; PKCε, protein kinase C ε-isoform; TAE, tetraethylammonium ion.

biomarkers for distinguishing AD from age-matched controls and non-AD dementias (Khan and Alkon, 2006, 2010; Chirila et al., 2013; Khan et al., 2015). Fibroblast-based AD biomarker assays require a punch biopsy for collection of the cell sample; therefore, this method would avoid lumbar puncture required for measuring CSF-based biomarkers.

7.5.1 Bradykinin-Induced Erk1/2 Phosphorylation in Skin Fibroblasts

Bradykinin receptors in cultured fibroblasts from AD patients were found to stimulate tau phosphorylation (Jong et al., 2003), and bradykinin was found to induce higher Erk1 and Erk2 phosphorylation in cultured skin fibroblasts from AD patients compared to age-matched controls, when normalized to untreated cells (Zhao et al., 2002). An internally controlled comparison of bradykinin-induced changes in Erk1 and Erk2 phosphorylation in AD and non-AD fibroblasts resulted in the derivation of an Alzheimer's disease index (AD-Index) that accurately distinguishes between fibroblasts from AD patients and those from age-matched controls and non-AD dementia patients (Khan and Alkon, 2006, 2010). The AD-Index was calculated as: AD-Index = [p-Erk1/p-Erk2]$^{BK+}$ − [p-Erk1/p-Erk2]$^{BK-}$ (bradykinin induction = BK+ and control = BK−). The assay procedure and a typical assay result are presented in Fig. 7.4. Importantly, the AD-Index assay

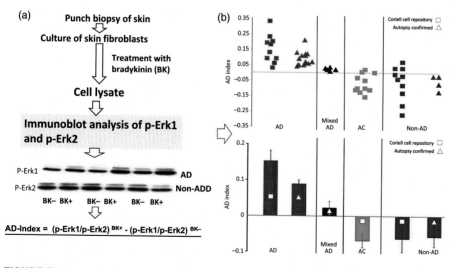

FIGURE 7.4

Bradykinin-induced Erk1/2 phosphorylation in skin fibroblasts. (a) Schematic flow diagram of the Erk1/2 AD-Index biomarker assay. (b) Performance of the AD-Index assay to distinguish between AD (*red square* and *red triangle*), mixed AD pathology with non-AD dementias (non-ADD) at autopsy (*black triangle*), AC (age-matched nondemented control) (*green square*), and pure non-ADD (*blue square* and *blue triangle*) (samples from the Coriell cell repository are denoted with *square* and autopsy registry are denoted with *triangle*). *(Part of the figure was reprinted from Khan and Alkon (2010); Neurobiol. Aging 31, 889–900; with permission from Elsevier press. Part of the figure was adapted from Khan and Alkon (2006); Proc. Natl. Acad. Sci. USA 103, 13203–13207 with permission).*

was autopsy confirmed and the accuracy of Erk1 and Erk2 AD-Index values was inversely correlated with disease duration, suggesting a maximal efficacy of the AD-Index bioassay in early diagnosis (ie, can be used for prospective preclinical AD evaluation). The AD-Index accurately distinguished AD from age-matched control cases within the first 4 years of disease symptoms, where clinical diagnosis usually fails (Khan and Alkon, 2010). The AD-Index has a remarkably high accuracy (96%) for autopsy confirmed cases, and in the absence of autopsy validation (ie, clinical diagnosis only), the accuracy of the Erk1/2 AD-Index biomarker for diagnosis of AD was 82% compared to a 67% accuracy of clinical diagnosis. When the AD-Index agrees with a clinical diagnosis of AD, there is a high probability of accuracy on the basis of autopsy validation. Furthermore, the AD-Index assay can be used for differential diagnosis of AD and non-AD dementias and AD can be distinguished in patients with mixed pathology of AD and non-AD dementias (comorbidity). The performance of AD-Index biomarker was very high (100% accuracy) for patients who had mixed AD dementias in autopsy-confirmed cases. In a separate cohort, the specificity of the Erk1/2 AD-Index biomarker was found to be quite high for ruling out AD for a subgroup of healthy age-matched controls. In that study, the healthy control cases were selected on the basis of the following criteria: no cancer, heart disease, arthritis, stroke, or family history of AD and a mini mental score examination (MMSE) of more than 27 individuals (Khan and Alkon, 2010).

As a potential test for AD therapeutic efficacy, the Erk1/2 AD-Index assay was able to detect changes in fibroblast signaling in response to bryostatin (a nontumorigenic potent PKC activator) and picolog (a synthetic analog of bryostatin with a similar potency with respect to PKC activation) treatment (Khan et al., 2009). These results suggest that the AD-Index measured in human fibroblasts could be used to evaluate AD therapeutic efficacy in cellular systems. PKC signaling deficits may contribute to the initial pathology of AD, and screening for PKC activators based on the AD-Index may provide an approach to rapid drug screening, at least for therapeutics targeting the early stages of AD. The Erk1/2 AD-Index biomarker assay still needs to be validated in mild cognitive impairment (MCI) and in longitudinal cohorts to determine its ability to track AD progression.

7.5.2 Ca^{2+} Imaging in Skin Fibroblasts

Disturbed Ca^{2+} homeostasis in AD is mainly due to dysregulation of the Ca^{2+} binding protein calpain, dysfunctional Ca^{2+} permeable receptor channels, and toxic oligomeric Aβ-triggered disturbance of internal Ca^{2+} buffering capacity. Calpain activation is tightly regulated to prevent massive proteolytic activity in the cell, and calpain dysfunction is seen in AD pathology. Oxidative stress, toxic Aβ, and leaky receptor channels contribute to the loss of buffer capacity of the Ca^{2+} pool inside cells. Oxidative stress is one of the risk factors of AD and altered Ca^{2+} homeostasis contributes to mitochondrial oxidative processes, such as glucose and glutamine oxidation in AD patients. PSEN1 and PSEN2 mutation also cause leaky Ca^{2+} channels in the endoplasmic reticulum of familial AD patients, as well as disturbed Ca^{2+} homeostasis (Tu et al., 2006). The calcium homeostasis modulator 1 (CALHM1) gene was found to be related to cytosolic Ca^{2+} and levels of Aβ in AD patients (Dreses-Werringloer et al., 2008). Altered Ca^{2+} homeostasis in AD brains and its manifestation in peripheral tissues (skin fibroblasts) were first reported by Peterson et al. (1985). This study found that Ca^{2+} uptake in cultured skin fibroblasts followed the order AD < age-matched control < young control. Other studies also support the finding of disturbed Ca^{2+} homeostasis in cultured skin fibroblasts of AD patients, linking the neuropathophysiology to the systemic pathophysiology (Peterson et al., 1986; Peterson and Goldman (1986). Treatment of skin fibroblasts with various Ca^{2+} sensitive drugs (3, 4-diaminopyridine, serum, N-formyl-methionyl-leucyl-phenylalanine, and bradykinin) has been shown to increase cytosolic free Ca^{2+} transiently, with the rate of the increase slower and the magnitude of the rise less pronounced in cells from age-matched controls and AD patients when compared to young controls (Peterson et al., 1988). Mitochondria in skin fibroblasts isolated and cultured from autopsy-confirmed AD patients showed decreased Ca^{2+} uptake and increased sensitivity toward free radical treatment (Kumar et al., 1994).

Borden et al. (1992) explored the possibility of developing a peripheral diagnostic biomarker for AD based on abnormal Ca^{2+} processing in cultured skin fibroblasts; however, they found that cytoplasmic ionic Ca^{2+} levels were not diagnostic and could not distinguish AD patients from non-AD cases. An alternative, though more complex, approach was subsequently investigated (Ito et al., 1994; Hirashima et al., 1996) to improve the diagnostic accuracy of Ca^{2+} abnormality in cultured skin fibroblasts. Intracellular Ca^{2+} release occurs in response to IP3 generation by low-dose (nanomolar range) bradykinin (BK) (Hirashima et al., 1996). On the other hand, the K^{+} channel blocker tetraethylammonium (TEA) ion increases intracellular Ca^{2+} in normal skin fibroblasts, whereas the response of TEA is lower in AD fibroblasts (Etcheberrigaray et al., 1994). The ratio of cytosolic Ca^{2+} induced by TAE and Ca^{2+} induced by bradykinin, as measured by fluorescence imaging, was investigated for its ability to distinguish fibroblasts from AD patients, age-matched controls, and patients with non-AD dementias. The ratio of differential responses to the TAE challenge and bradykinin challenge in terms of Ca^{2+} concentration measured by calcium imaging provided moderate to low specificity and sensitivity, though they were higher than in previous studies (Table 7.3). Several factors may contribute to the observed low accuracy of Ca^{2+}-based bioassays for AD, including (1) cell cycle dependence for the AD Ca^{2+} response, (2) skin fibroblast morphology, (3) cellular motility and cytoskeleton dynamics, and (4) suboptimal culture conditions. Future validation studies may improve the accuracy of Ca^{2+}-based AD biomarker assays by controlling the aforementioned conditions.

7.5.3 Deficit of PKCε in AD Cultured Skin Fibroblasts

Among the PKC isozymes, PKCε is predominantly expressed in the brain (Wetsel et al., 1992). Reduced PKC levels in the AD brain have been hypothesized to contribute to early AD pathology, including synaptic loss (Alkon et al., 2007). Activation of PKCε by bryostatin1 was found to prevent synaptotoxic Aβ-oligomer elevation, PKCε deficits, early synaptic loss, cognitive deficits, and amyloid plaque formation in AD transgenic mice (Hongpaisan et al., 2011). Activation of PKCε also attenuates activation of α-secretase to generate the synaptogenic nontoxic sAPPα (Etcheberrigaray et al., 2004), promotes synaptogenesis and the degradation of Aβ via the endothelin converting enzyme (ECE) (Nelson et al., 2009), and reduces GSK3-β activity (Takashima, 2006), and hence decreases hyperphosphorylation of tau (Kanno et al., 2015). In autopsy brains, PKCε levels were lower compared to age-matched autopsy-confirmed control cases and found to be inversely related with the Braak score, such that PKCε concentration decreased as the Braak score increased (Khan et al., 2015). The Braak score is a measure of severity of AD.

Skin fibroblasts also express PKCε. An early study found lower basal and activated levels of PKC in human cultured skin fibroblasts from AD patients (van Huynh et al., 1989). Degradation of PKC isozymes by soluble Aβ (Favit et al., 1998) and reduced levels of PKC isozymes in cultured AD skin fibroblasts (van Huynh et al., 1989) have been reported. This led to the development of an AD biomarker assay on the basis of Aβ-induced effects on PKC in cultured fibroblasts. In a proof-of-concept study, the concentration of PKCε in skin fibroblasts was assessed before and after treatment with 500 nM toxic amylospheroids (ASPDs) derived from soluble oligomeric $Aβ_{1-42}$ using an enzyme-linked immunosorbent assay (ELISA) specific to PKCε (Khan et al., 2015). The actual bioassay procedure that was developed is presented in Fig. 7.5. The quantitative output of this assay is the rate of change of PKCε levels in skin fibroblasts of AD patients and non-AD controls with increasing concentration of externally added

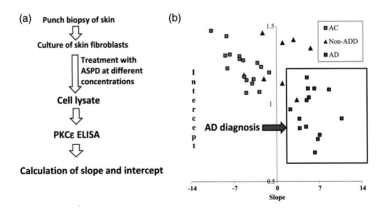

FIGURE 7.5

PKCε as a biomarker of Alzheimer's disease (AD) in cultured skin fibroblasts.
(a) Schematic flow diagram of the skin fibroblast-based PKCε biomarker assay for AD.
(b) Performance of the assay for distinguishing between age-matched controls (AC) (*green square*); non-AD dementia (non-ADD) (*blue triangle*); and Alzheimer's disease (AD) (*red square*). *(Part of the figure was adapted from Khan, T.K., Sen, A., Hongpaisan, J., Lim, C.S., Nelson, T.J., Alkon, D.L., 2015. J. Alzheimers Dis. 43, 491–509; with copyright permission from IOS press).*

toxic ASPDs. PKCε levels were found to be inversely related to the concentration of externally added toxic ASPDs for fibroblasts from age-matched control and non-AD dementias (Huntington's disease, FTD, PD, VaD). Basal levels of PKCε were found to be low in cultured skin fibroblasts isolated from AD patients and increased with increasing ASPD treatment. A subset of the patient population in the study had autopsy confirmation. The inverse response of PKCε to ASPD in skin fibroblasts from AD versus non-AD patients provided the basis for further development of a skin fibroblast-based PKCε diagnostic assay (Table 7.3) (Khan et al., 2015). Treatment with toxic ASPDs reduces PKCε levels in fibroblasts from non-AD individuals, on the basis of the recognized reciprocal relationship between Aβ and PKCε. In contrast, treatment with toxic ASPDs increases PKCε levels in fibroblasts from AD patients, consistent with evidence suggesting abnormal de novo synthesis of PKCε in response to Aβ in AD (Khan et al., 2015). The resulting PKCε fibroblast-based assay showed high specificity and sensitivity, although the study sample size was low and the results need to be validated in larger cohorts and in patients with MCI and non-AD dementias.

7.5.4 Fibroblasts Network Morphology Assay

A growing body of literature suggests that extracellular matrix production (Bellucci et al., 2007), metabolic properties (Sims et al., 1987), and dynamics of cytoskeleton proteins (Brandan et al., 1996; Takeda et al., 1992; Uéda et al., 1989) are different in skin fibroblasts from AD patients compared to those from age-matched non-AD cases. These abnormalities might be visualized and quantified in terms of the dynamics of network formation and cell–cell contacts during skin fibroblast culture. The formation of cellular networks was found to be altered for cultured AD skin fibroblasts (Chirila et al., 2013). The difference in the dynamics of cell network formation during three-dimensional culture of skin fibroblasts from AD or non-AD dementia patients and age-matched controls was quantified in terms of the number and area of the network aggregates. The total number of cellular aggregates was expected to be lower in AD cases than age-matched control and non-AD dementia cases. In the assay, AD fibroblasts formed a few large isolated aggregates, whereas fibroblasts from age-matched controls and non-AD dementia cases formed a greater number of aggregates that were smaller in size. As a result, the ratio of the area of each aggregate to the number of aggregates was very high for AD cases compared to age-matched controls and non-AD dementia cases (Fig. 7.6). A subset of the patient population had autopsy confirmation, and the accuracy, sensitivity, and specificity were very high, although the sample size in this study was low. This assay has been cross-validated with the Erk1/2 AD-Index assay and PKCε assay described previously (Chirila et al., 2014). Similar to the PKCε biomarker, the cell network morphology biomarker needs to be validated in larger cohorts and in patients with other modalities such as MCI, as well as by other laboratories, before it can be translated to the clinical setting. Longitudinal cohort studies are also necessary for validating the fibroblast network morphology assay and for studying the ability of the assay to follow AD progression.

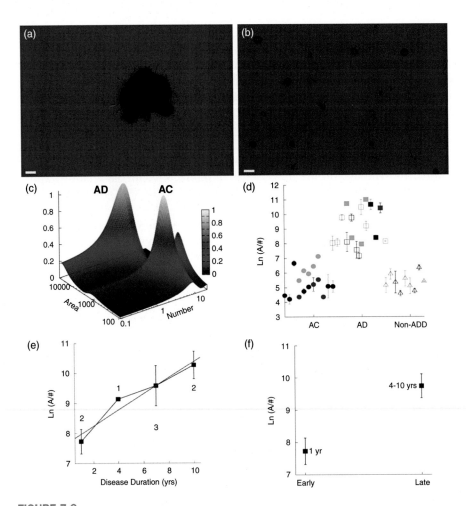

FIGURE 7.6

Fibroblast network morphology assay of Alzheimer's disease (AD). (a) Typical micrograph of aggregated cultured skin fibroblasts from AD patients after 48 h in a specially designed, three-dimensional matrix. (b) Aggregated cultured skin fibroblasts from age-matched control (AC) cases after 48 h. (c) Probability distribution plot of the area and number of aggregates. (d) Scatter plot of the natural log of the ratios of aggregate areas to aggregate number, Ln(Area/#), for AC (*circles*), AD (*squares*), non-AD dementia (non-ADD) (*triangle*); *red*, experiment conducted under double-blind conditions; *black*, samples from cell banks; *green*, samples obtained directly from subjects by punch biopsy. Empty symbols represent autopsy- or genetic-validated patients. (e) Ln(Area/#) increases proportionally to AD disease duration. (f) Ln(Area/#) value measures the disease severity. Error bars are the standard error of the mean (SEM). Scale bar is 10 μm. *(Adapted from Chirila, F.V., Khan, T.K., Alkon, D.L., 2013. J. Alzheimers Dis. 33, 165–176; with copyright permission from IOS press).*

7.5.5 Fibroblast-Based p53 Biomarkers for Alzheimer's Disease

p53 plays an important role in neuronal death (Morrison et al., 2003), and impaired p53 activity has been documented in skin fibroblasts (Uberti et al., 2006; Lanni et al., 2007) and blood cells (Lanni et al., 2008) of patients with AD. p53 is an apoptosis mediator, and its conformation is altered in cultured cells by treatment with nanomolar concentrations of Aβ peptide. In cultured skin fibroblasts from patients with early-stage AD or non-AD controls, Aβ induced transcriptional and conformational changes of p53, leading to further exploration as a potential peripheral biomarker of AD (Lanni et al., 2007). After Aβ treatment, the proportion of p53 in an unfolded conformation is higher in AD cases and the proportion in a folded conformation is higher in age-matched control cases (Lanni et al., 2007, 2008) (Table 7.3). Conformational changes in p53 occur with normal aging, which has presented a challenge to the diagnostic potential of this approach (Lanni et al., 2008). Gene sequencing of the entire p53 gene failed to find any point mutation of p53 in patients with AD; therefore, the conformational differences in p53 in AD and control fibroblasts was due to changes in tertiary structure of p53 (Uberti et al., 2006). In these studies, the sample size was low and the skin fibroblast p53 conformational change needs to be validated as a biomarker in larger cohorts as well as in MCI and non-AD dementias.

7.6 OCULAR BIOMARKERS IN ALZHEIMER'S DISEASE

It is well recognized that patients with AD have impaired vision, and difficulty with motion realization and color distinction. Evidence of AD pathology in the ocular system was first reported in 1994 (Scinto et al., 1994). The eye is the only part of the human body where neuronal tissues are accessible to external examination without the need for invasive procedures. The retina contains nerve cells that can be studied externally, and various biomarkers of AD within the nerve cells of the retina have been studied. The different ocular AD biomarkers and their diagnostic performance are summarized in Table 7.4. The cholinergic hypothesis of AD proposed in the last century is on the basis of dysfunction of the cholinergic system. Pupil response to cholinergic stimulation was the first attempt to develop an AD peripheral ocular biomarker on the basis of the deficit in cholinergic function in AD (Scinto et al., 1994). Those studies found a larger pupillary response to tropicamide (a cholinergic antagonist) in AD cases compared to non-AD controls (Scinto et al., 1994; Gomez-Tortosa et al., 1996; Grunberger et al., 1999; Arai et al., 1996; Kalman et al., 1997). However, some studies also suggested that the response of the pupil to tropicamide was not a reliable test for AD (FitzSimon et al., 1997; Kardon, 1998; Growdon et al., 1997). Later, a re-evaluation study using a lower concentration of tropicamide stimulation showed better results and found that the pupil response was able to distinguish AD versus VaD cases (Iijima et al., 2003).

Table 7.4	Ocular Biomarkers of Alzheimer's Disease	
Biomarker/ Measurement Techniques	**Patient Population/Assay Performance/Comments**	**References**
Aβ compound: fluorescence imaging of aftobetin-HCl (molecular weight = 489)	*The Cognoptix SAPPHIRE II eye test* Phase II clinical trial: sensitivity = 85%, specificity = 95% AD ($n = 20$), age-matched control ($n = 20$), no non-AD dementia patients were included	http://www. cognoptix.com/ news/releases/ 051115.htm
Thinning of retinal nerve fiber layer (RNFL) by optical coherence tomography evaluation of retina	AD ($n = 10$), DLB ($n = 10$), PD ($n = 10$), age-matched control ($n = 10$). Statistically significant ($P < 0.001$) difference between age-matched control and all dementia patients. No statistically significant difference between AD and all other non-AD dementia patients	Moreno-Ramos et al. (2013)
A set of 13 different vascular parameters derived from retinal high-resolution photographs analyzed by specialized software called Singapore I vessel assessment	*AIBL Flagship Study of Aging*: AD ($n = 25$), healthy control ($n = 125$); sensitivity = 81.2%, specificity = 75.7% and accuracy = 87.7%	Frost et al. (2013)
Cytosolic Aβ deposition in eye lenses	Patient groups: AD ($n = 9$); control ($n = 8$). Moderate specificity and sensitivity. Low sample size.	Goldstein et al. (2003)
Pupil response to cholinergic stimulation measured in dilation rate	AD ($n = 14$), VaD ($n = 14$) and young control ($n = 16$). Dilation rate after cholinergic stimulation is AD was higher than non-AD cases ($P < 0.001$)	Iijima et al. (2003)
Flashing stimulation in pupil: a set of 10 parameter derived from pupillary light reflex	AD ($n = 23$), age-matched control ($n = 23$) AD versus age-matched control, $P < 0.005$	Fotiou et al. (2007)

AD, Alzheimer's disease; Aβ, beta-amyloid; AIBL, Australian Imaging Biomarkers and Lifestyle; DLB, Lewy body dementia; PD, Parkinson's disease; VaD, vascular dementia.

Decrease in retinal nerve fiber layer (RNFL) thickness, as measured by optical coherent tomography, was also found to be associated with AD (Moreno-Ramos et al., 2013). This study found a significant difference in RNFL thickness between AD (n = 10) and age-matched control groups (n = 10) (P < 0.001); however, there was no significant difference between AD and non-AD dementias such as PD (n = 10) and DLB (n = 10). A decrease in RNFL thickness was found to be significantly correlated with cognitive scales such as MMSE and the Mattis Dementia Rating. Another ocular biomarker of AD under investigation is Aβ deposition in the human lens (Goldstein et al., 2003). Aβ deposition in the eye lens was higher in autopsy-confirmed AD patients than in age-matched control cases. This study was limited by a low sample size (AD = 9, age-matched control cases n = 8) and moderate diagnostic specificity and sensitivity. Nevertheless, Cognoptix (Acton, MA), a Massachusetts-based company, has been engaged in developing a noninvasive method for Aβ measurement in the eye using fluorescence imaging (http://www.cognoptix.com/index.htm). Their technique uses an Aβ-binding compound aftobetin-HCl. The company published data from their Phase II clinical trial in AD patients (n = 20) and age-matched controls (n = 20), reporting a specificity of 95% and a sensitivity of 85%. This clinical trial did not include any non-AD dementia patients for differential AD diagnosis.

Visual deficits and retinal nerve fiber thinning are also associated with other age-related eye diseases such as glaucoma, which confounds their utility as diagnostic biomarkers of AD. Other ocular AD biomarkers may be better able to distinguish between AD, age-matched controls, and young controls. Those biomarkers may also be useful for disease monitoring. The performance of existing ocular biomarkers is poor in the differential diagnosis of AD and non-AD dementias.

7.7 OTHER PERIPHERAL CELL-BASED BIOMARKERS FOR ALZHEIMER'S DISEASE

7.7.1 Buccal Cells

Buccal cell-based AD biomarkers are recent development (François et al., 2014; Thomas et al. 2007) (Table 7.5). Similar to skin cells, the buccal mucosa and brain are derived from ectoderm during embryogenesis. As a result, they share several common AD-specific pathologies, such as higher expression of tau protein (Hattori et al., 2002). Buccal cells exhibit cytological and nuclear morphological changes related to AD pathology; specifically, DNA damage (Thomas et al. 2007), significantly shorter telomere length (Thomas et al., 2008), neutral lipid content (François et al., 2014), and altered nuclear shape (François et al., 2014) have been identified in buccal cells of AD patients. Similar to other peripheral cell sources, buccal cells are easily collected by painless, inexpensive procedures, including cytobrushing (Richards et al., 1993; Garcia-Closas et al., 2001; King et al., 2002), cotton swabbing (Richards et al., 1993), and the "swish and spit" method (Hayney et al., 1996; Lum and Le Marchand, 1998; Feigelson et al., 2001). Among all buccal cell-based AD biomarkers, the most

Table 7.5	Peripheral Cells Other Than Blood Cells and Skin Fibroblasts With Abnormalities in Alzheimer's Disease		
Cell Types/ Measurement Technique	Biomarker	Patient Population and Assay Performance	References
Buccal cells; laser scanning cytometry	DNA content, neural lipid content, nuclear shape	Patient groups: AD (*n* = 13); MCI (*n* = 13); control (*n* = 26). Moderate specificity and sensitivity	François et al. (2014)
Nasal cells	Higher Aβ and phosphorylated tau	Patient groups: AD (*n* = 79); non-AD dementia (*n* = 63); control (*n* = 25) Low specificity and sensitivity	Arnold et al. (2010)

AD, Alzheimer's disease; *Aβ*, beta-amyloid; *MCI*, mild cognitive impairment.

advanced is an automated laser-scanning cytometric measure of nuclear DNA content and shape, neutral lipid content, and cell type ratio (François et al., 2014) (Table 7.5).

7.7.2 Nasal Cells

Olfactory dysfunction and loss of smell are common in AD and other neurodegenerative diseases. Increased oxidative damage in the olfactory system in AD manifests as higher levels of lipid peroxidation and hemeoxygenase-1. Increased Aβ production and higher levels of phosphorylated tau (Arnold et al., 2010), HNE-derived 2-pentylpyrrole (4-hydroxy-2-nonenal, HNE; "HNE-pyrrole"), and hemeoxygenase-1 (Perry et al., 2003) have been observed in nasal cells from AD patients. HNE-pyrrole is a product of lipid oxidation.

7.8 FUTURE APPROACHES TO CELL-BASED ALZHEIMER'S DISEASE PERIPHERAL BIOMARKER DISCOVERY

7.8.1 Multigenic Genetic Approaches

It is well-established fact that AD is not caused by dysfunction in a single protein or mutation of a single gene. Sporadic AD is a multifactorial, heterogeneous degenerative condition resulting from many interrelated mutations occurring in multiple genes by the combined effects of genetic/epigenetic alterations and other risk factors (eg, environment, head injury, diet, educational background, social system, etc.). Familial AD is linked to alterations in the APP, PS1, and PS2 genes; by contrast, only the Apo lipoprotein E4 (APOE4) and to some extent the sortilin-related receptor (SORL1) genes have been associated with sporadic AD.

Next-generation sequencing techniques may help define additional molecular factors that underlie sporadic AD.

7.8.2 Stimulated Signaling Biomarkers in AD Cellular Models in Blood Cells

As we understand more about AD-related dysfunctional cellular signaling that occurs outside of brain and how these processes are manifested in peripheral systems, new peripheral AD biomarkers may be discovered and validated in noninvasive tests to diagnose preclinical and clinical AD with high accuracy. Several stimulated signaling biomarkers in AD cellular models in blood cells have been reported (Table 7.6). Aβ-induced stimulation of cytokine production showed partial increase of IL-6, and a significantly decreased production of IL-10 in AD versus age-matched control blood cells ($P = 0.023$) (Arosio et al., 2004). Cell-cycle regulatory deficits and poor survival are not limited to the neurons of AD patients only, but are also seen in peripheral cells such as lymphocytes and fibroblasts (Tatebayashi et al., 1995; Nagy et al., 2002). Immortalized lymphocytes (lymphoblasts) derived from AD patient samples were found to have a greater cell survival rate compared to those from ALS and control cases after incubation in serum-free conditions for 72 h (Bartolomé et al., 2009). Minimizing static variation by subtracting nonstimulated cells as an internal control is an advantage to using stimulated signaling biomarkers.

The G1/S phase check point failure in AD lymphocytes was discussed in Section 7.3.3 of this chapter. Assessment of the G1/S phase of activated cultured blood lymphocytes by PHA followed by induction with rapamycin found a significantly elevated number of cells in S phase ($P = 0.001$) and lower number in G1 phase ($P = 0.01$) for AD lymphocytes compared to control cells by flow cytometry (Song et al., 2012b). AD lymphocytes were less responsive to the G1/S-phase-transition inhibitor rapamycin. On the basis of these studies, Amarantus Bioscience Holding, Inc. (http://www.amarantus.com; San Francisco, CA) is developing a test for AD (The Lymphocyte Proliferation (LymPro) Test) that measures the proliferative response of cultured blood lymphocytes to mitogenic induction. This test is based on differential cell cycle response of cultured lymphocytes from AD patients compared to those from non-AD controls. A UK-based diagnostic company, Cytox Ltd (http://www.cytoxgroup.com), is also focused on cell cycle dysregulation as a diagnostic biomarker of AD.

7.9 ADVANTAGES AND DISADVANTAGES OF CELL-BASED ALZHEIMER'S DISEASE BIOMARKERS

Collection of peripheral blood cells is easy to execute in a noninvasive manner. Blood contains a liquid component (serum or plasma) and different types of blood cells (mononuclear cells, erythrocytes, and platelets), thus providing a unique opportunity to study both peripheral fluid-based AD biomarkers (Chapter 6) as well as cell-based AD biomarkers in a single sample. Yet, while isolation and manipulation of blood cells are routine procedures, cell-based

Table 7.6	Stimulated Peripheral Cell-Based Signaling Biomarkers in Alzheimer's Disease		
Alzheimer's Disease Cellular Models	**Stimulated Molecular Signaling**	**Method/Comments**	**References**
PBMCs Patient groups: AD ($n = 65$); control ($n = 65$).	Aβ-peptide induced cytokine production IL-6 was higher, but nonsignificant IL-10 was significantly lower ($P = 0.023$)	In vitro cytokine production can be monitored easily	Arosio et al. (2004)
Lymphoblasts Patient groups: AD ($n = 20$); control ($n = 20$); ALS ($n = 10$)	Cell survival following 72 h serum starvation. In AD case cell survival rate was significantly higher ($P < 0.01$)	Immortalization of isolated lymphocytes to lymphoblasts requires extra step	Bartolomé et al. (2009)
Lymphocytes Patient groups: AD ($n = 32$); healthy control ($n = 30$); PD ($n = 26$)	Multivariate analysis of stimulated and unstimulated cell surface markers; CD69, CD28. Positive predictive value = 91%; negative predictive value = 92%	Flow cytometry with and without mitogenic stimulation by PHA, PWM	Stieler et al. (2012)
Lymphocytes Patient groups: Severe AD ($n = 15$), mild AD ($n = 15$), age-matched control ($n = 15$) All clinically confirmed by NINCDS-ADRDA criteria and MMSE score	G1/S: non-AD = 4.30 ± 0.30, mild AD = 0.94 ± 0.13, severe AD = 0.96 ± 0.11	Assessment of G1/S phase checkpoint of blood lymphocytes activated by PHA followed by Rapamycin induction. Flow cytometry measurement	Zhou and Jia (2010)
Lymphocytes Mild AD ($n = 26$); control ($n = 28$). All clinically confirmed by NINCDS-ADRDA criteria and K-MMSE score	Assessment of G1/S phase checkpoint: G1: AD: 70.29 ± 6.32, control: 76.03 ± 9.05 S: AD: 12.45 ± 6.09, control: 6.03 ± 5.11	Assessment of G1/S phase checkpoint of blood lymphocytes activated by PHA followed by Rapamycin induction. Flow cytometry measurement	Song et al. (2012b)

AD, Alzheimer's disease; ALS, amyotrophic lateral sclerosis; IL, CD28, cluster of differentiation 28; CD69, cluster of differentiation 69; interleukin; K-MMSE, Korean version of the mini mental score examination; MMSE, mini mental score examination; NINCDS-ADRDA, National Institute of Neurological Communicative Disorders and Stroke/Alzheimer's Disease and Related Disorders Association Work Group criteria of AD; PBMC, peripheral blood mononuclear cells; PD, Parkinson's disease; PHA, phytohaemagglutinin; PWM, pokeweed mitogen.

bioassays require specialized cell biology techniques such as cell separation, cell culture, and specialized manipulation, stimulation, and imaging procedures. Each of these procedures is time consuming and requires skilled personnel. Transportation and stimulation/manipulation of samples must be done with extreme care.

There are several advantages of skin fibroblast-based diagnostic assays. The assay procedures are simple, and require inexpensive sample collection that can be performed in the primary care setting. Unlike CSF-based AD biomarkers, multiple samples can be taken over time to track disease or treatment efficacy. Performing multiple lumbar punctures in the elderly patient is simply not practical. Punch biopsies for collecting skin samples are less painful compared to lumbar puncture, and are routine procedures performed in the elderly population for assessing chronic skin disorders, inflammatory lesions, and cutaneous neoplasms. It is a very simple and straightforward technique, is fast (10–15 min), and can be performed by a nurse practitioner without the supervision of a physician. However, punch biopsies are less comfortable than peripheral blood draws. Technically, it is easy to culture fibroblasts from skin biopsies without contamination of other cell types such as keratinocytes. Fibroblasts in culture are fairly homogeneous and can generate a greater signal to noise ratio compared to mixed tissue samples, and the proliferative nature of primary fibroblasts allows repeat experiments with low passage numbers. A comparison of the RNA quality from lymphocytes and fibroblasts from the same patient suggests that blood samples are more susceptible to external conditions such as acute stimuli, nutritional status, fever, infections, and drug treatment. Cultured fibroblasts are ideal for manipulation and for stimulation of biomarkers compared to blood cells. Cell–cell contact morphologies can be assessed in adhering fibroblasts but not in nonadhering blood cells or buccal cells.

AD is a progressive and irreversible disease. Molecular signaling deficits in peripheral systems may occur before Aβ deposition and tau changes in the brain. Probing signaling events in peripheral cells may make it possible to detect AD in its preclinical stages. Changes in specific molecular signatures have also been found to differentiate between AD and non-AD dementias. Overlap in the clinical features of AD, LBD, VaD, FTD, and tauopathies are very common, and mixed (comorbid) dementia cases can be very difficult to distinguish from AD. The skin fibroblast-based AD-Index assay was found to be able to distinguish AD from non-AD dementias in patients with comorbid dementias at autopsy (Khan and Alkon, 2010).

Skin fibroblast-based biomarkers have been investigated for several diseases as reported in the medical literature. For example, skin fibroblasts have been used in assays of metabolic abnormalities linked to neurological disease, such as Refsum disease (Steinberg, 1983) and Lesch–Nyhan syndrome (Zoref and Sperling, 1979). Skin biopsies have been used to diagnose several neurometabolic and neurodegenerative diseases (Dolman, 1984; Idoate Gastearena and Vega, 1997). Fibroblast-based diagnostic laboratory tests are commonly used to detect several congenital metabolic and neurodegenerative diseases with specific genetic causes (Mayo Medical Laboratory, http://www.mayomedicallaboratories.com/index.html; and The NP-C Guidelines Working Group).

Skin fibroblast-based bioassays do have some disadvantages. There is a lag time between biopsy and test results. It takes several weeks to complete the assays due to the slow growth of skin fibroblasts in culture. Unlike blood collection, the punch biopsy is not completely noninvasive in nature. Nevertheless, the opportunities afforded by our growing understanding of the cellular changes that occur in AD continue to drive innovations that can overcome these technical challenges, and the development of skin fibroblast-based assays for AD remains an active area of research. Blood-based AD biomarkers also have some disadvantages. For example, the degree of fluctuation of proteins/peptides in the blood is influenced by the physiological state of AD patients, such as the sleep cycle, food intake, and other age-related factors (eg, elevated blood pressure, elevated blood glucose, age-related inflammation).

7.10   CONCLUSIONS

Amyloid pathogenesis and tau metabolic pathways are not limited to the brain, but are ubiquitous in the human body and found in blood, skin, saliva, and other peripheral tissues such as eye lenses. There is a strong need to develop easily measurable, inexpensive, minimally invasive diagnostic tests for AD to facilitate early (preclinical) detection of the disease, to predict the likelihood of developing AD, and to monitor disease progression and therapeutic efficacy. Use of widely researched neuroimaging and CSF biomarkers hold promise, but continue to face challenges related to invasiveness of sample collection, cost, interlaboratory variation, and inability to distinguish AD from non-AD dementias. Peripheral fluid-based AD biomarkers (eg, detected in blood plasma, saliva) have not achieved sufficiently high levels of specificity and sensitivity. Cellular biomarkers may be an attractive option for diagnosis of preclinical AD because some early-stage AD-related pathological abnormalities are seen in both neurons and peripheral cells. Some of the cellular biomarkers of AD represent multimodal aspects of the disease, such as abnormalities in Aβ metabolism, tau/microtubules, inflammation effectors, calcium homeostasis, and oxidative stress regulators, as well as neuronal loss and mitochondrial dysfunction. Factors that have systemic impact, such as genetics, hypoxia, ischemia, and metabolic dysfunction could, therefore, be critically important in the etiology of AD and be detected in peripheral cells at preclinical stages. For example, G1/S transition check point in AD lymphocytes is not related to disease severity, and differences in the regulation of the G1/S transition may be used as preclinical peripheral AD biomarker in the future. Taken together, these studies not only strengthen support for the idea that AD pathophysiology is reflected in the peripheral tissues, but also provide hope for the development of diagnostic assays that can detect the disease in its earliest stages. The discovery of AD biomarkers may also lead to the identification of new therapeutic targets and approaches.

Peripheral cell-based AD biomarkers hold great promise for detecting dysfunctional molecular signaling occurring in the early stages of AD. Yet most of the cell-based AD peripheral biomarkers are in discovery stages and their development remains underfunded. Among the cell-based AD biomarkers, those detected in blood lymphocytes, platelets, and cultured skin fibroblasts have been studied in multiple laboratories. More validation studies are required in large, multicenter cohorts consisting of heterogeneous and multiethnic populations. More longitudinal and autopsy-validation studies are also urgently needed for cell-based AD biomarkers. A critical evaluation and standardization of analytical methodologies and more technological development approaches are necessary before new peripheral cell-based AD bioassays can be translated into clinical trials.

Bibliography

Akiyama, H., Barger, S., Barnum, S., Bradt, B., Bauer, J., Cole, G.M., Cooper, N.R., Eikelenboom, P., Emmerling, M., Fiebich, B.L., Finch, C.E., Frautschy, S., Griffin, W.S., Hampel, H., Hull, M., Landreth, G., Lue, L., Mrak, R., Mackenzie, I.R., McGeer, P.L., O'Banion, M.K., Pachter, J., Pasinetti, G., Plata-Salaman, C., Rogers, J., Rydel, R., Shen, Y., Streit, W., Strohmeyer, R., Tooyoma, I., Van Muiswinkel, F.L., Veerhuis, R., Walker, D., Webster, S., Wegrzyniak, B., Wenk, G., Wyss-Coray, T, 2000. Inflammation and Alzheimer's disease. Neurobiol. Aging 21, 383–421.

Alkon, D.L., Sun, M.K., Nelson, T.J., 2007. PKC signaling deficits: a mechanistic hypothesis for the origins of Alzheimer's disease. Trends Pharmacol. Sci. 28, 51–60.

Ambasudhan, R., Talantova, M., Coleman, R., Yuan, X., Zhu, S., Lipton, S.A., Ding, S., 2011. Direct reprogramming of adult human fibroblasts to functional neurons under defined conditions. Cell Stem Cell 9, 113–118.

Arai, H., Terajima, M., Nakagawa, T., Higuchi, S., Mochizuki, H., Sasaki, H., 1996. Pupil dilatation assay by tropicamide is modulated by apolipoprotein E epsilon 4 allele dosage in Alzheimer's disease. Neuroreport 7, 918–920.

Arendt, T., 2012. Cell cycle activation and aneuploid neurons in Alzheimer's disease. Mol. Neurobiol. 46, 125–135.

Arendt, T., Holzer, M., Gärtner, U., Brückner, M.K., 1998. Aberrancies in signal transduction and cell cycle related events in Alzheimer's disease. J. Neural. Transm. Suppl. 54, 147–158.

Armentero, M.T., Sinforiani, E., Ghezzi, C., Bazzini, E., Levandis, G., Ambrosi, G., Zangaglia, R., Pacchetti, C., Cereda, C., Cova, E., Basso, E., Celi, D., Martignoni, E., Nappi, G., Blandini, F., 2011. Peripheral expression of key regulatory kinases in Alzheimer's disease and Parkinson's disease. Neurobiol. Aging 32, 2142–2151.

Arnold, S.E., Lee, E.B., Moberg, P.J., Stutzbach, L., Kazi, H., Han, L.Y., Lee, V.M., Trojanowski, J.Q., 2010. Olfactory epithelium amyloid-beta and paired helical filament-tau pathology in Alzheimer disease. Ann. Neurol. 67, 462–469.

Arosio, B., Trabattoni, D., Galimberti, L., Bucciarelli, P., Fasano, F., Calabresi, C., Cazzullo, C.L., Vergani, C., Annoni, G., Clerici, M., 2004. Interleukin-10 and interleukin-6 gene polymorphisms as risk factors for Alzheimer's disease. Neurobiol. Aging 25, 1009–1015.

Bartolomé, F., Muñoz, U., Esteras, N., Esteban, J., Bermejo-Pareja, F., Martín-Requero, A., 2009. Distinct regulation of cell cycle and survival in lymphocytes from patients with Alzheimer's disease and amyotrophic lateral sclerosis. Int. J. Clin. Exp. Pathol. 2, 390–398.

Baskin, F., Rosenberg, R.N., Iyer, L., Hynan, L., Cullum, C.M., 2000. Platelet APP isoform ratios correlate with declining cognition in AD. Neurology 54, 1907–1909.

Bellucci, C., Lilli, C., Baroni, T., Parnetti, L., Sorbi, S., Emiliani, C., Lumare, E., Calabresi, P., Balloni, S., Bodo, M., 2007. Differences in extracellular matrix production and basic fibroblast growth factor response in skin fibroblasts from sporadic and familial Alzheimer's disease. Mol. Med. 13, 542–550.

Borden, L.A., Maxfield, F.R., Goldman, J.E., Shelanski, M.L., 1992. Resting $[Ca^{2+}]_i$ and $[Ca^{2+}]_i$ transients are similar in fibroblasts from normal and Alzheimer's donors. Neurobiol. Aging 13, 33–38.

Brandan, E., Melo, F., García, M., Contreras, M., 1996. Significantly reduced expression of the proteoglycan decorin in Alzheimer's disease fibroblasts. Clin. Mol. Pathol. 49, M351–M356.

Bruel, A., Cherqui, G., Columelli, S., Margelin, D., Roudier, M., Sinet, P.M., Prieur, M., Pérignon, J.L., Delabar, J., 1991. Reduced protein kinase C activity in sporadic Alzheimer's disease fibroblasts. Neurosc. Lett. 133, 89–92.

Caiazzo, M., Dell'Anno, M.T., Dvoretskova, E., Lazarevic, D., Taverna, S., Leo, D., Sotnikova, T.D., Menegon, A., Roncaglia, P., Colciago, G., Russo, G., Carninci, P., Pezzoli, G., Gainetdinov, R.R., Gustincich, S., Dityatev, A., Broccoli, V., 2011. Direct generation of functional dopaminergic neurons from mouse and human fibroblasts. Nature 476, 224–227.

Casoli, T., Di Stefano, G., Balietti, M., Solazzi, M., Giorgetti, B., Fattoretti, P., 2010. Peripheral inflammatory biomarkers of Alzheimer's disease: the role of platelets. Biogerontology 11, 627–633.

Cazzaniga, E., Bulbarelli, A., Lonati, E., Re, F., Galimberti, G., Gatti, E., Pitto, M., Ferrarese, C., Masserini, M., 2008. Enhanced folate binding of cultured fibroblasts from Alzheimer's disease patients. Neurosci. Lett. 436, 317–320.

Chirila, F.V., Khan, T.K., Alkon, D.L., 2013. Spatiotemporal complexity of fibroblast networks screens for Alzheimer's disease. J. Alzheimers Dis. 33, 165–176.

Chirila, F.V., Khan, T.K., Alkon, D.L., 2014. Fibroblast aggregation rate converges with validated peripheral biomarkers for Alzheimer's disease. J. Alzheimers Dis. 42, 1279–1294.

Citron, M., Vigo-Pelfrey, C., Teplow, D.B., Miller, C., Schenk, D., Johnston, J., Winblad, B., Venizelos, N., Lannfelt, L., Selkoe, D.J., 1994. Excessive production of amyloid beta-protein by peripheral cells of symptomatic and presymptomatic patients carrying the Swedish familial Alzheimer disease mutation. Proc. Natl. Acad. Sci. USA 91, 11993–11997.

Cupello, A., Favale, E., Audenino, D., Scarrone, S., Gastaldi, S., Albano, C., 2005. Decrease of serotonin transporters in blood platelets after epileptic seizures. Neurochem. Res. 30, 425–428.

de Barry, J., Liégeois, C.M., Janoshazi, A., 2010. Protein kinase C as a peripheral biomarker for Alzheimer's disease. Exp. Gerontol. 45, 64–69.

Di Luca, M., Pastorino, L., Bianchetti, A., Perez, J., Vignolo, L.A., Lenzi, G.L., Trabucchi, M., Cattabeni, F., Padovani, A., 1998. Differential level of platelet amyloid beta precursor protein isoforms: an early marker for Alzheimer disease. Arch. Neurol. 55, 1195–1200.

Dolman, C.L., 1984. Diagnosis of neurometabolic disorders by examination of skin biopsies and lymphocytes. Semin. Diagn. Pathol. 1, 82–97.

Dreses-Werringloer, U., Lambert, J.C., Vingtdeux, V., Zhao, H., Vais, H., Siebert, A., Jain, A., Koppel, J., Rovelet-Lecrux, A., Hannequin, D., Pasquier, F., Galimberti, D., Scarpini, E., Mann, D., Lendon, C., Campion, D., Amouyel, P., Davies, P., Foskett, J.K., Campagne, F., Marambaud, P., 2008. A polymorphism in CALHM1 influences Ca^{2+} homeostasis, Abeta levels, and Alzheimer's disease risk. Cell 133, 1149–1161.

Esteras, N., Alquézar, C., de la Encarnación, A., Villarejo, A., Bermejo-Pareja, F., Martín-Requero, A., 2013. Calmodulin levels in blood cells as a potential biomarker of Alzheimer's disease. Alzheimers Res. Ther. 5, 55.

Etcheberrigaray, R., Ito, E., Oka, K., Tofel-Grehl, B., Gibson, G.E., Alkon, D.L., 1993. Potassium channel dysfunction in fibroblasts identifies patients with Alzheimer disease. Proc. Natl. Acad. Sci. USA 90, 8209–8213.

Etcheberrigaray, R., Ito, E., Kim, C.S., Alkon, D.L., 1994. Soluble beta-amyloid induction of Alzheimer's phenotype for human fibroblast K+ channels. Science 264, 276–279.

Etcheberrigaray, R., Tan, M., Dewachter, I., Kuipéri, C., Van der Auwera, I., Wera, S., Qiao, L., Bank, B., Nelson, T.J., Kozikowski, A.P., Van Leuven, F., Alkon, D.L., 2004. Therapeutic effects of PKC activators in Alzheimer's disease transgenic mice. Proc. Natl. Acad. Sci. USA 101, 11141–11146.

Evin, G., Zhu, A., Holsinger, R.M., Masters, C.L., Li, Q.X., 2003. Proteolytic processing of the Alzheimer's disease amyloid precursor protein in brain and platelets. J. Neurosci. Res. 74, 386–392.

Faux, N.G., Ellis, K.A., Porter, L., Fowler, C.J., Laws, S.M., Martins, R.N., Pertile, K.K., Rembach, A., Rowe, C.C., Rumble, R.L., Szoeke, C., Taddei, K., Taddei, T., Trounson, B.O., Villemagne, V.L., Ward, V., Ames, D., Masters, C.L., Bush, A.I., 2011. Homocysteine, vitamin B12, and folic acid levels in Alzheimer's disease, mild cognitive impairment, and healthy elderly: baseline characteristics in subjects of the Australian Imaging Biomarker Lifestyle study. J. Alzheimers Dis. 27, 909–922.

Favit, A., Grimaldi, M., Nelson, T.J., Alkon, D.L., 1998. Alzheimer's-specific effects of soluble beta-amyloid on protein kinase C-alpha and -gamma degradation in human fibroblasts. Proc. Natl. Acad. Sci. USA 95, 5562–5567.

Feigelson, H.S., Rodriguez, C., Robertson, A.S., Jacobs, E.J., Calle, E.E., Reid, Y.A., Thun, M.J., 2001. Determinants of DNA yield and quality from buccal cell samples collected with mouthwash. Cancer Epidemiol. Biomarkers. Prev. 10, 1005–1008.

FitzSimon, J.S., Waring, S.C., Kokmen, E., McLaren, J.W., Brubaker, R.F., 1997. Response of the pupil to tropicamide is not a reliable test for Alzheimer disease. Arch. Neurol. 54, 155–159.

Fotiou, D.F., Brozou, C.G., Haidich, A.B., Tsiptsios, D., Nakou, M., Kabitsi, A., Giantselidis, C., Fotiou, F., 2007. Pupil reaction to light in Alzheimer's disease: evaluation of pupil size changes and mobility. Aging Clin. Exp. Res. 19, 364–371.

François, M., Leifert, W., Hecker, J., Faunt, J., Martins, R., Thomas, P., Fenech, M., 2014. Altered cytological parameters in buccal cells from individuals with mild cognitive impairment and Alzheimer's disease. Cytometry A 85, 698–708.

Frost, S., Kanagasingam, Y., Sohrabi, H., Vignarajan, J., Bourgeat, P., Salvado, O., Villemagne, V., Rowe, C.C., Macaulay, S.L., Szoeke, C., Ellis, K.A., Ames, D., Masters, C.L., Rainey-Smith, S., Martins, R.N., AIBL Research Group, 2013. Retinal vascular biomarkers for early detection and monitoring of Alzheimer's disease. Transl. Psychiatry 3, e233.

Garcia-Closas, M., Egan, K.M., Abruzzo, J., Newcomb, P.A., Titus-Ernstoff, L., Franklin, T., et al., 2001. Collection of genomic DNA from wash. Cancer Epidemiol Biomarkers Prev. 10, 687–696.

Gasparini, L., Racchi, M., Binetti, G., Trabucchi, M., Solerte, S.B., Alkon, D.L., Etcheberrigaray, R., Gibson, G., Blass, J., Paoletti, R., Govoni, S, 1998. Peripheral markers in testing pathophysiological hypotheses and diagnosing Alzheimer's disease. FASEB J. 12, 17–34.

Goldstein, L.E., Muffat, J.A., Cherny, R.A., Moir, R.D., Ericsson, M.H., Huang, X., Mavros, C., Coccia, J.A., Faget, K.Y., Fitch, K.A., Masters, C.L., Tanzi, R.E., Chylack, Jr., L.T., Bush, A.I., 2003. Cytosolic beta-amyloid deposition and supranuclear cataracts in lenses from people with Alzheimer's disease. Lancet 361, 1258–1265.

Gomez-Tortosa, E., del Barrio, A., JimenezAlfaro, I., 1996. Pupil response to tropicamide in Alzheimer's disease and other neurodegenerative disorders. Acta Neurol. Scand. 94, 104–109.

Govoni, S., Bergamaschi, S., Racchi, M., Battaini, F., Binetti, G., Bianchetti, A., Trabucchi, M., 1993. Cytosol protein kinase C downregulation in fibroblasts from Alzheimer's disease patients. Neurology 43, 2581–2586.

Growdon, J.H., Graefe, K., Tennis, M., Hayden, D., Schoenfeld, D., Wray, S.H., 1997. Pupil dilation to tropicamide is not specific for Alzheimer disease. Arch. Neurol. 54, 841–844.

Grunberger, J., Linzmayer, L., Walter, H., Rainer, M., Masching, A., Pezawas, L., SaletuZyhlarz, G., Stohr, H., Grunberger, M., 1999. Receptor test (pupillary dilatation after application of 0.01% tropicamide solution) and determination of central nervous activation (Fourier analysis of pupillary oscillations) in patients with Alzheimer's disease. Neuropsychobiology 40, 40–46.

Hattori, H., Matsumoto, M., Iwai, K., Tsuchiya, H., Miyauchi, E., Takasaki, M., Kamino, K., Munehira, J., Kimura, Y., Kawanishi, K., Hoshino, T., Murai, H., Ogata, H., Maruyama, H., Yoshida, H., 2002. The tau protein of oral epithelium increases in Alzheimer's disease. J. Gerontol. A Biol. Sci. Med. Sci. 57, M64–M70.

Hayney, M.S., Poland, G.A., Lipsky, J.J., 1996. A noninvasive 'swish and spit' method for collecting nucleated cells for HLA typing by PCR in population studies. Hum. Hered. 46, 108–111.

Herrup, K., 2012. The contributions of unscheduled neuronal cell cycle events to the death of neurons in Alzheimer's disease. Front Biosci (Elite Ed) 4, 2101–2109.

Herrup, K., Arendt, T., 2002. Re-expression of cell cycle proteins induces neuronal cell death during Alzheimer's disease. J. Alzheimers Dis. 4, 243–247.

Herrup, K., Neve, R., Ackerman, S.L., Copani, A., 2004. Divide and die: cell cycle events as triggers of nerve cell death. J. Neurosci. 24, 9232–9239.

Hirashima, N., Etcheberrigaray, R., Bergamaschi, S., Racchi, M., Battaini, F., Binetti, G., Govoni, S., Alkon, D.L., 1996. Calcium responses in human fibroblasts: a diagnostic molecular profile for Alzheimer's disease. Neurobiol. Aging 17, 549–555.

Hongpaisan, J., Sun, M.K., Alkon, D.L., 2011. PKCε activation prevents synaptic loss, Aβ elevation, and cognitive deficits in Alzheimer's disease transgenic mice. J. Neurosci. 31, 630–643.

Honig, L.S., Schupf, N., Lee, J.H., Tang, M.X., Mayeux, R., 2006. Shorter telomeres are associated with mortality in those with APOE epsilon4 and dementia. Ann. Neurol. 60, 181–187.

Hye, A., Kerr, F., Archer, N., Foy, C., Poppe, M., Brown, R., Hamilton, G., Powell, J., Anderton, B., Lovestone, S., 2005. Glycogen synthase kinase-3 is increased in white cells early in Alzheimer's disease. Neurosci. Lett. 373, 1–4.

Idoate Gastearena, M.A., Vega, V.F., 1997. Diagnosis of neurometabolic and neurodegenerative diseases by cutaneous biopsy. Rev. Neurol. 25, S269–S280.

Iijima, A., Haida, M., Ishikawa, N., Ueno, A., Minamitani, H., Shinohara, Y., 2003. Re-evaluation of tropicamide in the pupillary response test for Alzheimer's disease. Neurobiol. Aging 24, 789–796.

Ito, E., Oka, K., Etcheberrigaray, R., Nelson, T.J., McPhie, D.L., Tofel-Grehl, B., Gibson, G.E., Alkon, D.L., 1994. Internal Ca²⁺ mobilization is altered in fibroblasts from patients with Alzheimer disease. Proc. Natl. Acad. Sci. USA 91, 534–538.

Janoshazi, A., de Barry, J., 1999. Rapid in vitro conformational changes of the catalytic site of PKC alpha assessed by FIM-1 fluorescence. Biochem 38, 13316–13327.

Janoshazi, A., Sellal, F., Marescaux, C., Danion, J.M., Warter, J.M., de Barry, J., 2006. Alteration of protein kinase C conformation in red blood cells: a potential marker for Alzheimer's disease but not for Parkinson's disease. Neurobiol. Aging 27, 245–251.

Jenkins, E.C., Velinov, M.T., Ye, L., Gu, H., Li, S., Jenkins, Jr., E.C., et al., 2006. Telomere shortening in T lymphocytes of older individuals with Down syndrome and dementia. Neurobiol. Aging 27, 941–945.

Johnston, J.A., Cowburn, R.F., Norgren, S., Wiehager, B., Venizelos, N., Winblad, B., Vigo-Pelfrey, C., Schenk, D., Lannfelt, L., O'Neill, C., 1994. Increased beta-amyloid release and levels of amyloid precursor protein (APP) in fibroblast cell lines from family members with the Swedish Alzheimer's disease APP670/671 mutation. FEBS Lett. 354, 274–2748.

Jong, Y.J., Ford, S.R., Seehra, K., Malave, V.B., Baenziger, N.L., 2003. Alzheimer's disease skin fibroblasts selectively express a bradykinin signaling pathway mediating tau protein Ser phosphorylation. FASEB J. 17, 2319–2321.

Kalman, J., Kanka, A., Magloczky, E., Szoke, A., Jardanhazy, T., Janka, Z., 1997. Increased mydriatic response to tropicamide is a sign of cholinergic hypersensitivity but not specific to late-on-set sporadic type of Alzheimer's dementia. Biol. Psychiatry 41, 909–911.

Kanno, T., Tsuchiya, A., Tanaka, A., Nishizaki, T., 2015. Combination of PKCε activation and PTP1B inhibition effectively suppresses Aβ-induced GSK-3β activation and tau phosphorylation. Mol. Neurobiol. doi: 10.1007/s12035-015-9405-x.

Kardon, R.H., 1998. Drop the Alzheimer's drop test. Neurology 50, 588–591.

Keeney, J.T., Swomley, A.M., Harris, J.L., Fiorini, A., Mitov, M.I., Perluigi, M., Sultana, R., Butterfield, D.A., 2012. Cell cycle proteins in brain in mild cognitive impairment: insights into progression to Alzheimer disease. Neurotox. Res. 22, 220–230.

Khan, T., Alkon, D.L., 2006. An internally controlled peripheral biomarker for Alzheimer's disease: Erk1 and Erk2 responses to the inflammatory signal bradykinin. Proc. Natl. Acad. Sci. USA 103, 13203–13207.

Khan, T.K., Alkon, D.L., 2010. Early diagnostic accuracy and pathophysiologic relevance of an autopsy-confirmed Alzheimer's disease peripheral biomarker. Neurobiol. Aging 31, 889–900.

Khan, T.K., Nelson, T.J., Verma, V.A., Wender, P.A., Alkon, D.L., 2009. A cellular model of Alzheimer's disease therapeutic efficacy: PKC activation reverses Abeta-induced biomarker abnormality on cultured fibroblasts. Neurobiol. Dis. 34, 332–339.

Khan, T.K., Sen, A., Hongpaisan, J., Lim, C.S., Nelson, T.J., Alkon, D.L., 2015. PKCε deficits in Alzheimer's disease brains and skin fibroblasts. J. Alzheimers Dis. 43, 491–509.

King, I.B., Satia-Abouta, J., Thornquist, M.D., Bigler, J., Patterson, R.E., Kristal, A.R., Shattuck, A.L., Potter, J.D., White, E., 2002. Buccal cell DNA yield, quality, and collection costs: comparison of methods for large-scale studies. Cancer Epidemiol Biomarkers Prev. 11, 1130–1133.

Kumar, U., Dunlop, D.M., Richardson, J.S., 1994. Mitochondria from Alzheimer's fibroblasts show decreased uptake of calcium and increased sensitivity to free radicals. Life Sci. 54, 1855–1860.

Lanni, C., Uberti, D., Racchi, M., Govoni, S., Memo, M., 2007. Unfolded p53: a potential biomarker for Alzheimer's disease. J. Alzheimers Dis. 12, 93–99.

Lanni, C., Racchi, M., Mazzini, G., Ranzenigo, A., Polotti, R., Sinforiani, E., Olivari, L., Barcikowska, M., Styczynska, M., Kuznicki, J., Szybinska, A., Govoni, S., Memo, M., Uberti, D., 2008. Conformationally altered p53: a novel Alzheimer's disease marker? Mol. Psychiatry 13, 641–647.

Leidinger, P., Backes, C., Deutscher, S., Schmitt, K., Mueller, S.C., Frese, K., Haas, J., Ruprecht, K., Paul, F., Stähler, C., Lang, C.J., Meder, B., Bartfai, T., Meese, E., Keller, A., 2013. A blood based 12-miRNA signature of Alzheimer disease patients. Genome Biol. 14, R78.

Lopes, J.P., Oliveira, C.R., Agostinho, P., 2009. Cell cycle re-entry in Alzheimer's disease: a major neuropathological characteristic? Curr. Alzheimer Res. 6, 205–212.

Lukens, J.N., Van Deerlin, V., Clark, C.M., Xie, S.X., Johnson, F.B., 2009. Comparisons of telomere lengths in peripheral blood and cerebellum in Alzheimer's disease. Alzheimers Dement. 5, 463–469.

Lum, A., Le Marchand, L., 1998. A simple mouthwash method for obtaining genomic DNA in molecular epidemiological studies. Cancer Epidemiol Biomarkers Prev. 7, 719–724.

Matsushima, H., Shimohama, S., Fujimoto, S., Takenawa, T., Kimura, J., 1995. Reduction of platelet phospholipase C activity in patients with Alzheimer disease. Alzheimer Dis. Assoc. Disord. 9, 213–217.

Migliore, L., Botto, N., Scarpato, R., Petrozzi, L., Cipriani, G., Bonuccelli, U., 1999. Preferential occurrence of chromosome 21 malsegregation in peripheral blood lymphocytes of Alzheimer disease patients. Cytogenet. Cell Genet. 87, 41–46.

Migliore, L., Fontana, I., Trippi, F., Colognato, R., Coppedè, F., Tognoni, G., Nucciarone, B., Siciliano, G., 2005. Oxidative DNA damage in peripheral leukocytes of mild cognitive impairment and AD patients. Neurobiol. Aging 26, 567–573.

Moreno-Ramos, T., Benito-León, J., Villarejo, A., Bermejo-Pareja, F., 2013. Retinal nerve fiber layer thinning in dementia associated with Parkinson's disease, dementia with Lewy bodies, and Alzheimer's disease. J. Alzheimers Dis. 34, 659–664.

Morrison, R.S., Kinoshita, Y., Johnson, M.D., Guo, W., Garden, G.A., 2003. p53-dependent cell death signaling in neurons. Neurochem. Res. 28, 15–27.

Nagasaka, Y., Dillner, K., Ebise, H., Teramoto, R., Nakagawa, H., Lilius, L., Axelman, K., Forsell, C., Ito, A., Winblad, B., Kimura, T., Graff, C., 2005. A unique gene expression signature discriminates familial Alzheimer's disease mutation carriers from their wild-type siblings. Proc. Natl. Acad. Sci. USA 102, 14854–148549.

Nagy, Z., 2000. Cell cycle regulatory failure in neurones: causes and consequences. Neurobiol. Aging 21, 761–769.

Nagy, Z., 2007. The dysregulation of the cell cycle and the diagnosis of Alzheimer's disease. Biochim. Biophys. Acta 1772, 402–408.

Nagy, Z., Esiri, M.M., Smith, A.D., 1997a. Expression of cell division markers in the hippocampus in Alzheimer's disease and other neurodegenerative conditions. Acta Neuropathol. 93, 294–300.

Nagy, Z., Esiri, M.M., Cato, A.M., Smith, A.D., 1997b. Cell cycle markers in the hippocampus in Alzheimer's disease. Acta Neuropathol. 94, 6–15.

Nagy, Z., Esiri, M.M., Smith, A.D., 1998. The cell division cycle and the pathophysiology of Alzheimer's disease. Neuroscience 87, 731–739.

Nagy, Z.S., Smith, M.Z., Esiri, M.M., Barnetson, L., Smith, A.D., 2000. Hyperhomocysteinaemia in Alzheimer's disease and expression of cell cycle markers in the brain. J. Neurol. Neurosurg. Psychiatry 69, 565–566.

Nagy, Z., Combrinck, M., Budge, M., McShane, R., 2002. Cell cycle kinesis in lymphocytes in the diagnosis of Alzheimer's disease. Neurosci. Lett. 317, 81–84.

Nelson, T.J., Cui, C., Luo, Y., Alkon, D.L., 2009. Reduction of beta-amyloid levels by novel protein kinase C (epsilon) activators. J. Biol. Chem. 284, 34514–34521.

Pang, Z.P., Yang, N., Vierbuchen, T., Ostermeier, A., Fuentes, D.R., Yang, T.Q., Citri, A., Sebastiano, V., Marro, S., Südhof, T.C., Wernig, M., 2011. Induction of human neuronal cells by defined transcription factors. Nature 476, 220–223.

Pani, A., Mandas, A., Diaz, G., Abete, C., Cocco, P.L., Angius, F., Brundu, A., Muçaka, N., Pais, M.E., Saba, A., Barberini, L., Zaru, C., Palmas, M., Putzu, P.F., Mocali, A., Paoletti, F., La Colla, P., Dessì, S., 2009a. Accumulation of neutral lipids in peripheral blood mononuclear cells as a distinctive trait of Alzheimer patients and asymptomatic subjects at risk of disease. BMC Med. 7, 66.

Pani, A., Dessì, S., Diaz, G., La Colla, P., Abete, C., Mulas, C., Angius, F., Cannas, M.D., Orru, C.D., Cocco, P.L., Mandas, A., Putzu, P., Laurenzana, A., Cellai, C., Costanza, A.M., Bavazzano, A., Mocali, A., Paoletti, 2009b. Altered cholesterol ester cycle in skin fibroblasts from patients with Alzheimer's disease. J. Alzheimers Dis. 18, 829–841.

Panossian, L.A., Porter, V.R., Valenzuela, H.F., Zhu, X., Reback, E., Masterman, D., Cummings, J.L., Effros, R.B., 2003. Telomere shortening in T cells correlates with Alzheimer's disease status. Neurobiol. Aging 24, 77–84.

Perry, G., Castellani, R.J., Smith, M.A., Harris, P.L., Kubat, Z., Ghanbari, K., Jones, P.K., Cordone, G., Tabaton, M., Wolozin, B., Ghanbari, H., 2003. Oxidative damage in the olfactory system in Alzheimer's disease. Acta Neuropathol. 106, 552–556.

Peterson, C., Goldman, J.E., 1986. Alterations in calcium content and biochemical processes in cultured skin fibroblasts from aged and Alzheimer donors. Proc. Natl. Acad. Sci. USA 83, 2758–2762.

Peterson, C.P., Gibson, G.E., Blass, J.P., 1985. Altered calcium uptake in cultured skin fibroblasts from patients with Alzheimer's disease. New Engl. J. Med. 312, 1063–1065.

Peterson, C., Ratan, R.R., Shelanski, M.L., Goldman, J.E., 1986. Cytosolic free calcium and cell spreading decrease in fibroblasts from aged and Alzheimer donors. Proc. Natl. Acad. Sci. USA 83, 7999–8001.

Peterson, C., Ratan, R.R., Shelanski, M.L., Goldman, J.E., 1988. Altered response of fibroblasts from aged and Alzheimer donors to drugs that elevate cytosolic free calcium. Neurobiol. Aging 9, 261–266.

Pfisterer, U.1., Wood, J., Nihlberg, K., Hallgren, O., Bjermer, L., Westergren-Thorsson, G., Lindvall, O., Parmar, M., 2011. Efficient induction of functional neurons from adult human fibroblasts. Cell Cycle 10, 3311–3316.

Rainesalo, S., Keränen, T., Saransaari, P., Honkaniemi, J., 2005. GABA and glutamate transporters are expressed in human platelets. Brain Res. Mol. Brain. Res. 141, 161–165.

Richards, B., Skoletsky, J., Shuber, A.P., Balfour, R., Stern, R.C., Dorkin, H.L., et al., 1993. Multiplex PCR amplification from the CFTR gene using DNA prepared from buccal brushes/swabs. Hum. Mol. Genet. 2, 159–163.

Scheuner, D., Eckman, C., Jensen, M., Song, X., Citron, M., Suzuki, N., Bird, T.D., Hardy, J., Hutton, M., Kukull, W., Larson, E., Levy-Lahad, E., Viitanen, M., Peskind, E., Poorkaj, P., Schellenberg, G., Tanzi, R., Wasco, W., Lannfelt, L., Selkoe, D., Younkin, S., 1996. Secreted amyloid beta-protein similar to that in the senile plaques of Alzheimer's disease is increased in vivo by the presenilin 1 and 2 and APP mutations linked to familial Alzheimer's disease. Nat. Med. 2, 864–870.

Schreml, S., Kaiser, E., Landthaler, M., Szeimies, R.M., Babilas, P., 2010. Amyloid in skin and brain: what's the link? Exp. Dermatol. 19, 953–957.

Schuitemaker, A., Dik, M.G., Veerhuis, R., Scheltens, P., Schoonenboom, N.S., Hack, C.E., Blankenstein, M.A., Jonker, C., 2009. Inflammatory markers in AD and MCI patients with different biomarker profiles. Neurobiol. Aging 30, 1885–1889.

Scinto, L.F., Daffner, K.R., Dressler, D., Ransil, B.I., Rentz, D., Weintraub, S., Mesulam, M., Potter, H., 1994. A potential noninvasive neurobiological test for Alzheimer's disease. Science 266, 1051–1054.

Sims, N.R., Finegan, J.M., Blass, J.P., 1987. Altered metabolic properties of cultured skin fibroblasts in Alzheimer's disease. Ann. Neurol. 21, 451–457.

Smith, C.C., 1997. Stimulated release of the b-amyloid protein of Alzheimer's disease by normal human platelets. Neurosc. Lett. 235, 157–159.

Soininen, H., Syrjänen, S., Heinonen, O., Neittaanmäki, H., Miettinen, R., Paljärvi, L., Syrjänen, K., Beyreuther, K., Riekkinen, P., 1992. Amyloid beta-protein deposition in skin of patients with dementia. Lancet 339, 245–1245.

Song, J., Wang, S., Tan, M., Jia, J., 2012a. G1/S checkpoint proteins in peripheral blood lymphocytes are potentially diagnostic biomarkers for Alzheimer's disease. Neurosci. Lett. 526, 144–149.

Song, M., Kwon, Y.A., Lee, Y., Kim, H., Yun, J.H., Kim, S., Kim, D.K., 2012b. G1/S cell cycle checkpoint defect in lymphocytes from patients with Alzheimer's disease. Psychiatry Investig. 9, 413–417.

Steinberg, D., 1983. Phytanic acid storage disease (Refsum's disease). In: Stanbury, J.B., Wyngaarden, J.B., Fredrickson, D.S., Goldstein, J.L., Brown, M.S. (Eds.), The Metabolic Basis of Inherited Disease. fifth ed. McGrow Hill, New York, pp. 731–747.

Stieler, J., Grimes, R., Weber, D., Gartner, W., Sabbagh, M., Arendt, T., 2012. Multivariate analysis of differential lymphocyte cell cycle activity in Alzheimer's disease. Neurobiol. Aging 33, 234–241.

Takashima, A., 2006. GSK-3 is essential in the pathogenesis of Alzheimer's disease. J. Alzheimers Dis. 9 (3 Suppl), 309–317.

Takeda, M., Tatebayashi, Y., Nishimura, T., 1992. Change in the cytoskeletal system in fibroblasts from patients with familial Alzheimer's disease. Prog. Neuropsychopharmacol. Biol. Psychiatry 16, 317–328.

Tan, J., Town, T., Abdullah, L., Wu, Y., Placzek, A., Small, B., Kroeger, J., Crawford, F., Richards, D., Mullan, M., 2002. CD45 isoform alteration in CD4+ T cells as a potential diagnostic marker of Alzheimer's disease. J. Neuroimmunol. 132, 164–172.

Tang, K., Hynan, L.S., Baskin, F., Rosenberg, R.N., 2006. Platelet amyloid precursor protein processing: a bio-marker for Alzheimer's disease. J. Neurol. Sci. 240, 53–58.

Tatebayashi, Y., Takeda, M., Kashiwagi, Y., Okochi, M., Kurumadani, T., Sekiyama, A., Kanayama, G., Hariguchi, S., Nishimura, T., 1995. Cell-cycle-dependent abnormal calcium response in fibroblasts from patients with familial Alzheimer's disease. Dementia 6, 9–16.

Thomas, P., Hecker, J., Faunt, J., Fenech, M., 2007. Buccal micronucleus cytome biomarkers may be associated with Alzheimer's disease. Mutagenesis 22, 371–379.

Thomas, P., O' Callaghan, N.J., Fenech, M., 2008. Telomere length in white blood cells, buccal cells and brain tissue and its variation with ageing and Alzheimer's disease. Mech. Ageing Dev. 129, 183–190.

Tu, H., Nelson, O., Bezprozvanny, A., Wang, Z., Lee, S.F., Hao, Y.H., Serneels, L., De Strooper, B., Yu, G., Bezprozvanny, I., 2006. Presenilins form ER Ca^{2+} leak channels, a function disrupted by familial Alzheimer's disease-linked mutations. Cell 126, 981–993.

Uberti, D., Lanni, C., Carsana, T., Francisconi, S., Missale, C., Racchi, M., Govoni, S., Memo, M., 2006. Identification of a mutant-like conformation of p53 in fibroblasts from sporadic Alzheimer's disease patients. Neurobiol. Aging 27, 1193–1201.

Uéda, K., Cole, G., Sundsmo, M., Katzman, R., Saitoh, T., 1989. Decreased adhesiveness of Alzheimer's disease fibroblasts: is amyloid beta-protein precursor involved? Ann. Neurol. 25, 246–251.

van Huynh, T., Cole, G., Katzman, R., Huang, K.P., Saitoh, T., 1989. Reduced protein kinase C immunoreactivity and altered protein phosphorylation in Alzheimer's disease fibroblasts. Arch. Neurol. 46, 1195–1199.

Veerhuis, R., Van Breemen, M.J., Hoozemans, J.M., Morbin, M., Ouladhadj, J., Tagliavini, F., Eikelenboom, P., 2003. Amyloid beta plaque-associated proteins C1q and SAP enhance the Abeta1–42 peptide-induced cytokine secretion by adult human microglia in vitro. Acta Neuropathol. 105, 135–144.

Viel, T.A., Buck, H.S., 2011. Kallikrein-kinin system mediated inflammation in Alzheimer's disease in vivo. Curr. Alzheimer Res. 8, 59–66.

Vierbuchen, T., Ostermeier, A., Pang, Z.P., Kokubu, Y., Südhof, T.C., Wernig, M., 2010. Direct conversion of fibroblasts to functional neurons by defined factors. Nature 463, 1035–1041.

Vignini, A., Giusti, L., Raffaelli, F., Giulietti, A., Salvolini, E., Luzzi, S., Provinciali, L., Mazzanti, L., Nanetti, L., 2013. Impact of gender on platelet membrane functions of Alzheimer's disease patients. Exp. Gerontol. 48, 319–335.

Vincent, I., Jicha, G., Rosado, M., Dickson, D.W., 1997. Aberrant expression of mitotic cdc2/cyclin B1 kinase in degenerating neurons of Alzheimer's disease brain. J. Neurosci. 17, 3588–3598.

Webber, K.M., Raina, A.K., Marlatt, M.W., Zhu, X., Prat, M.I., Morelli, L., Casadesus, G., Perry, G., Smith, M.A., 2005. The cell cycle in Alzheimer disease: a unique target for neuropharmacology. Mech. Ageing Dev. 126, 1019–1025.

Wetsel, W.C., Khan, W.A., Merchenthaler, I., Rivera, H., Halpern, A.E., Phung, H.M., Negro-Vilar, A., Hannun, Y.A., 1992. Tissue and cellular distribution of the extended family of protein kinase C isoenzymes. J. Cell Biol. 117, 121–133.

Yang, Y., Geldmacher, D.S., Herrup, K., 2001. DNA replication precedes neuronal cell death in Alzheimer's disease. J. Neurosci. 21, 2661–2668.

Yang, Y., Mufson, E.J., Herrup, K., 2003. Neuronal cell death is preceded by cell cycle events at all stages of Alzheimer's disease. J. Neurosci. 23, 2557–2563.

Yoo, A.S., Sun, A.X., Li, L., Shcheglovitov, A., Portmann, T., Li, Y., Lee-Messer, C., Dolmetsch, R.E., Tsien, R.W., Crabtree, G.R., 2011. MicroRNA-mediated conversion of human fibroblasts to neurons. Nature 476, 228–231.

Zhao, W.Q., Ravindranath, L., Mohamed, A.S., Zohar, O., Chen, G.H., Lyketsos, C.G., Etcheberrigaray, R., Alkon, D.L., 2002. MAP kinase signaling cascade dysfunction specific to Alzheimer's disease in fibroblasts. Neurobiol. Dis. 11, 166–183.

Zhou, X., Jia, J., 2010. P53-mediated G1/S checkpoint dysfunction in lymphocytes from Alzheimer's disease patients. Neurosci. Lett. 468, 320–325.

Zoref, E., Sperling, O., 1979. Increased de novo purine synthesis in cultured skin fibroblasts from heterozygotes for the Lesch-Nyhan syndrome. A sensitive marker for carrier detection. Hum. Hered. 29, 64–68.

Zoumakis, E., Kalantaridou, S.N., Chrousos, G.P., 2007. The "brain-skin connection": nerve growth factor-dependent pathways for stress-induced skin disorders. J. Mol. Med. 85, 1347–1349.

Index